Inequality – The Unbeatable Challenge

RIVER PUBLISHERS SERIES IN CHEMICAL, ENVIRONMENTAL, AND ENERGY ENGINEERING

The "River Publishers Series in Chemical, Environmental, and Energy Engineering" is a series of comprehensive academic and professional books which focus on Environmental and Energy Engineering subjects. The series focuses on topics ranging from theory to policy and technology to applications.

Books published in the series include research monographs, edited volumes, handbooks and textbooks. The books provide professionals, researchers, educators, and advanced students in the field with an invaluable insight into the latest research and developments.

Topics covered in the series include, but are by no means restricted to the following:

- Energy and Energy Policy
- Chemical Engineering
- Water Management
- Sustainable Development
- Climate Change Mitigation
- Environmental Engineering
- Environmental System Monitoring and Analysis
- Sustainability: Greening the World Economy

For a list of other books in this series, visit www.riverpublishers.com

Inequality – The Unbeatable Challenge

Editors

Medani P. Bhandari

Gandaki University, Nepal
Akamai University, USA and Sumy State University, Ukraine

Shvindina Hanna

Sumy State University, Ukraine

LONDON AND NEW YORK

Published 2021 by River Publishers
River Publishers
Alsbjergvej 10, 9260 Gistrup, Denmark
www.riverpublishers.com

Distributed exclusively by Routledge
4 Park Square, Milton Park, Abingdon, Oxon OX14 4RN
605 Third Avenue, New York, NY 10158

First published in paperback 2024

Inequality – The Unbeatable Challenge / by Medani P. Bhandari, Shvindina Hanna.

Routledge is an imprint of the Taylor & Francis Group, an informa business

Publisher's Note
The publisher has gone to great lengths to ensure the quality of this reprint but points out that some imperfections in the original copies may be apparent.

While every effort is made to provide dependable information, the publisher, authors, and editors cannot be held responsible for any errors or omissions.

ISBN: 978-87-7022-623-3 (hbk)
ISBN: 978-87-7004-291-8 (pbk)
ISBN: 978-1-003-33854-3 (ebk)

DOI: 10.1201/9781003338543

Dedication

We dedicate this book to all loved ones Worldwide who lost their lives due to Corona pandemic from the beginning to date; we acknowledge that some of you were victimized due to inequality in healthcare systems. In your memory, in this book, we have raised the questions of why the world was not able to save you. The world will remember and salute you all, and we hope that future science will be able to control such pandemic.

Contents

Preface

We may have no doubts that inequality is an unsolved problem, but now we need to find out—is it unbeatable? There is still lack of knowledge around how inequality has been grounded throughout human civilization, why society is stratified and classified, economically, politically, socially, and religiously, and why the discrimination due to gender, sexual orientation, country of origin, language differences, immigration status, caste, race, and ethnicity? This book addresses these issues in a holistic way as well as including case studies of various countries. It tries to find out why inequality has been unbeatable and what would be the best policies to overcome this challenge.

This edited book presents some unexplored issues of economic inequality, including case studies of various countries. Inequality is a chronic and divisive factor of society. Inequality exists as an integral attribute of human development. Communities, nations, and systems are not evolving at the same speed and rate and thus require different resources in different amounts. However, the distribution of winnings is also uneven due to the multidimensionality of influencing factors.

When we talk about inequality, it is not just inequality of income or wealth; it is first, inequality in access to priorities and human needs—to shelter, to clean water, air, health care, and also to appropriate vaccination systems and assistance, security systems and safety guarantees for the future. Past financial crises and the current pandemic shock has revealed bugs in the system, shaking it and changing our perception of the norms.

There are different types of inequalities, such as disparities in the development of regions or neighboring countries, where there is a constant cross-border exchange of resources not in favor of a more impoverished neighbor. A specific type of inequality is associated with the globalization of the social and economic life of humankind, which manifests hunger in certain countries, the inability of low-income people to access health care services, including vaccination, diagnosis, and prevention of serious diseases. Particular attention needs to be devoted to the inequality, which arises as a result of permanent causal links between phenomena, for instance, the so-called wealth-related inequality in access to education, and as a consequence, the possible growth or loss of individual wealth. Different communities have different access to technologies that could accelerate their development; therefore, some nations are evolving at a cosmic rate, while others in the early 21st century are trying to move away from traditional crafting.

The issue of gender inequality and age inequality is one of the oldest types of population distribution in access to resources and decision-making. Institutional weakness in addressing inequalities and finding a sustainable path for the development of different communities in a multicultural, multi-plane reality also emphasizes the relevance of inequality. Labor migration is one of the markers, by the way, of the difference in the quality of life between different countries, communities, and systems. And migration itself is not only a consequence but also an accelerator of the inequalities increases in the donor system.

Will the issue of inequality ever be resolved? We are talking about policies developed in several areas of social institutions—education and science, economics, public administration, health, law, religion, family, and individual development. That is why researchers from different background contribute to the problem solving, trying to find the ways out the unbearable inequality in global and local contexts. Book covers verities of topics, which basically, concentrate on why, Inequality is the problems of all times, and treated as unbeatable challenge; the problems and consequences of inequality—how it divides society, hampers the social harmony, leads toward the conflicts and even wars; Inequality due to gender, age, origin, ethnicity, disability, sexual orientation, class, refugee status, and religion; how Covid 19 is hitting the poor and how it is playing the triggering factor to widen inequality; Inequalities in Education / Healthcare / Food Supply and the impact of social exclusion in consumption patterns—and economic patterns, within the titles: Inequality—Social Inclusion, Sustainable Development and Asta-Ja; Forecasting Inequality—The Innovation Implementation; Inequality and Financial Development; Green Investment as an Alternative to Minimize Inequality; Effect of Shadow Economy on Social Inequality; Cultural Dynamics and Inequality; Social Inequality in the Globalization Era;

The Luxury Lockdown: Tackle to Covid-19—a factor of widening inequality; Rural Agricultural Transformation and Inequality; Income Inequality and Poverty Status.

Social Responsibility as a Tool for the Human Resources Policy Development to Overcome Inequality and Pathways to Address the Challenges of Inequality, etc.

First, book, begins with the global scenarios of inequality, largely economic inequality, provides the current stage of inequality with the varieties of argumentations (references) and states that to overcome this challenge, primarily, there is a need of humanitarian approach like "Bashudaiva Kutumbakam"—"The entire world is our home, and all living beings are our relatives" or any other approaches, which give the equal value to all humans, without any discrimination. The documentary evidence shows that, the major cause of widening inequality is due to the lack of holistic, inclusive approaches in the political authority. Most of the prominent scholars give the arguments that "The main barrier to tackling inequalities is lack of political will" (United Nations 2020). Why there has been a lack of political will to minimize the inequality? Basically, those are in power, have not actually felt the pain of being marginalized, poor, live in poverty. Even though, the people who might have faced such problem; once they are in power, their concentration is to remain in power but not to support, include or empower who are in the stage of haves not.

Specifically, book, not only analyzes the socio-political spectrum the inequality; however, goes, beyond and highlights and proposes new aspects and essence of social inclusion, related to sustainable development and elaborates how the proper utilization of the ASTA-JA framework (which means eight resources [(*Ja—Jal* (water), *Jamin* (land), *Jungle* (forest), *Jadibuti* (medicinal and aromatic plants), *Janashakti* (manpower), *Janawar* (animal), *Jarajuri* (crop plants), and *Jalabayu* (climate)] can be a pathway to minimize inequality through the social justice. The book points that, there is a need of factual knowledge and practices which can bridge the gap in the divided society. Awareness needs to come from individual to individuals, society to societies, nation to nations. To overcome the inequality problems the concept of "Bashudaiva Kutumbakam"—"The entire world is our home, and all living beings are our relatives" and "Live and let other live"—"the harmony within, community, nation and global" is needed. There is a need of trust, love, and respect to each other.

This edited book, follows theoretical frameworks, and raises the questions, pinpoints problems, provides recommendations and fulfills the knowledge gap of complex social discourse in reducing the global, regional, national, and local inequality. This book tries to provide some of the basic answers of why inequality is growing in general and especially, how Covid-19 is playing to widen inequality. As known fact that, the world is already facing various challenges: political, social, religious, etc., inequalities, ageing, AIDS, atomic energy, children, climate change, economic colonization, democracy fights, poverty, food insecurity, gender inequality, lack of access to healthcare, human rights, questions on international law and justice, migration, challenges on peace and security, population growth, refugees challenges, scarcity of water, misuse of technology, drugs, growing individualistic approach among youth, deviant behavior to attend temporary pleasure, corruption, wars, violation of social norms and values, and so on. In addition to that, the inequality is growing and the discriminations on the basis of race, caste, color, country of origin, religion, migratory status, gender, and sexual orientations are still in practice. The economic inequality is widening and widening, and the Covid-19 is playing a favorable role to increase inequality. At the same time, the climate change–induced challenges are also playing important role in widening the inequality, where the poor are the main victims.

In this book, we; including chapter authors, have tried to remind how inequality has been growing in each and every sector and how the Covid 19, has been playing a crucial role to support the title of this book "Inequality—The Unbeatable Challenge." The authors of this book argue that, the inequality challenges are beatable, inequality cannot be eliminated; however, can be reduced if all concerned stakeholders intent on changing it. In general notion, it seems that no one wants to bear the pain, see the pain; however, there is no true visible commitments, drive and seriousness to beat the challenges of inequality. Yes, in paper, policy, or in directives of United Nations to governments, and all development agencies, academic domain, there is raised concern; however, the concern or publications itself has nothing to do, if there is no practical action. This book aims to make aware of the reality by showing that, inequality is almost omnipresent in all sectors of society and the victims are those who has been marginalized through generation to generation, due to the dominant nature of haves' circle, almost in the entire world. The discrimination on the basis of race, ethnicity, gender, origin, sexual orientation, refugee status, language, color, religious belief, and even physical outlook, etc., is helping to continuously widen the social inequality. Similarly, the inequality in economic, political, religious, knowledge, etc., are due to the greediness of power, money, self-profit by dominating the have nots population. This situation is severe in the developing world in comparison to the developed world. This book tries to give some examples through the case studies. Hopefully, these case studies will help to pinpoint the problems and consequences of inequality.

Medani P. Bhandari, PhD.
Gandani University, Nepal
Akamai University, USA and Sumy State University, Ukraine

Forewords

Inequality—The Unbeatable Challenge, editors, Prof. Medani P. Bhandari and Associate Prof. Shvindina Hanna

Based in this text, the goal of significant reduction of economic and social inequalities in developing countries, remains unachieved. Amelioration of human suffering in the third world continues as an important goal for the United Nations. The authors of this important book address these issues and lay out the concerns in an effective manner, with an important focus upon equivalency of the global moral standard. The title of the book, "Reducing Intra- and International Inequalities and Disproportions Towards Sustainable Development Goals" must be viewed among the premier missions of the global community. Prof. Dr. Medani Bhandari and his colleague, Associate Prof. Shvindina Hanna, in their editing, have addressed a path forward toward the final cure for human inequality.

The hope of the book parallels the philosophies of many of the great leaders of the international humanitarian movement. For instance, the American human rights hero, Dr. Martin Luther King, Jr. once stated: *Everybody can be great. . . . because everybody can serve. . . . .You only need a heart full of grace and a soul generated by love.* He said: *The time is always right . . . . . . to do what is right.* The great Filipino Benigno Ninoy Aquino once said: *We should not depend on one man, we should depend on all of us. All of us in the cause for freedom must stand up now and be a leader, and when all of us are leaders, we will expedite the cause of freedom. Overcoming major human challenges is the reality of your world, as remembered in the words of the great Vietnam leader, Ho Chi Minh: "Remember that the storm is a good opportunity for the pine and the cypress to show their strength and their stability. After the rain . . . good weather In the wink of an eye, the universe throws off its muddy clothes. Further, Mahatma Gandhi stated: As human beings, our greatness lies not so much in being able to remake the world, rather our greatness rests— in being able to remake ourselves. A man is but the product of his thoughts ... what he thinks he becomes."* Further, he said: *My life is my message... You must be the change you wish to see in the world.*

These men were great humanitarians because they had a true mission, and they pursued it with every ounce of humanity. As is a foundation of this book, the global community must establish human equality and betterment of the human condition as primary goals for all humanity. World leaders then, must achieve this goal with unity.

Another basis for betterment of the human condition is founded in a premise of the major world religions. The holy teachings of Jesus of Nazareth states: *whatever you want others to do to you, do also to them....* This is widely referred to as the Golden Rule, and internationally, across the major religions, this teaching underscores law and community across much of the human race. For Buddhism, the same teaching states: *Hurt not others in ways that you yourself would find hurtful.* For Hinduism, we find written: *This is the sum of duty; do nothing unto others ... what you would not have them do unto you.* In Islam, it is believed that: *No one of you is a believer until he desires for his brother that which* he desires for himself.

In the preface of the book, Prof. Bhandari states that "The book points that, there is a need of factual knowledge and practices which can bridge the gap in the divided society. Awareness needs to come from individual to individuals, society to societies, nation to nations. To overcome the inequality problems the concept of "Bashudaiva Kutumbakam"—"The entire world is our home, and all living beings are our relatives" and "Live and let other live"—"the harmony within, community, nation and global" is needed. There is a need of trust, love, and respect to each other." I agree with Prof. Bhandari, and hope that, the readers of this book also understand the essence of humanity and contribute to overcome the challenges of inequality. I say inequality, may not be eliminated but for sure it can be minimized if everyone of us takes this as my problem, our problem and it is my responsibility, our responsibility to minimize.

These foundations then support the premise of this book, that the nations united in policy and effort must move toward this goal of global human equality. I thank the authors for pursuing such a grand undertaking.

Prof. Douglass Lee Capogrossi,
Ph.D. President EmeritusAkamai University, USA

Throughout history, the topic of inequality has successfully been buried due to lack of knowledge, lack of transparency, and/or fear of uncovering grave injustices that stay hidden and blinded from view to look deeper, create solutions, and act as guided. The word emphasized here is "guided."

Truths, solutions, and actions are best uncovered through an open heart, asking for courage and protection, and wisdom to move forward. This effort is not to be taken by the timid or lazy rather requires a bold, inward search that inspires you and courageously points a direction for you, for me.

In my personal journey of life as a nurse, educator, practitioner, and researcher, I have always considered myself to be open, guided and treat others equally. I searched to care about those with less than what I had and serve them. I did not come across as the bossy type, do what I say or believe what I consider to be "the truth." So, I patted myself on the back and said, ok Bulbrook, you did a good job.

On reading this newest book on Inequalities by Prof. Bhandari and Associate Prof. Hanna, I uncovered some areas that I could do more and was inspired to do just that. So, if you are willing to have your eyes opened, see the unseen, and be guided and inspired, read on. Learn about economic, social, political, cultural and I would add religious inequalities rather than group it under cultural.

What gets triggered for you will set you free from a lame duck position of life to a soar with the eagles or fly with angels' activating your inner state of being filled with curiosity, compassion, and concern to right wrongs and thus inequalities.

At Akamai University, the topic of inequality comes center and front stage as one of the three core colleges identified in this historic transition of our innovative university dedicated to human and planetary health. We take the risks to uncover hidden truths, to educate and inspire others to tackle the difficult challenges facing society worldwide. Inequality is one of the most hidden dynamics of life worldwide that cannot be ignored any longer.

It will take each of us to look deeper, harder inside and around us. Team up with others in this search that seems daunting and out of reach. What is asked is to take one step at a time and the path will enfold. But you have to first look inside, notice what rings your bell, piques' your curiosity, or challenges your humanity to evolve into an introspective, caring concerned citizen of the world dedicated to do your part however large or small, will make a change setting in motion a cause, a solution, and or a calling to service.

Do not be afraid. Do not hesitate. Rather say, "count me in." "Please guide and help me to identify and do my part addressing inequality of gender, age, origin, nationality, ethnicity, disability, sexual orientation, class, religion, social justice, dominance, education, power, social exclusion, economic, political, cultural, etc. At the end of your life enjoy being able to say: "I did my part and am proud and happy to have made a contribution to address inequality!"

Prof. Mary Jo Bulbrook,
BSN, RN, Med, EdD, CEMP/S/I, HTCP, HTP/I, ICF, President Akamai University

I'd like to express my pleasure about a new book production edited by Professor Medani P. Bhandari and Associate Professor Hanna Shvindina "Unbeatable Challenge of Inequality," one in a set of book series devoted to the problem of social injustice. There are many types of inequalities spread in the world, and many contributors joined the book to present their ideas and successful cases in a sphere of overcoming the inequality challenge. However, the scale and scope of inequality issues made the editors of the book point out the question about the possibility of solving the problem. The book unites the chapters elaborated by different authors from different countries focused on finding out how to accelerate the development in diverse communities toward justice and equality. Besides the appreciation to the editors of the book, I'd like to share my gratefulness to the project's contributors, authors of the chapters, reviewers, and publishers who raise public awareness about the inequality as one of the main challenges of the modern world.

Prof. Vasyl Karpusha,
Rector, Sumy State University, Sumy, Ukraine

Economic, social, political, and other inequalities are a complex and still unexplored issue primarily relating to unequal distribution of income, wealth, and opportunity between different groups in a society. Although inequality as a phenomenon is an element accompanying the development of every society and is an integral attribute of human development, strong disproportions in the development of societies are an unfavorable phenomenon. They intensify the already disturbed balance in communities, nations, and systems evolving, their speed and rate. This phenomenon is deeply rooted in historical, cultural, political, and economic processes, because living conditions in different places or different countries vary, so the differences are often staggering.

Although inequality is often associated with the concept of economic inequality, it is a phenomenon with a much broader basis, relating to inequality in access to priority human needs, access to water, health care, education, work, and decent living conditions. However, the issue of economic inequality cannot be ignored. This phenomenon intensifies and is visible in periods of special challenges for countries, economies or social systems related to global economic crises (e.g., the Great World Crisis, 2007–2011) or the crisis related to the SARS-CoV2 virus pandemic. Albeit there are various methods of measuring economic inequality, the obtained results indicate possible causes related to diversified access to material goods leading to the economic inequalities within a given socio-political system, which to a large extent impacts the functioning of individuals and influences the quality of their lives. Economic inequalities and social inequalities are closely connected and interact with each other.

The already complicated image of possible sources of inequality is also overlapped by the phenomenon of technological inequality and the resulting social and professional exclusions. The SARS-CoV2 virus pandemic highlighted the problem of technological exclusion of elderly people who do not have sufficient competences to use information technology in everyday life, such as the use of a computer for bank transfers, registration for vaccinations, participation in classes via educational platforms. IT technology, which has become an element of everyday lives for the younger society, has become an inadequate obstacle for the older age group.

So, Inequality is a complex issue felt by every modern society. Inequalities are found in the aforementioned levels of income and consumption, access to information, employment, culture and the degree of participation in public life. As a consequence, large groups of people suffer economic and social exclusion which, in turn, triggers changes in values of a given society. The literature on the subject also indicates that the basis of economic inequalities that may turn into social inequalities lie in the personal characteristics of individuals and their tendency to compete for their property status. This tendency to compete along with the progressing technological exclusion has a significant impact on the progressive stratification of societies.

The issue of inequalities is the subject of numerous scientific considerations, activities undertaken by public institutions, public debates—concentrated on their elimination or limitation. The book by Professor Medani P. Bhandari and Associate Prof. Shvindina Hanna refers to the timeless issues related to unbeatable challenge of inequality. The book covers important issues related to, inter alia, consequences of inequality, gender inequality, Covid pandemic as an accelerator of inequality, problems with access to education and healthcare, social exclusion, financialization as a determinant of income inequality, poverty and discrepancies in access to housing and poverty status of households. The authors of the book attempt to answer the question of what needs to be done to reduce the scale of inequality in the world. This book can be recommended for those readers who are interested in the subject of inequality and / or for those involved in the elimination of strong economic and social imbalances.

Prof. Jacek Piotr Binda,
PhD Rector of the International Affairs, Bielsko-Biala School of Finance and Law, Poland

The themes of social inequality discussed in this book take on particular significance at the time of its publication, i.e., during the global pandemic of SAR-CoV-2 coronavirus. The power of the phenomenon of social inequality has as its essence the isolation of human beings in their multidimensional existence. This means that the cause-and-effect relationship will touch upon the issues of alienation, loneliness, rejection, lack of perspective. However, all these barriers have their origin in man, in his mentality and in his upbringing. The chapters of the book you are about to read fully reflect the depth of the issues raised, but they also encourage the reader to reflect further: where is my place in this process?

The chapters presented here illustrate, in significant part, a certain path followed by modern man in various fields of life and in different parts of the world. The wide representation of views gives the reader the opportunity to understand that the problem of inequality is a transnational issue relating not only to differences in economic and social existence. In addition to the general theoretical views accepted in science, each article shows the influence of cultural realities in shaping the problem of social inequality. Authors from different countries often unknowingly grasp the meaning of given scientific considerations in relation to the background in which they themselves live or which they are familiar with. This has its charm and beauty, because it enriches global theories and gives them a meaning that can be found by readers in every corner of the world. The solutions presented in most of the chapters also outline the power of creation over progressive social decay, which stems not from biological factors but from human behavior motivated by negative emotions.

Finding new opportunities in theories that have already been described hundreds of times gives hope that intergenerational change will bring not only new technological, medical or natural solutions, but also social ones. For the reason is that even in the phenomenon of social inequality or the related phenomenon of social exclusion we can see great opportunities for the development of the labor market, new forms of work, new forms of family development, entrepreneurship, agriculture, NGOs, or CSR. The very fact of the creation of this book, whose authors are from various countries of the world, gives great hope for a change in social attitudes, which will lay the foundations for reducing social inequalities and building a society that will see exclusion as an opportunity rather than a problem.

I would like to thank the editors of the book, Professor Medani P. Bhandari from the United States and Dr Hanna Shvindina from Ukraine, for compiling all the material and editing such a wonderful edition. Despite the distance between your countries, you have managed to edit a wonderful work that will do much good to the deliberations on social inequality.

Dr. Stanisław Ciupka,
Dean of Faculty of Law and Social Science, Bielsko-Biała School of Finance and Law, Poland

The book "Inequality—the Unbeatable Challeges" edited by Prof. Medani Bhandari and Prof. Shvindina Hanna makes important contributions in the study of the various forms of inequalities in the world and towards formulating policies to mitigate the problem.

Especially, in the past two centuries, the world has witnessed an astonishing level of economic advancement. But the growing economic inequality has remained a stubborn problem for the humanity. According to data from the International Monetary Fund (IMF), the average global per capita annual income in 2021 was around $11,770 in current US dollars. While this average figure represents an amazingly high level of income, many countries and population groups are still abjectly poor. For example, according to the IMF data for 2021, Burundi had a per capita annual income of a mere $267 and South Sudan $322 compared to $126,000 for Belgium for the same year. Dozens of other countries have annual per capita income of less than US $1000.

There remains a stark and painful inequality of income among the people in both poor and rich countries. For example, according to the 2019 US census data, there were 34 million people (10.5% of the total population) in the United States who lived below the poverty line. The Bloomberg organization estimates this number to be 40 million in 2020.

While the world is witnessing an unprecedented level of prosperity where billions of people are enjoying an unprecedented high standard of living, poverty is also equally pervasive in all countries. The dramatic levels of wealth equalities can be seen in Oxfam International's report of 2020. The report notes: "There were 2,153 billionaires in 2019, and together they have the same amount of wealth as of the poorest 4.6 billion people in the world." The World Bank estimates that in 2015, almost 690 million people globally lived on less than $1.90/day, the updated measure of extreme poverty.

The current Covid-19 pandemic has also more blatantly exposed the deep-rooted inequalities in the world. During the current difficult public health crisis, the mega rich individuals have continued to prosper while the poor have fallen even deeper into the poverty. The Oxfam report (2021), "The Inequality Virus" states "The increase in the wealth of the 10 richest billionaires since the crisis began is more than enough to prevent anyone on Earth from falling into poverty because of the virus and to pay for a COVID-19 vaccine for all." While many poor individuals in the developing world and also in rich countries have been facing existential crisis imposed by the pandemic, many businesses and rich individuals have made a fortune during the same period.

To look at another example, as of April 2021, while many in the rich countries are getting vaccines for Covid-19, most poor countries are still to see a single dose of vaccine into the arms of their citizens. Several possible strategies would have reduced such a dramatic inequality in access to vaccine. For example, the big pharma companies which created the Covid vaccine could have waived the patents and intellectual property rights on the vaccine at least for several years during the pandemic. Although the vaccine companies that invested in the creation of the vaccine should be rewarded, there are several ways to maintain the incentives for companies to invest in research while helping to eradicate one of the worst public health crises in our lifetime.

The World Health Organization (WHO), and several advocacy groups for low-income countries have asked the big pharma companies and rich countries to suspend the intellectual property rights related to the Covid vaccines temporarily, and let the manufacturers in the developing world help make the vaccines rapidly. Beyond the temporary lifting of the patents, developing countries are also asking that rich countries share the vaccine doses with them. There is also a bottleneck in the supply chain of ingredients and tools related to the manufacture of vaccines in the developing world. For example, when the United States put an embargo in exporting ingredients and supporting items such as such as specialized bags and filters the Indian companies had to drastically reduce the amount of vaccine, they were able to produce. The United States government invoked the Defense Production Act to halt the export of items used in vaccine manufacturing. This is yet another example of the acute asymmetry among the countries in power, resources, and know-how.

A declaration led by the WHO and signed by hundreds of multilateral aid, non-profit organizations, universities, and UN-related agencies organizations have specifically called for several actions by the national governments, companies, and global leaders to facilitate the distribution and availability of vaccine worldwide. Recommended actions included support for the COVAX facility, a joint vaccine manufacturing facility funded and supported by dozens of rich countries. The WHO also asked vaccine manufacturers to share know-how to help drastically increase the global supply of vaccines. It requested governmental agencies to expedite the approval processes, and to provide Covid vaccine free of cost to the consumer. These measures would help mitigate vaccine inequalities up to some extent.

Traditionally, political and economic systems have also been considered as major factors that contribute towards an increase or reduction in inequalities. The conventional wisdom has been that capitalism is built to intrinsically create inequality to provide impetus for individuals to work hard, innovate and be enterprising to become richer than others. Socialism and communism were considered as systems that would bring about equality and fairness in the society. However, the results of these systems in actual practice have been mixed. Many erstwhile socialistic and communist countries often proved to be even more unequal than the well-developed capitalist societies. In the socialist societies, power was concentrated in the hands of a few leaders who monopolized the state authority, and who were often corrupt and prone to abuse their power to personalize the gains from the labor of the workers. This took place while the vast majority of the citizens in those countries remained impoverished and powerless.

On the other hand, the capitalist or market-driven economies often implemented minimum wage for the workers, and the workers' unions there frequently negotiated better salaries and working conditions for them. Many market economies also instituted social protection mechanism, free school level education and universal medical care. More importantly, the generally increased national wealth lifted many individuals from the situation of abject poverty like a tide lifting all the boats. However, there remains a wide gulf between the rich and poor in the capitalist and market economies too. In some countries, the gap between the rich and poor continues to expand. For example, in the United States, which is generally considered the richest and the most capitalist country in the world, a large proportion of the citizens are poor and millions face a hopeless situation. The Population Reference Bureau (PRB), a reputed US-based non-profit organization, estimated that in 2019, there were between 600,000 and 1.5 million homeless individuals in the United States.

Inequality is not only related to income, wealth, and political power, but also to access to know-how, information, and technology. These factors often follow the inequality of income and wealth and create a positive feedback loop to widen inequalities among countries and social groups. In addition, the world continues to witness social inequalities in the form of discrimination based on race, caste, and creed. Even in well-functioning democracies and advanced economies, such social inequalities persist.

Here, I would like to share some personal story. I have experienced first-hand the impact of poverty on human well-being while growing up in the remote hills of eastern Nepal in the mid-twentieth century. The hand-to-mouth existence, the need for back-breaking physical labor and lack of any opportunities for upward mobility left people in a state of perpetual poverty and despair. A poor person did not get enough to eat, or ate poor quality food. The frequent malnutrition, perpetual physical weakness, and poor sanitation-related pervasive health issues such as hookworms, body lice and bed bugs deprived one of any sense of well-being. People who were sunk deep in poverty routinely experienced major and chronic health issues such as tuberculosis, chronic gastritis, various forms of skin diseases, low vision, and even blindness. Class and caste discrimination made the situation even worse, and robbed many of their personal dignity.

The feudalism existing in those days in Nepal was the driving force for the pervasive equality in the society. Just a handful of families controlled the meager resources of the country, and exploited the common persons by extracting their labor and services for a very low cost.

Compared to the mid-twentieth century period, the economic and social situation in Nepal has dramatically improved today. Now, people are better fed; they wear decent clothes and shoes, and have access to basic health care most of the time. However, the systemic inequality of income, wealth and social class not only continues until today, but has even worsened in some cases such as in remote areas, and some urban slums. The old feudal system seems to be sometimes just replaced by the new system of employers and employees. In many instances, the daily wage-earners and informal employees receive the short end of the bargain and do not have many opportunities to negotiate for salaries or benefits.

The successful democratic changes brought about in Nepal first in 1951 and later in 1990, have greatly helped the society prosper and individuals improve their socio-economic status. With a relatively free press and informed citizenry, the working-class today is better prepared to bargain for rights and fair wages. Trade unions with their collective bargaining abilities have helped in the improvement of wages, working conditions and benefits for most workers. Democracy greatly helps toward reducing inequality and improving individual well-being, while economic development should also help in making people freer. In his seminal book "Development as Freedom (1999)" Amartya Sen states "Freedoms are not only the primary ends of development, but are also among its principal the means."

The story of changes in Nepal is similar to the transformations happening in many other developing countries. Most developing countries are gradually climbing up into the economic ladder enabling their citizens to live better and more prosperous lives than their parents or grandparents did. However, many countries in the lowest rung of economic development continue to be stuck in poverty traps and internal inequalities.

Poverty and wide social inequalities adversely impact human dignity and health and impart a sense of hopelessness for the poor. A person in entrenched poverty loses self-confidence, has a difficult time taking care of oneself and one's family, and misses out on many activities that can bring joy and fulfilment in life. Such a person cannot live up to his or her full potential. With relative low life-expectancy, their lives are often short-lived. Inequality can bring extreme differences in income, power, and status in the society. A conflicted, polarized, and divided society is not good for any of its members: rich, poor, or middle class. Everyone feels unsafe and exposed to external threats and lives in constant feeling of insecurity and vulnerability.

Prof. Medani Bhandari and Prof. Shvindina Hanna, as editors, have excellent background as educators and practitioners in social, environmental, and economic development arena to create this useful book as co-editors.

Bhandari began his education and career in humble settings in rural Nepal, where he created a non-profit organization to mobilize the community in the fields of environmental improvement and social upliftment. During this period, I had the opportunity to work with Bhandari, and got to know him and appreciate his passion for community development and environmental management. After several years of field work in Nepal, he moved to the United States for higher education and professional pursuits. His experience in a number different countries has provided him with a comparative perspective on how societies function and how social inequalities are a big challenge to the societies. He was also able to observe how social and economic inequalities can impact people in their abilities to realize the full potential as members of the society.

Hanna grew up and educated in Ukraine, and later spent time in France and USA pursuing higher education. In co-editing this book, she has utilized her education and training in economics, management, and international development. Her collaboration with Bhandari has brought about several other books related to the environment and inequalities.

In this book, many prominent authors from various countries have provided a rich variety of stories and case studies from their experience, study, and research. With their inputs, the book offers a rich compendium of comparative knowledge and insights into the issues of social, political, and economic inequalities in different settings.

Students, academics, researchers, policymakers, and laypeople who are interested in the topic of inequality and development can greatly benefit from this book.

Dr. Ambika Prasad Adhikari, DDes.
Urban Planner, and International Development Professional Phoenix, Arizona, USA, April, 2021

The 13-chaptered book entitled "Inequality—The Unbeatable Challenge" edited by Profs. Medani P. Bhandari and Shvindina Hanna includes contributions from 25 authors of various disciplines. Concerned with the widening income differences, this book presents various facets of inequalities.

The book states that inequality was, is, and will remain as an unsolved and unbeatable problem. There is still lack of knowledge how, inequality has been widening throughout the human civilization, why society is stratified economically, politically, socially, and religiously and why the discrimination due to gender orientations, country of origin, caste, race, and ethnically dividing society still prevail. This book addresses these issues in a wholistic manner and then presents various case studies exploring why inequality has been unbeatable and what would be the best policies to overcome this challenge.

The book starts how inequalities prevails from social inclusion and sustainable development citing examples from Nepal. Although various slogans are postulated by political parties in Nepal, government leaders have been unable to address inequalities even in their high-sounding agricultural transformation where gender issues influence wage rates. Other issues discussed in this book are inequality on financial development and green investment. Discussing financial inequalities, concerns are expressed on the growing shadow economy that has directly impacted cultural, social, educational, and health services to name a few. Cultural inequalities include disparities in the recognition and standing of the language, religion, customs, norms, and practices of different groups. It presents examples from Turkey where the Syrian refugees are facing cultural inequality emerging from the social and religious discrimination even in the age of globalization.

To make this argument clearer, the book brings the issues of Covid-19 lockdown, how this has become a luxury to someone while it has become a misery for others who must live on their day-to-day blue-collar work-based incomes. The book links this inequality with the emerging technology and brings an issue of "haves" and "havenots" where the "haves" groups enjoy the technology and the "havenots" group either suffer from the lack of access to advance technology, and thus, get infected due to the direct contact from Covid-19 at workplaces, and their income and health deteriorate. Dwindling economies obviously bring disparities in education and health. With the "haves" and "havenots" scenarios, the book brings the dilemma faced by the households living in the vicinities of national parks in Nigeria. Though the households living nearby national parks have plenty of natural resources and beautiful sceneries, their survival needs are different from the elite groups. The well-to-do /elites people enjoy the beauty of nature whereas the people living nearby national parks become the true sufferers of animal raiding on their crops. The book agues how despite people involving in the conservation and management of natural resources are facing the brunt of inequality where tourist entrepreneurs are taking economic advantages at the suffering of poor. The book argues that only sound policy that incorporates the reality of ecosystem services can address such issues to ameliorate inequality.

The book overviews the role of the United Nations, and it suggests that the UN can act as a key stakeholder to energize the governments, the World Bank, the Regional Development Banks, governmental and nongovernmental national and international organizations, and academic institutions. It presents scenarios how technological accesses among "have" can create inequalities in health and economies if policies are not designed properly. The 21st century technology is available to "haves" group whereas the "havenots" groups are suffering from the lack of its access. Obviously, this has created disparity in education, manufacturing, and getting into white collar jobs, and the emerging pandemics such as covid-19 has created vicious cycles where "havenots" groups are facing the brunt under bad government policies.

This book suggests that the governments should address the social inclusion component in their developmental projects targeting the marginalized and poverty-stricken population in their capacity-building if they are serious to address the inequality, social justice, and marginalization. The arguments are that despite the United Nations (2015) clearly acknowledging the existence of various forms of inequality and its Goal 10 explicitly dealing with inequalities, it has become pervasive, thus, individual governments must involve in mitigating inequalities through policy instruments. The UN aims to achieve universal and equitable access to safe and affordable drinking water for all by 2030 with adequate and equitable sanitation and hygiene for all and end open defecation by then. Goal 7 of the UN aims to ensure affordable, reliable, sustainable, and modern energy for all. Overall, the UN goals are aimed to clearly identify the major indicators of inequality based on gender, age, disability, race, ethnicity, origin, religion, and economy to eliminate any discriminatory laws, policies, and practices. Despite high sounding goals of the UN and the world being trapped under covid-19, the actions should be at the local levels, thus, the governments must be actively involved to minimize the traumatic state of widening inequality.

Other forms of inequality this book discusses is the lack of access to financial ownership and natural resource-based incomes, social capitals, and political inequalities. In political inequalities, it discusses the distribution of political opportunities and power among groups, such as control over local, regional, and national institutions of governance, the army, and the police. The book further explains how political inequality has created limited people's capabilities to participate politically and express their genuine needs.

Pervasive discriminations, social exclusions and poverty, the economic growth, community resiliency, and socio-economic transformation must be dealt with from the multidimensional socio-economic developmental perspectives while improving infrastructure that directly touch the living conditions of everyone in a day-to-day life. While forecasting the inequality, the book suggests a radical tax approach and suggest expediting the roles of local organizations to mitigate inequality. The book makes attempt to justify this approach citing examples Ukraine. The book talks about the nexus of development status among countries and argues how development and poverty spillover through spatial associations. The book critically presents the current model of profit making how everyone is racing for profit. As profit making motive continues, resource poor persons always fall behind and resource rich people take benefit of opportunity cost which widens inequality. Profit making tendency further exacerbates even in the globalization era thus, the government must involve to curb inequalities through policy instruments.

The book also argues the Green Investment as an economic instrument to achieve sustainable development that not only utilizes untapped resources but also helps in global climate mitigation citing examples from Ukraine. The book is equally concerned with the emerging shadow economy where wrong policy instruments will gravitate the control of green economy at the hands of elite group and inequality widens further as resources are skewed at the hands of elites. The book is much concerned with the pain of Covid 19, how it is triggering inequality that will have long-lasting effects in multisectoral areas of present and future society. Readers and policy makers interested in exploring the root causes of inequality may find this book very useful.

April 17, 2021

Keshav Bhattarai,
Ph.D. Professor of Geography, School of Geoscience, Physical & Safety
University of Central Missouri, Humphreys 223CWarrensburg, MO
64093bhattarai@ucmo.edu,
Phone: 660-543-8805, Fax: 660-543-8142

Endorsements

This book "Inequality—The Unbeatable Challenge" introduces the reader to the problems of inequality in the world, and the possibilities of solving unsolvable problems. All sections are lined up in a certain sequence, are a reflection of the developments of different authors' teams, answer difficult questions and raise new ones. This book is the result of many years of collaboration of volunteers, activists, sociologists, environmentalists, and experts in policymaking aimed at a solution to the problem of inequality. This book is a continuation of discussions at the global level, and I think it has answers for some local challenges. In continuation, I'd like to say that I have had experience working with the editors of this book, and I believe that their work and efforts should be appreciated at the highest level.

Professor **Yevgen KHLOBYSTOV** National University of Kyiv-Mohyla Academy/University of Economics and Humanities (Poland)
Ievgen Khlobystov
Contacts: National University of Kyiv-Mohyla Academy 2 Skovorody vul., Faculty of Natural Sciences, Department of Environmental Studies, Kyiv 04070, Ukraine
E-mail: ievgen.khlobystov@ukr.net

The book "Inequality—The Unbeatable Challenge" edited by Professor Medani P. Bhandari and Associate Professor Hanna Shvindina should be acknowledged among the scholars, practitioners, policy-makers, environmentalists, and sociologists. This book brought the readers closer to an understanding of inequality disaster and the devastation that could certainly be less if the sustainable development goals were taken into account. It is my pleaser to right an endorsement for the book that united so many authors from different fields of study and location of research. The book is the result of excellent work and coordination of editors and deep investigations of the scholars who joined the project. The ideas presented in the book and discussed phenomena interlinks between causes and casualties of social injustice and inequality may become a source of valuable insights for investigators in the field in the future.

Liubov Mykhailova
Professor of Department of Management, Sumy National Agrarian University

Acknowledgements

First, I would like to thank my colleague, co-editor of this volume Associate Professor Shvindina Hanna and all members of her family (parents, Olexandr and Valentina, and best friend and supporter Arseniy Prokhasko) and Prof. Oleksandr Telizhenko, Prof. Volodymyr Boronos, Prof. Tetyana Vasilyeva, Prof. Oleg Balatskyi (her mentor 1999–2012). I should acknowledge that, she is the key manager of this project—making initial book plan to establishing contacts with authors—collecting manuscripts, and other editorial tasks as needed.

Equally, I would also like to thank all authors who contributed to this volume—Drs./ Profs. Durga D. Poudel; Anna Rosohata and Liubov Syhyda; Olha Kuzmenko, Anton Boyko and Victoria Bozhenko; Olena Chygryn, Tetyana Pimonenko, Oleksii Lyulyov; Sergij Lyeonov, Tatiana Vasylieva, Inna Tiutiunyk and Iana Kobushko; Kemal YILDIRIM; Teletov Aleksandr Sergeevych, and Teletovà Svetlana Grigorievna; Man Bahadur Bk; Prem B. Bhandari; Daniel Etim Jacob and Imaobong Ufot Nelson; Aleksander Sapiński, Sabina Sanetra-Półgrabi, Serhii Y. Kasian, Nataliya Rozhko for your contribution in the book. Each of you have done an excellent job, your scholarly work bridges the knowledge gap in the complex field—inequality with reference to sustainable development goals.

I would also like to thank Prof. Douglas Capogrossi (my mentor, Emeritus, President of Akamai University, Hawaii), Dr. Mary Jo Bulbrook, (President, Akamai University), Dr. Ambika Adhikari (mentor for environment conservation), Dr. Krishna Prasad Oli (Member of National Planning Commission, Government of Nepal); Prof. Keshav Bhattarai (Arizona State University), Mr. Kedar Neupane (former staff of UNHCR), Dr. Jacek Piotr Binda and Dr. Stanisław Ciupka, (Rector, and Dean, of Bielsko-Biała School of Finance and Law, Poland respectively), Prof. Yevgen Khlobystov, National University of Kyiv-Mohyla Academy, Ukriane, Prof. Liubov Mykhailova, Sumy National Agrarian University, Ukraine and other for their reviews, forwards, and endorsements. Your togetherness with us adds value in this book project.

I would like to remember Prof. Bishnu Paudel (Guru of all of us, who passed away due to heart attract in early 2020, when Corona pandemic was just at the beginning stage. His understanding of Bashudhiva Kutumbakam (all living beings are our relatives and neighbors). He used to say that the essence of Bashudhiva Kutumbakam is to see everyone as you see yourself, love everyone as you love yourself. He always used to say that any contribution towards addressing the challenges like inequalities, climate change, sustainability, pandemic, and other social, economic, political, and natural challenges are the pathway to change human behavior towards Bashudhiva Kutumbakam (all living being are our relatives and neighbors). If we can understand the real meaning of Bashudhiva Kutumbakam and could make others aware about this philosophy, the challenges of inequality could turn the opportunity to service of mankind.

I would also like to thank my wife Prajita Bhandari for her encouragements and for giving insightful information on how women are victims of inequality—at home to work environment. Without your support, this project was impossible. I would also like to thank our son Prameya, daughter Manaslu, and daughter-in-law Kelsey for helping me to find relevant resources in the field of inequality, social division, stratifications, and gender issues. I would also like to thank to our granddaughter for giving us a joyful environment, which always helps to concentrate in the work I am tuned in. I would also like to thank our son-in-law Abhimanyu Iyer, and his parents Mahesh and Uma Iyer for insightful thoughts on inequality issues. I would also like to thank my mother Hema Devi, brothers Krishna, Hari, sisters Kali, Bhakti, Radha, Bindu, Sita, and their families for encouraging me by providing a peaceful environment. I would also like to thank my friends Kshitij Prasai, Prem Bhandari, Man Bahadur Khari, Rajan Adhikari, Govinda Luitel, Medini Adhikari, Guna Raj Luitel, Parshuram Bhandari, Prof. Binod Pokhrel, Prof. Gopi Uprety, Tirtha Koirala, Prof. Sanjay Mishra, Bijay Kattel (my mentor for wildlife conservation in Nepal), Dr. Aleksander Sapinski (Poland), Dhir Prasad Bhandari and all who have been always encouraging to us to give back to the society through knowledge sharing. Thank you to Rajeev Prasad, Junko, and all friends of River Publishers, for encouraging and empowering us to complete this book project on time. Thank you to you all, who have given their inputs for this book project directly or indirectly.

Thank you,
Prof. **Medani P. Bhandari**, Ph.D.

List of Figures

List of Tables

List of Contributors

Bahadur Bk, Man, *Former Chief Secretory, Bagmati Province, Government of Nepal, Nepal*

Bhandari, Medani P., *Professor and Advisor of Gandaki University, Pokhara, Nepal, Prof. Akamai University, USA and Sumy State University, Ukraine*

Bhandari, Prem B., *Managing Director of South Asia Research Consult, Inc. Michigan, USA*

Boyko, Anton, *Assistant Professor at the Department of Economic Cybernetics at Sumy State University, Sumy, Ukraine*

Bozhenko, Victoria, *Assistant Professor at the Department of Economic Cybernetics at Sumy State University, Sumy, Ukraine*

Chygryn, Olena, *Associate Professor, Economics, Entrepreneurship and Business Administration Department, Sumy State University, Sumy, Ukraine*

Grigorievna, Teletova Svetlana, *Associate Professor of the Department of Russian Language, Foreign Literature and Methods of their Teaching, Sumy State A. S. Makarenko Pedagogical University, Sumy, Ukraine*

Jacob, Daniel Etim, *Forestry and Wildlife Department, University of Uyo, Nigeria; E-mail: danieljacob@uniuyo.edu.ng*

Kasian, Serhii Y., *Associate Professor, Head of Marketing Department Dnipro University of Technology, Ukraine; E-mail:kasian.s.ya@nmu.one*

Kobushko, Iana, *Senior Lecturer of the Management Department, Sumy State University, Sumy, Ukraine*

Kuzmenko, Olha, *Head of the Department of Economic Cybernetics at Sumy State University, Sumy, Ukraine*

Lyeonov, Sergij, *Doctor of Economics, Professor of the Economic Cybernetics Department, Sumy State University, Sumy, Ukraine*

Lyulyov, Oleksii, *Associate Professor, Economics, Entrepreneurship and Business Administration Department, Sumy State University, Sumy, Ukraine*

Nelson, Imaobong Ufot, *Biodiversity Preservation Center, Uyo, Nigeria*

Pimonenko, Tetyana, *Associate Professor, Economics, Entrepreneurship and Business Administration Department, Sumy State University, Sumy, Ukraine*

Poudel, Durga D., *The Founder of Asta-Ja Framework, Professor, University of Louisiana at Lafayette, Lafayette, Louisiana, USA; E-mail: durga.poudel@louisiana.edu*

Rosohata, Anna, *Senior Lecturer, Department of Marketing, Sumy State University, Sumy, Ukraine*

Rozhko, Nataliya, *Associate Professor, Ternopil Ivan Puluj National Technical University, Ukraine*

Sanetra-Półgrabi, Sabina, *Pedagogical University of Krakow, Poland; E-mail: olek.sapinski@interia.pl*

Sapiński, Aleksander, *Lecturer at Bielsko-Biala School of Finance and Law, Poland; E-mail: olek.sapinski@interia.pl*

Sergeevych, Teletov Aleksandr, *Professor of the Department of Public Management and Administration, Sumy National Agrarian University, Sumy, Ukraine*

Shvindina, Hanna, *Sumy State University, Ukraine*

Syhyda, Liubov, *Senior Lecturer, Department of Marketing, Sumy State University, Sumy, Ukraine*

Tiutiunyk, Inna, *Associate Professor, Senior Lecturer of the Finance and Entrepreneurship Department, Sumy State University, Sumy, Ukraine*

Vasylieva, Tatiana, *Doctor of Economics, Professor of the Finance and Entrepreneurship Department, Director of Oleg Balatskyi Academic and Research Institute of Finance, Economics and Management, Sumy State University, Sumy, Ukraine*

Yildirim, Kemal, *European School of Law and Governance, Kosovo; E-mail: conflictresearch@yahoo.com*

List of Abbreviations

CAMP	Coastal Area Management Program
COVID-19	Infectious disease caused by the strain of coronavirus SARS-CoV-2
CERF	UN Office for the Coordination of Humanitarian Affairs
EU	European Union
FAO	Food and Agriculture Organization
FIVIMS	Food Insecurity and Vulnerability Information and Mapping Systems
HIV	Human Immunodeficiency Virus
IBRD	International Bank for Reconstruction and Development (World Bank Group)
ISSC	International Social Science Council
IISD	International Institute for Sustainable Development Sustainable Development
ILO	International Labor Organization
IPCC	Intergovernmental Panel on Climate Change
IUCN	International Union for Conservation of Nature and Natural Resources
LGBTQ+	Lesbian, Gay, Bisexual, Transgender, and Queer and/or Questioning
MDG	Millennium Development Goal
MEA	Millennium Ecosystem Assessment
OCHA	UN Office for the Coordination of Humanitarian Affairs
ODI	Overseas Development Institute
PAP	Priority Actions Program
RAC	Regional Activity Centre
SA	South Asia
SD21	Sustainable Development in the 21st Century
SDGs	Sustainable Development Goals
UN	United Nations

UN DESA	United Nations Department of Economic and Social Affairs
UN ESCAP	Economic and Social Commission for Asia and the Pacific
UNAIDS	Joint United Nations Programme on HIV/AIDS
UNCCD	United Nations Convention to Combat Desertification
UNDP	United Nations Environmental Program
UNEP	United Nations Environment Programme
UNESCO	United Nations Educational, Scientific and Cultural Organization
UNHCR	Office of the United Nations High Commissioner for Refugees
UNHCR	UN High Commissioner for Refugees
UNICEF	United Nations Children's Fund
Vasudhaiva Kutumbakam	The entire world is my home and all living being are my relatives
WB	World Bank
WFP	World Food Program
WHO	World Health Organization
WWF	World Wildlife Fund

Word Phrase Definitions

20:20 Ratio—It compares the ratio of the average income of the richest 20 per cent of the population to the average income of the poorest 20 per cent of the population. Used by the United Nations Development Programme Human Development Report (called "income quintile ratio") (United Nations 2015).

ASTTA-JA—framework (interconnectedness of eight natural resources) *Ja* in Nepali letter, *Jal* (water), *Jamin* (land), *Jungle* (forest), *Jadibuti* (medicinal and aromatic plants), *Janshakti* (manpower), *Janawar* (animal), *Jarajuri* (crop plants), and *Jalabayu* (climate).

Atkinson's Inequality Measure (or Atkinson's Index)—This is the most popular welfare-based measure of inequality. It presents the percentage of total income that a given society would have to forego in order to have more equal shares of income between its citizens. This measure depends on the degree of society aversion to inequality (a theoretical parameter decided by the researcher), where a higher value entails greater social utility or willingness by individuals to accept smaller incomes in exchange for a more equal distribution. An important feature of the Atkinson index is that it can be decomposed into within and between-group inequality. Moreover, unlike other indices, it can provide welfare implications of alternative policies and allows the researcher to include some normative content to the analysis (Bellù, 2006) (United Nations 2015).

Beijing Declaration and Platform for Action (1995)—Adopted at the Fourth World Conference on Women in September 1995, comprehensive commitments to women are called for under 12 critical areas of concern: poverty, education and training, health, violence against women, armed conflict, the economy, power and decision-making, institutional mechanisms, human rights, media, environment, and the girl child (UNICEF 2017).

Climate Change—(a) The Inter-governmental Panel on Climate Change (IPCC) defines climate change as "a change in the state of the climate that can be identified (e.g., by using statistical tests) by changes in the mean and/or the variability of its properties, and that persists for an extended period, typically decades or longer. Climate change may be due to natural internal processes or external forcing, or to persistent anthropogenic changes in the composition of the atmosphere or in land use" (IPCC).

Climate Change—Includes both global warming and its effects, such as changes to precipitation, rising sea levels, and impacts that differ by region (IPCC).

Coefficient of Variation—The coefficient of variation is the square root of the variance of the incomes divided by the mean income. It has the advantages of being mathematically tractable and its square is subgroup decomposable, but it is not bounded from above United Nations 2015).

Commission on the Status of Women (CSW) (1946)—The main global intergovernmental body exclusively dedicated to the promotion of gender equality and the empowerment of women. At its sixtieth session, in 2016, the Commission passed resolution 60/2, on women, the girl child and HIV and AIDS. At its sixty-first session, in 2017, the Commission urged governments to mainstream gender perspectives in education and training, including science, technology, engineering and math (STEM), develop gender-sensitive curricula, eradicate female illiteracy and facilitate girls' and women's effective transition to work. Girls with disabilities and their right to education were highlighted in conclusions adopted at both sessions (UNICEF 2017).

Convention on the Elimination of All Forms of Discrimination Against Women (CEDAW) (1979)—Adopted in 1979 by the United Nations General Assembly, "CEDAW is often described as an international bill of rights for women." Consisting of a preamble and 30 articles, it defines what constitutes discrimination against women and sets up an agenda for national action to end such discrimination. The Convention defines discrimination against women as "...any distinction, exclusion or restriction made on the basis of sex which has the effect or purpose of impairing or nullifying the recognition, enjoyment or exercise by women, irrespective of their marital status, on a basis of equality of men and women, of human rights and fundamental freedoms in the political, economic, social, cultural, civil or any other field." (UNICEF 2017).

CORONOMICS—The Economic Analysis Based on change scenario due to Covid-19.

COVID-19 death—Defined for surveillance purposes as a death resulting from a clinically compatible illness in a probable or confirmed COVID-19 case, unless there is a clear alternative cause of death that cannot be related to COVID-19 disease (e.g., trauma). There should be no period of complete recovery between the illness and death. Further guidance for certification and classification (coding) of COVID-19 as cause as cause of death is available in WHO (2020e) (WHO, 2020).

COVID-19—Infectious disease caused by the strain of coronavirus SARS-CoV-2 discovered in December 2019. Coronaviruses are a large family of viruses which may cause illness in animals or humans. In humans, several coronaviruses are known to cause respiratory infections ranging from the common cold to more severe diseases such as Middle East Respiratory Syndrome (MERS) and Severe Acute Respiratory Syndrome (SARS). The most recently discovered coronavirus causes coronavirus disease COVID-19 (WHO, 2020).

Cultural Inequality—Discriminations based on gender, ethnicity and race, religion, disability, and other group identities. (ISSC, IDS and UNESCO 2016).

Decile Dispersion Ratio (or Inter-decile Ratio)—It is the ratio of the average income of the richest x per cent of the population to the average income of the poorest x per cent. It expresses the income (or income share) of the rich as a multiple of that of the poor. However, it is vulnerable to extreme values and outliers. Common decile ratios include: D9/D1: ratio of the income of the 10 per cent richest to that of the 10 per cent poorest; D9/D5: ratio of the income of the 10 per cent richest to the income of those at the median of the earnings distribution; D5/ D1: ratio of the income of those at the median of the earnings distribution to the 10 per cent poorest. The Palma ratio and the 20/20 ratio are other examples of decile dispersion ratios (United Nations 2015).

Discrimination (Gender Discrimination)—"Any distinction, exclusion or restriction made on the basis of sex which has the effect or purpose of impairing or nullifying the recognition, enjoyment or exercise by women, irrespective of their marital status, on the basis of equality of men and women, of human rights and fundamental freedoms in the political, economic, social, cultural, civil or any other field" [United Nations, 1979. 'Convention on the Elimination of all forms of Discrimination Against Women,' Article 1] (UNICEF 2017).

Economic Inequality—Differences between levels of incomes, assets, wealth, and capital, living standards and employment. (ISSC, IDS and UNESCO 2016)

Empowerment—Refers to increasing the personal, political, social, or economic strength of individuals and communities. Empowerment of women and girls concerns women and girls gaining power and control over their own lives. It involves awareness-raising, building self-confidence, expansion of choices, increased access to and control over resources and actions to transform the structures and institutions which reinforce and perpetuate gender discrimination and inequality (UNICEF 2017).

Environmental Inequality—Unevenness in access to natural resources and benefits from their exploitation; exposure to pollution and risks; and differences in the agency needed to adapt to such threats; (ISSC, IDS and UNESCO 2016).

Food Insecurity—A situation that exists when people lack secure access to enough safe and nutritious food for normal growth and development and an active and healthy life. It may be caused by the unavailability of food, insufficient purchasing power, inappropriate distribution, or inadequate use of food at the household level. Food insecurity, poor conditions of health and sanitation, and inappropriate care and feeding practices are the major causes of poor nutritional status. Food insecurity may be chronic, seasonal, or transitory. (FIVIMS).

Food Security—A situation that exists when all people, at all times, have physical, social and economic access to sufficient, safe and nutritious food that meets their dietary needs and food preferences for an active and healthy life. (FIVIMS)

Galt Score—It measures the inequality of wage among workers. The Galt score is a simple ratio of a company's CEO pay to the pay of that company's Median worker. A company which pays its CEO many times more than its median employee will have a high Galt score (UNDP).

Gender—A social and cultural construct, which distinguishes differences in the attributes of men and women, girls, and boys, and accordingly refers to the roles and responsibilities of men and women. Gender-based roles and other attributes, therefore, change over time and vary with different cultural contexts. The concept of gender includes the expectations held about the characteristics, aptitudes and likely behaviors of both women and men (femininity and masculinity). This concept is useful in analyzing how commonly shared practices legitimize discrepancies between sexes (UNICEF 2017).

Gender Analysis—A critical examination of how differences in gender roles, activities, needs, opportunities, and rights/entitlements affect men, women, girls, and boys in certain situations or contexts. Gender analysis examines the relationships between females and males and their access to and control of resources and the constraints they face relative to each other (UNICEF 2017).

Gender and Development (GAD)—Gender and Development (GAD) came into being as a response to the perceived shortcomings of women in development (WID) programs. GAD-centered approaches are essentially based on three premises: (1) Gender relations are fundamentally power relations; (2) Gender is a socio-cultural construction rather than a biological given; and (3) Structural changes in gender roles and relations are possible. Central to GAD is the belief that transforming unequal power relations between men and women is a prerequisite for achieving sustainable improvements in women's lives. The onus is on women and men to address and re-shape the problematic aspects of gender relations. The conceptual shift from "women" to "gender" created an opportunity to include a focus on men and boys (UNICEF 2017).

Gender Balance—This is a human resource issue calling for equal participation of women and men in all areas of work (international and national staff at all levels, including at senior positions) and in programs that agencies initiate or support (e.g., food distribution programs). Achieving a balance in staffing patterns and creating a working environment that is conducive to a diverse workforce improves the overall effectiveness of our policies and programs and will enhance agencies' capacity to better serve the entire population (UNICEF 2017).

Gender Development Index (GDI)—The Gender Development Index (GDI) measures gender gaps in human development achievements in three basic dimensions of human development: (1) health (measured by female and male life expectancy at birth); (2) education (measured by female and male expected years of schooling for children and female and male mean years of schooling for adults ages 25 and older); and (3) command over economic resources (measured by female and male estimated earned income) (UNICEF 2017).

Gender Empowerment Measure (GEM)—Developed by the United Nations system in 1995, GEM measures inequalities between men's and women's opportunities in a country. An annually updated tool, it is used in formulating and applying gender equality indicators in programs. It provides a trends-tracking mechanism for comparison between countries, as well as for one country over time (UNICEF 2017).

Gender Equality—The concept that women and men, girls and boys have equal conditions, treatment, and opportunities for realizing their full potential, human rights, and dignity, and for contributing to (and benefitting from) economic, social, cultural, and political development. Gender equality is, therefore, the equal valuing by society of the similarities and the differences of men and women, and the roles they play. It is based on women and men being full partners in the home, community, and society. Equality does not mean that women and men will become the same but that women's and men's rights, responsibilities and opportunities will not depend on whether they are born male or female (UNICEF 2017).

Gender Gap—Disproportionate difference between men and women and boys and girls, particularly as reflected in attainment of development goals, access to resources and levels of participation. A gender gap indicates gender inequality (UNICEF 2017).

Gender Inequality Index—A composite measure which shows the loss in human development due to inequality between female and male achievements in three dimensions: reproductive health, empowerment, and the labor market. The index ranges from zero, which indicates that women and men fare equally, to one, which indicates that women fare as poorly as possible in all measured dimensions (Gender Inequality Index (GII), UNDP, 2015).

Gini Coefficient—Measure of the deviation of the distribution of income among individuals or households within a country from a perfectly equal distribution. A value of 0 represents absolute equality, a value of 100 absolute inequality (World Bank 2013).

Global Gender Gap Index—Index to measure one important aspect of gender equality—the relative gaps between women and men across four key areas: health, education, economy, and politics (The Global Gender Gap Report 2016).

Hoover Index—Hoover index (also known as the Robin Hood index, Schutz index, or Pietra ratio) It shows the proportion of all income which would have to be redistributed to achieve a state of perfect equality. In other words, the value of the index approximates the share of total income that has to be transferred from households above the mean to those below the mean to achieve equality in the distribution of incomes. Higher values indicate more inequality, and that more redistribution is needed to achieve income equality. It can be graphically represented as the maximum vertical distance between the Lorenz curve and the 45-degree line that represents perfect equality of incomes (United Nations 2015).

Human Immunodeficiency Virus—The virus infects cells of the immune system, destroying or impairing their function. Infection with the virus results in progressive deterioration of the immune system, leading to "immune deficiency." The immune system is considered deficient when it can no longer fulfill its role of fighting infection and disease. Infections associated with severe immunodeficiency are known as "opportunistic infections", because they take advantage of a weakened immune system (World Health Organization, WHO, 2017).

Human Rights—Agreed international standards that recognize and protect the inherent dignity and the equal and inalienable rights of every individual, without any distinction as to race, color, sex, language, religion, political or other opinion, national or social origins, property, birth, or other status (The Universal Declaration of Human Rights).

Inequality—Unfair situation in society when some people have more opportunities, money, etc., than other people. (Cambridge Dictionary, and FAO, 2014).

Inequality-Adjusted Human Development Index (IHDI)—The IHDI combines a country's average achievements in health, education, and income with how those achievements are distributed among country's population by "discounting" each dimension's average value according to its level of inequality. Thus, the IHDI is distribution-sensitive average level of human development. Two countries with different distributions of achievements can have the same average HDI value. Under perfect equality the IHDI is equal to the HDI but falls below the HDI when inequality rises (UNDP).

Knowledge-Based Inequality—differences in access and contribution to different sources and types of knowledge, as well as the consequences of these disparities. (ISSC, IDS and UNESCO 2016)

Labor—Binding labor law instruments include the 1973 Minimum Age Convention (No. 138), the 1999 Worst Forms of Child Labor Convention (No. 182), the 2011 Domestic Workers Convention (No. 189) of the International Labor Organization and the Protocol of 2014 to the Forced Labor Convention, 1930 (UNICEF 2017).

LGBTQ+—Umbrella term for all persons who have a non-normative gender or sexuality. LGBTQ stands for lesbian, gay, bisexual, transgender, and queer and/or questioning. Sometimes a + at the end is added to be more inclusive. A UNICEF position paper, "Eliminating Discrimination Against Children and Parents Based on Sexual Orientation and/or Sexual Identity (November 2014)," states all children, irrespective of their actual or perceived sexual orientation or gender identity, have the right to a safe and healthy childhood that is free from discrimination (UNICEF 2020).

Maternal Mortality—Death of a woman while pregnant or within 42 days of termination of pregnancy, irrespective of the duration and site of the pregnancy, from any cause related to or aggravated by the pregnancy or its management but not from accidental or incidental causes. World Health Organization (UN, 2014).

Palma Ratio—It is the ratio of national income shares of the top 10 per cent of households to the bottom 40 per cent. It is based on economist José Gabriel Palma's empirical observation that difference in the income distribution of different countries (or over time) is largely the result of changes in the "tails" of the distribution (the poorest and the richest) as there tends to be relative stability in the share of income that goes to the "middle" (Cobham, 2015) (as in United Nations 2015).

Patriarchy—Social system in which men hold the greatest power, leadership roles, privilege, moral authority and access to resources and land, including in the family. Most modern societies are patriarchies (UNICEF 2020).

Political Inequality—The differentiated capacity for individuals and groups to influence political decision-making processes and to benefit from those decisions, and to enter into political action (ISSC, IDS and UNESCO 2016).

Sexual Exploitation—Any actual or attempted abuse of a position of vulnerability, differential power, or trust, for sexual purposes. (UN, 2014). Sexual violence under international law encompasses rape, sexual slavery, enforced prostitution, forced pregnancy, enforced sterilization, trafficking and any other form of sexual violence of comparable gravity, which may, depending on the circumstances, include situations of indecent assault, trafficking, inappropriate medical examinations, and strip searches (ONU, 2014).

Social Inequality—Differences between the social status of different population groups and imbalances in the functioning of education, health, justice, and social protection systems (ISSC, IDS and UNESCO 2016).

Social Protection, or Social Security—is a human right and is defined as the set of policies and programs designed to reduce and prevent poverty and vulnerability throughout the life cycle. Social protection includes benefits for children and families, maternity, unemployment, employment injury, sickness, old age, disability, survivors, as well as health protection. Social protection systems address all these policy areas by a mix of contributory schemes (International Labor Organization, ILO 2020).

Son Preference—The practice of preferring male offspring over female offspring, most often in poor communities, that view girl children as liabilities and boy children as assets to the family (UNICEF 2017).

Spatial Inequality—Spatial and regional disparities between centers and peripheries, urban and rural areas, and regions with more or less diverse resources (ISSC, IDS and UNESCO 2016).

Substantive Equality—This focuses on the outcomes and impacts of laws and policies. Substantive equality goes far beyond creating formal legal equality for women (where all are equal under the law) and means that governments are responsible for the impact of laws. This requires governments to tailor legislation to respond to the realities of women's lives. Striving for substantive equality also places a responsibility on governments to implement laws, through gender-responsive governance and functioning justice systems that meet women's needs. Substantive equality is a concept expressed in the Convention on the Elimination of All Forms of Discrimination against Women (CEDAW). It recognizes that because of historic discrimination, women do not start on an equal footing to men (UNICEF 2017).

Theil Index—Theil index and General Entropy (GE) measures the values of the GE class of measures vary between zero (perfect equality) and infinity (or one, if normalized). A key feature of these measures is that they are fully decomposable, i.e., inequality may be broken down by population groups or income sources or using other dimensions, which can prove useful to policy makers (United Nations 2015).

United Nations Girls' Education Initiative (UNGEI)—A multi-stakeholder partnership committed to improving the quality and availability of girls' education and contributing to the empowerment of girls and women through education. The UNGEI Secretariat is hosted by UNICEF in New York City (UNICEF 2017).

Wage Share—The labor share is defined as the share of value added which is paid out to workers. It is therefore often also called the wage share. Generally, it is assumed that value added is produced with capital and labor as input factors so that Y = F(K, L) where Y is value added or output1 , K the capital input, and L labor (Dorothee Schneider 2011).

Women in Development (WID)—A Women in Development (WID) approach is based on the concept that women are marginalized in development-oriented interventions, with the result that women are often excluded from the benefits of development. Hence, the overall objective is to ensure that resources and interventions for development are used to improve the condition and position of women (UNICEF 2017).

1

Introduction—the Unbeatable Challenges of Inequality, Growing Problems, and Consequences

Medani P. Bhandari[1] **and Hanna Shvindina**[2]

[1]Professor and Advisor of Gandaki University, Pokhara, Nepal, Prof. Akamai University, USA and Sumy State University, Ukraine
[2]Sumy State University, Ukraine

Abstract

The chapter introduces the title of the book and outlines the coverage and unveils the challenges of inequality and presents why inequality is widening and how Covid-19 can be a triggering factor to make inequality an "unbeatable challenge," from today to the coming decades. The authors of this book argue that the inequality challenges are beatable, inequality cannot be eliminated; however, it can be reduced if all concerned stakeholders want the necessary change.

1.1 Introduction

"*Inequality—the state of not being equal, especially in status, rights, and opportunities—is a concept very much at the heart of social justice theories. However, it is prone to confusion in public debate as it tends to mean different things to different people. Some distinctions are common though. Many authors distinguish "economic inequality," mostly meaning "income inequality," "monetary inequality," or more broadly, inequality in "living conditions." Others further distinguish a rights-based, legalistic approach to inequality—inequality of rights and associated obligations (e.g., when people are not equal before the law, or when people have unequal political power)*" (United Nation 2018:1).

The theory behind the inequality books is still to explore why society is running within the paradigms of discrimination, social injustice, and exclusion toward marginalized groups by the mainstream social elites. Most of us might have faced, seen, or experienced similar situations and got into discomfortable scenarios. However, the reality is reality, and until or unless we identify the root cause of the problems, we could not solve them. It is important to accept the fact that, to some extent we are and will also be responsible for the widening inequality if we remain silent, did not identify the causal impact of our behavior within family and community, and society, and change motive to want to want to give, get service to give service the society will remain unchanged. The context of Covid-19, its impact, and the examples of widening inequality can be good lessons and even act as pathways to minimize the inequality and create a socially inclusive environment in the society.

"Never before has inequality been so high on the agenda of policymakers worldwide, or such a hot topic for social science research. More journal articles are being published on the topic of inequality and social justice today than ever before" (ISSC, IDS and UNESCO 2016:back cover page). As inequality has been a black shadow, a dividing factor of human civilization; it has been an important discourse of discussion in the academic as well as development domains (Barker, 2004; Beall, Guha-Khasnobis and Kanbur, 2012; Bhandari, 2018; 2019; 2020; Bhandari and Shvindina, 2019; Brockmann and Delhey, 2010; Carson, 1962; Cheng, 2018; Colyvas and Powell, 2006; Crossan & Bedrow, 2003; Daly, 2007; Desa, 2013; Freistein and Mahlert, 2015; Holvino & Merrill-Sands, 2004; IISD, 2005; IUCN, UNDP & WWF, 1991; Jackson, 2009; Jan-Peter et al., 2007; Harris, 2000; Lélé, 1991; Marshall, 1961; Marx, 1953ff; Meadows, 1998; Meadows et al., 1972; Norgaard, 1988; North, 1981; Purvis et al., 2018; Rockström, 2009; Scott, 2003; Soares et al., 2014; Sterling, 2010; United Nations, 1972; 1997; 2002; 2010; 2012, 2015, 2015 a, b, c, d; UNEP, 2010; UNESCO, 2013; Vander-Merwe & Van-der-Merwe, 1999; Vos, 2007; Wals, 2009; World Bank, 2015; WCED, 1987; WID, 2018; Wright, 1994; World Economic Outlook, 2007). Furthermore, in recent history, the major international agencies such as all development banks, especially the World Bank and Regional Development Banks have been implementing various programs to reduce inequalities (World Bank, 2011; World Bank, 2020; United Nations, 2020; World Economic Forum, 2020; ILO, 2020; UNDP, 2020). All of the United Nations line agencies have a special focus on reducing inequalities (United Nations, 1972; 1997; 2002; 2010; 2012, 2015, 2015 a, b, c, d; UNDP 2019).

The academic scholars have published thousands of papers and books about the problems and consequences of various types of inequalities in the societal level to global level (Blau and Kahn, 2017; Goldin & Rouse, 2000; Neumark, Bank & Nort, 1996; Olivetti & Petrongolo, 2008; Ortiz-Ospina 2018; Jayachandran, 2015). However, instead of reducing, inequalities continue to increase globally (Christiansen and Jensen, 2019; Bhandari and Shvindina, 2019; United Nations 2020a, b; United Nations, 2020; Dehm, 2018; Chhabria, 2016).

All concerned stakeholders of the world—the academia, media, development agencies accept the fact that inequality has been a major problem of social development (World Bank, 2011; 2020; United Nations, 2020; World Economic Forum, 2020; ILO, 2020; UNDP, 2020). Among them, the United Nations has been insisting on implementing various programs to reduce inequality by member countries. As the acknowledgment of inequality, a major challenge of the contemporary world, United Nations (2015) has given a priority in its sustainable development goals. As an evidence, "Out of the 17 goals, eleven goals address various forms of inequities, (Goals 1, 3, 4, 5, 6, 7, 8, 9, 10, 11, 16, and 17), and one goal (Goal 10) explicitly proposes to reduce various forms of inequalities" (Freistein and Mahlert, 2015:7).

As such, all-research reports and academic publications reveal the true picture of inequality, recommend the way out to minimize it; however, why the inequality is still widening and why the program and policies which aim to minimize inequality are ineffective are still not clear. As such, there is nothing impossible if the concerned stakeholders (the governments, United Nations, World Bank, Development Banks, International Monetary Fund, governmental and nongovernmental agencies, governmental and nongovernmental organizations, media, social activists, etc.) really want to create the minimal equity environment for all people of the world. In other words, there is a possibility to change the title "Unbeatable Challenges of Inequality, Growing Problems, and Consequences" to "Pathways to Minimize Inequality—the Diminishing Problems and Measurable and Applicable Solutions." Due to Covid-19, the world is in a different crisis, and this crisis is playing a triggering role in widening the inequality in almost every sector and particularly in healthcare and economic sectors.

In this book, we, including chapter authors, have tried to remind how inequality has been growing in each and every sector and how Covid-19 has played a crucial role in supporting the title of this book, "Unbeatable Challenges of Inequality, Growing Problems, and Consequences." The authors of this book argue that the inequality challenges are beatable, inequality can not be eliminated; however, it can be reduced if all concerned stakeholders intent on making a change. In general notion, it seems that no one wants to bear the pain, see the pain; however, there are no true visible commitments, drive, and seriousness to beat the challenges of inequality. In papers, policies, or directives of United Nations to governments, and in all development agencies, academic domain, there are raised concerns; however, these concerns have no effect, if there are no practical actions. This book aims to make us aware of the reality by showing that inequality is almost omnipresent in all sectors of society. The victims are those who have been marginalized from generation to generation due to the dominant nature of haves' circle almost in the entire world. The discrimination on the basis of race, ethnicity, gender, origin, sexual orientation, refugee status, language, color, religious belief, and even physical outlook, etc., is helping to continuously widen the social inequality. Similarly, the inequalities in economic, political, religious, knowledge, etc., are due to the greediness of power, money, and self-profit by dominating the have nots population. This situation is severe in the developing world in comparison to the developed world. This book tries to give some examples through some case studies. Hopefully, these case studies will help to pinpoint the problems and consequences of inequality.

"The social sciences have long played a leading role in analyzing inequalities. But gaps remain in our understanding of inequalities and how to address them. The urgency of reducing inequality demands new kinds of research and knowledge and a robust role for social science in identifying and building transformative pathways to greater equality" (ISSC, IDS and UNESCO 2016:13). Will the issue of inequality ever be resolved? We are talking about policies developed in several areas of social institutions—education and science, economics, public administration, health, law, religion, family, and individual development. That is why researchers from different backgrounds contribute to the problem solving, trying to find the ways out of the unbearable inequality in global and local contexts.

As such, "Inequality exists as an integral attribute of human development. Communities, nations, and systems are not evolving with the same speed and rate and require different resources in different amounts. However, the distribution of winnings is also uneven due to the multidimensionality of influencing factors. When we talk about inequality, it is not just inequality of income or wealth; it is, first of all, inequality in access to priority human needs—to habitation, to clean water, air, health care, and accordingly—to appropriate vaccination systems and assistance, security systems and safety guarantees for the future. Past financial crises and the current pandemic shock revealed bugs in the system, shaking it and changing our perception of the norms" (ISSC, IDS and UNESCO (2016:5)).

There are different types of inequalities, such as disparities in the development of regions or neighboring countries, where there is a constant cross-border exchange of resources not in favor of a more impoverished neighbor.

As noted in ISSC, IDS, and UNESCO (2016:5), dimensions of inequality can be analyzed in the following contexts:

- Economic inequality: differences between levels of incomes, assets, wealth, and capital, living standards, and employment;
- Social inequality: differences between the social status of different population groups and imbalances in the functioning of education, health, justice, and social protection systems;
- Cultural inequality: discriminations based on gender, ethnicity and race, religion, disability, and other group identities;
- Political inequality: the differentiated capacity for individuals and groups to influence political decision-making processes and to benefit from those decisions, and to enter into political action;
- Spatial inequality: spatial and regional disparities between centers and peripheries, urban and rural areas, and regions with more or less diverse resources;
- Environmental inequality: unevenness in access to natural resources and benefits from their exploitation; exposure to pollution and risks; and differences in the agency needed to adapt to such threats; and
- Knowledge-based inequality: differences in access and contribution to different sources and types of knowledge, as well as the consequences of these disparities. (ISSC, IDS and UNESCO 2016:5).

There are many papers on evidence of different types of inequalities that may be interlinked; for instance, gender inequality leads to an economic and social one. The inequality between households may be argued through the lens of environmental inequality since location does matter—if the household is in the zone of possible environmental or climate disaster or in the area of accumulation resources for living. Social and cultural inequalities have a tremendous impact on living standards all around the world. One of the most destructive phenomena is the inequality in the access to health care service, depending on the origin, ethnicity, age, etc., disguised as wealth-related access. Spatial and political inequalities are highly intercorrelated as soon as housing wealth is defined by income, amount of investments, and housing market dynamics. That leads to the differences between social groups in cities, between cities and rural areas, and different communities with different capacities to accumulate capital gains. We may argue about the differences and inequalities, which are not the same. However, we raised a lot of questions about the justice in resource allocation, social policy adaptation, and spatial "reconfiguration" toward the world of equal opportunities.

There is a big question in knowledge sharing if we take into account the dilemma between "value for money" idea and the sharing-for-free concept. The best practices in business are produced, generalized, and accumulated at the best business schools that are highly competitive. To release the knowledge and make it accessible means devaluation of it, and at the same time, we should admit that education may become a powerful driving force for future changes.

A specific type of inequality is associated with the globalization of the social and economic life of humankind, which manifests hunger in certain countries, the inability of low-income people to access health care services including vaccination, diagnosis, and prevention of serious diseases. Particular attention needs to be devoted to the inequality, which arises as a result of permanent causal links between phenomena, for instance, the so-called wealth-related inequality in access to education, and consequently, the possible growth or loss of individual wealth. Different communities have different access to technologies that could accelerate their development; therefore, some nations are evolving at a cosmic rate, while others in the early 21st century are trying to move away from traditional crafting (ISSC, IDS, and UNESCO 2016:5).

The issue of gender inequality and age inequality is one of the oldest types of population distribution in access to resources and decision-making. Institutional weakness in addressing inequalities and finding a sustainable path for the development of different communities in a multicultural, multiplane reality also emphasizes the relevance of inequality. Labor migration is one of the markers, by the way, of the difference in the quality of life between different countries, communities, and systems. And migration itself is not only a consequence but also an accelerator of the inequalities increases in the donor system.

There is no doubt that inequality is an unsolved problem, and now we need to find out if it is unbeatable. There is still a lack of knowledge on how inequality has been grounded throughout the human civilization, why society is stratified, classified, economically, politically, socially, and religiously and why the discrimination due to gender, sexual orientations, country of origin, language differences, immigration status, caste, race, and ethnically divided inequal. This book addresses these issues in holism and the case studies of various countries and tries to find out why inequality has been unbeatable and the best policies to overcome this challenge—Inequality. Further, the book also argues that, Covid-19, has a devastating impact on all aspects of human lives and it is playing a triggering role in scaling up an inequality.

1.2 The Current Context—The Covid-19 and Inequality

Covid-19 is a global crisis and the crisis always hits the people who have no resources to alter with the situation. The Covid-19 regime is about a year old; however, it completely changed human civilization's global picture. Firstly, it remained with the blame culture, the so-called conspiracies within nations and even among scientists, and the world media—big and small—also played important roles to horrify the general public. The global economic situation and political or power greediness divided the world due to few tempered leaderships, whose role was basically to blame other nations, or in the worse cases, own scientists, and society. Many scholarly papers, organizational reports already began to alert that Covid-19 regime would have a strong hit and problems could be long lasting; Covid-19 has exacerbated the wealth gap (Institute of Employment Rights, 2020; Inequality in A Rapidly Changing World, World Social Report 2020 (United Nations, 2020); Coronavirus crisis "exacerbating inequalities as richer build savings faster" (The Independent, 2020); Low-Income Households Turn to Debt as Rich Save in Lockdown (Meakin, 2020); How Wealth Managers Helped Millionaire Clients Grow Richer During Lockdown (Williams, 2020); Poverty and Shared Prosperity (World Bank, 2020); Even It Up: Time to end extreme inequality (Oxfam, 2021); Due to the pandemic, working hours are estimated to have declined by 17.3% in the second quarter of 2020 (compared with the fourth quarter of 2019), which is equivalent to 495m full-time jobs. Working hour losses in the third and fourth quarters eased slightly, but the jobs deficit at the end of 2020 remains significant (ILO, 2020); How Pandemics Leave the Poor Even Farther Behind (Furceri, Loungani and Ostry, 2020); Precarity and the pandemic: COVID-19 and poverty incidence, intensity, and severity in developing countries (Sumner, Ortiz-Juarez and Hoy, 2020); Power, Profits and the Pandemic: From corporate extraction for the few to an economy that works for all (Gneiting, Lusiani and Tamir, 2020); The Inequality Virus, bringing together a world torn apart by coronavirus through a fair, just and sustainable economy (Oxfam, 2021); The hunger virus: how COVID-19 is fueling hunger in a hungry world (Oxfam, 2020); Who will Bear the Brunt of Lockdown Policies? Evidence from Tele-workability Measures Across Countries (Brussevich, Dabla-Norris and Khalid, 2020); Black workers face two of the most lethal preexisting conditions for coronavirus—racism and economic inequality (Gould and Wilson, 2020); How Much Does Reducing Inequality Matter for Global Poverty? (Lakner, Mahler, Negre and Prydz, 2020;Tackling the Inequality Pandemic (United Nations, 2020); COVID-19, neoliberalism and health systems in 30 European countries: relationship to deceases (Barrera-Algarín et al., 2020); Privatization and Pandemic: A cross-country analysis of COVID-19 rates and health-care financing structures (Assa and Calderon, 2020); Community vulnerability to epidemics in Nepal: A high-resolution spatial assessment amidst COVID-19 pandemic (Khanal et al., 2020); Early estimates of the indirect effects of the COVID-19 pandemic on maternal and child mortality in low-income and middle-income countries (Robertson et al., 2020); Social protection responses to the COVID-19 pandemic in developing countries: Strengthening resilience by building universal social protection (ILO, 2020); How are mothers and fathers balancing work and family under lockdown (Andrew et al., 2020); largest number of hungry people ever, as coronavirus devastates poor nations (ILO, 2020); From Insights to Action: Gender equality in the wake of COVID-19 (UN Women, 2020); Missing: Where are the migrants in pandemic influenza preparedness plans (Wickramage et al., 2018); Why LGBT People are Disproportionately Impacted by COVID-19 (LGBT Foundation, 2020); Gender Gaps in the Care Economy during the COVID-19 Pandemic in Turkey (UNDP, 2020); why should we worry, what can we do in Covid regime (Bhandari, 2020, 2020, 2020, 2020, a, b, c, d).

Conquering the Great Divide

"The pandemic has laid bare deep divisions, but it's not too late to change course"

"COVID-19 has not been an equal opportunity virus: it goes after people in poor health and those whose daily lives expose them to greater contact with others. And this means it goes disproportionately after the poor, especially in poor countries and in advanced economies like the United States where access to health care is not guaranteed. One of the reasons the United States has been afflicted with the highest number of cases and deaths (at least as this goes to press) is because it has among the poorest average health standards of major developed economies, exemplified by low life expectancy (lower now than it was even seven years ago) and the highest levels of health disparities........ Unfortunately, as bad as inequality had been before the pandemic, and as forcefully as the pandemic has exposed the inequalities in our society, the post-pandemic world could

(*Continued*)

experience even greater inequalities unless governments do something. The reason is simple: COVID-19 won't go away quickly. And the fear of another pandemic will linger. Now it is more likely that both the private and the public sectors will take the risks to heart. And that means certain activities, certain goods and services, and certain production processes will be viewed as riskier and costlier. While robots do get viruses, they are more easily managed. So, it is likely that robots will, where possible, at least at the margin, replace humans. "Zooming" will, at least at the margin, replace airline travel. The pandemic broadens the threat from automation to low-skilled, person-to-person services workers that the literature so far has seen as less affected—for example, in education and health. All of this will mean that the demand for certain types of labor will decrease. This shift will almost surely increase inequality—accelerating, in some ways, trends already in place.........................
COVID-19 has exposed and exacerbated inequalities between countries just as it has within countries. The least developed economies have poorer health conditions, health systems that are less prepared to deal with the pandemic, and people living in conditions that make them more vulnerable to contagion, and they simply do not have the resources that advanced economies have to respond to the economic aftermath.
The pandemic won't be controlled until it is controlled everywhere, and the economic downturn won't be tamed until there is a robust global recovery. That's why it's a matter of self-interest—as well as a humanitarian concern—for the developed economies to provide the assistance the developing economies and emerging markets need. Without it, the global pandemic will persist longer than it otherwise would, global inequalities will grow, and there will be global divergence." Prof. Joseph Stiglitz[1] (IMF 2020) (https://www.imf.org/external/pubs/ft/fandd/2020/09/COVID19-and-global-inequality-joseph-stiglitz.htm)

COVID-19, like any other global disaster, reveals the weaknesses of the systems and disrupts our perception about what is normal and what is tolerable, and what is absolute nonsense. It makes little sense to support the wealth-related difference in access to health care services if a pandemic affects all. The inequalities may be a consequence of differences of capacities but should be solved toward the equity of opportunities for further development.

As these few references above show, the globe already entered social devastation. There is a chance of frustrations among us, which could create deviant behavior. It could lead to system failure, both socially, economically, and culturally political arena. The trust in power could be altered, and there is a chance of an increase of famine with an epidemic. As we begin to notice, people are so freighted, people already start to strike back to the social and political norms "Necessity knows no laws" (see news related to lockdown).

[1] Prof. Joseph Stiglitz is a professor at Columbia University and a recipient of the Nobel Memorial Prize in Economic Sciences. These texts were taken from (IMF 2020), (https://www.imf.org/external/pubs/ft/fandd/2020/09/COVID19-and-global-inequality-joseph-stiglitz.htm)

This pandemic is breaking world order toward the negativism, and it is hard to foresee the degree of social breakdown. However, the symbols show that it will be the hardest harm in human civilization. Still, we could minimize the havoc impact if we disciplined ourselves as responsible citizens. It depends on how the leaderships of each country work collectively, mutually, and wisely. It also depends on how we as responsible citizens provide a positive environment to support the leadership. It is not a time of blame but is the time to contribute to saving humanity.

Now the people of the world are divided into two thoughts: Covid-19 prevention and control and proximity of hunger due to the prolonged lockdown. Especially the rich are more afraid of the disease because the hunger never touches them. But the oppressed-class people are not as worried about the disease as they are about hunger.

According to the report of World Bank Group (Update on IDA, 2021), Covid-19 hit poor and vulnerable countries the hardest, which increases the gap in capacities in the future. The developing countries become poorer, and existing inequalities—more obvious and dreadful. Those households that had a chance to escape poverty are left behind in economic development.

Figure 1.1 Going home by foot—Covid-19 crisis.
Source: Social media and Man BK collection—people are walking home for several days and taking rest in the public place during lockdown in Nepal.

Figure 1.2 Walking with belongings.
Source: Social media and Man BK collection—people are walking home for several days in the Highway during lockdown in Nepal.

Figure 1.3 People are waiting to get some supports to survive.
Source: Social media and Man BK collection—people are waiting for food support during lockdown in Nepal. Hunger is more dangerous than Covid-19, Covid 19 may not take the life but body will stop working without food.

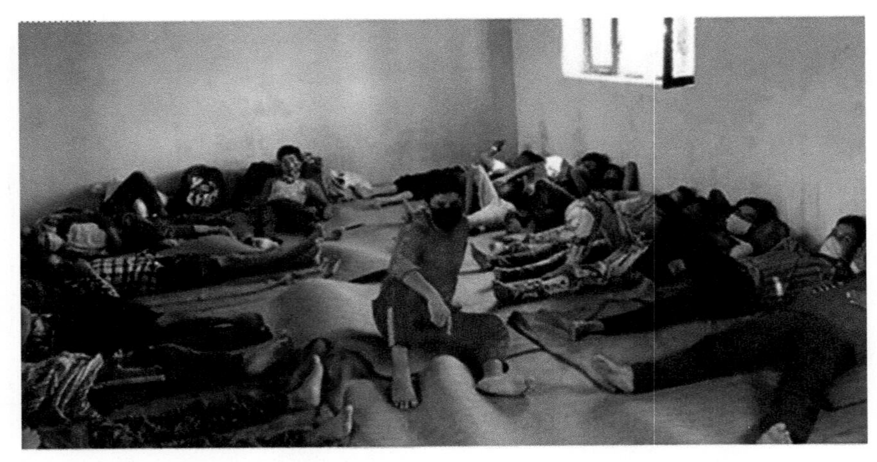

Figure 1.4 Even people in quarantine.
Source: social media and Man BK collection—people are in quarantine—had to sleep in floor in a large group, is this the right way to keep people in quarantine? Are they safe and healthy? The case of Nepal during lockdown.

Figure 1.5 A normal day-to-day life of poor.
Source: Social media and Man BK collection. This is a representative picture of general public of rural village of Nepal, where people still rely on minimum living standard. Still no access to drinking water, electricity, healthcare, and other bare requirements for survival. This picture was taken before the Covid-19; and we do not have information, about the pain of the lockdown; when the supply chain was completely disturbed and still continues. There are many people in the world, who are completely forgotten.

Figure 1.6 Cemetery in South Africa, deaths due to Covid-19 crisis.
Source: South Africa has suffered significantly more COVID-19–related deaths than any other African country. Available at https://www.dw.com/en/southern-africa-caught-in-covid-19-surge/a-56372105

Figure 1.7 Family in East Asia, left in poverty.
Source: Coronavirus: Millions will be left in poverty, World Bank warns (March 31, 2021). Available at https://www.bbc.com/news/business-52103666. According to the World Bank, the financial impact of Coronavirus will stop almost 24 million people from escaping poverty in East Asia and the Pacific.

In Ukraine, the regional disparities became more threatening. For instance, there are pieces of evidence of regional disparities in terms of capacity to respond to Covid-19 challenge, and therefore, certain regions were more or less vulnerable depending on the type of industry predominant in the country. In other words, if the residents of a certain region are employed at the chemical industry, for instance, it means that environmental pollution in the same area and professional specialization will be related to the high morbidity of lung diseases. That means that the same area will be more vulnerable to Covid-19 pandemic, and the mortality will be higher for the same residents. Some regions dependent on industries suffer more economically and lose the battle for the resources to respond to the crisis. The vicious circle can be interrupted only with proper social and economic policies.

As shown in few pictures, the walking people are seen on the highway; resting on the open area with hunger and tiredness, standing to get food from supporters, or even staying in quarantine in a group; sleeping in the floor; and the labor of the old man for survival show the same situation as there is no place to eat on the way due to the lockdown, neither they have affordability. Even now, with the emphasis on the prevention of Corona infection around the world, international organizations and nation–states do not seem to be as concerned about starvation as they seem to have been. People are dying of hunger three times a day more than that of Covid-19. According to the (United Nations, 2015), from farm to fork, the food supply chain accounts for 29% of global greenhouse gas emissions as it has been dealing by the big corporate houses. About a third of all food produced goes to waste. It would preserve enough food to feed 2 billion people—more than twice the number of undernourished people across the globe. It would also increase household income, improve food availability, reduce food imports, and improve the balance of trade. However, for that, it needs to restore the distribution chain intact; otherwise, the lockdown may further increase the food wastage as the farmers being not able to do the harvesting properly and bring products to the market.

Here, it is worthy to note that the poor have no access, not only the healthcare system but also have no excess of food and shelter because our social, political, economic, religious, traditional, and cultural systems are not favorable for the poor and marginalized population. The social structure is not inclusive and it is supportive of increasing the inequality.

As a known fact that, the world is already facing various challenges: political, social, religious, and other inequalities, ageing, aids, atomic energy, children, climate change, economic colonization, democracy fights, poverty, food insecurity, gender inequality, lack of access to healthcare, human rights, questions on international law and justice, migration, challenges on peace and security, population growth, refugees challenges, scarcity of water, misuse of technology, drugs, growing individualistic approach among youth, deviant behavior to attend temporary pleasure, corruption, wars, violation of social norms and values and so on. In addition to that, the inequality is growing and the discriminations on the basis of race, caste, color, country of origin, religion, migratory status, gender, and sexual orientations are still in practice. The economic inequality is broadly widening, and the Covid-19 is playing a favorable role in increasing the inequality. At the same time, climate change-induced challenges are also playing an important role in widening the inequality, where the poor are the main victims. There is no mechanism that empowers the victims to raise their voices against this social injustice. Therefore, scholarly world has an important role in overcoming these challenges. We hope that this book can play a role in making the appropriate policy to overcome the challenges of inequality.

1.3 The Scenarios of Economic Inequality—How It Is

The world society is mainly divided based on economic inequality followed by social inequality; cultural inequality; political inequality; spatial inequality; environmental inequality; knowledge-based inequality; and other parameters of inequalities. The following presented scenario shows the economic aspects of inequality based on Wid. World Data and Texts.

The scenarios of inequality based on open-source data and texts (https: //wid.world/news-article/2020-regional-updates/November10,2020;PreparedandWrittenBy~WID.WORLD)

Table 1.1 General overview of inequality worldwide.

Region	Situation
Latin America and the Middle East stand as the world's most unequal regions	Top 10% of the income distribution capturing respectively 54% and 56% of the average national income. In Latin America, amidst a decline of inequality levels in a handful of countries, inequality persisted, and even increased, in some others. In the Middle East, Gulf countries (Bahrain, Kuwait, Oman, Qatar, UAE, and Saudi Arabia) have been marked extreme inequality levels with little variation since the 1990s.

(*Continued*)

Table 1.1 (*Continued*)

Region	Situation
Africa comes next as one of the world's most unequal regions, with the top 10% capturing half of national income.	Contrary to widespread view, there is no equality Africa exceptionalism, its inequality levels are very close to those of Latin America or the Middle East. Extreme inequality levels can be found among nations which historically experienced white settlers' colonization and extreme forms of racial injustices (e.g., South Africa). The persistence of inequality in such countries is largely due to the lack of land ownership reforms, the absence of social security, and progressive taxation systems.
Russia	In 2019, the top 10% captured 46% of the national income, more than twice the share of the bottom 50%. Russia is a country which experienced, after the fall of the Soviet Union, one of the fastest increase in inequality. Its top 10% income share doubled in a few years after 1988 and it stands today at 46%. Lack of financial transparency, progressive taxation and a highly deregulated economy maintain current inequality rates at such high levels.
Asia, within-country inequality has been rising significantly since the 1990s.	In India, the top 10% income share grew from 30% in the 1980s to over 56% today, following deregulation and liberalization reforms. In contrast, the top 10% share in China grew from 28% to 41% in 2019. The lower rise of inequality in China was associated with stronger growth rates and much faster poverty eradication than in India, showing that more economic growth need not necessarily mean increased inequality.
USA	The United States shows a rise in the concentration of income unseen in other rich nations. The top 10% increased from 34% to 45% between 1980 and 2019. Half of the American population was shut from pretax economic growth.
Europe stands as the most equal of all regions	With the top 10% receiving 35% of income in 2019. This can be largely explained by public investments in education and health (i.e., by pre-distribution policies), financed by a fair amount of taxes (redistribution mechanisms).
Global inequality data shows that rising inequality is not a fatality and that countries with strong investments in public services and welfare policies have the lowest inequality levels. Tackling inequality is a matter of political choice. … The countries most affected by the Covid-19 pandemic were mostly those that display high and rising levels of inequality. For countries to best address this issue, governments' commitment to the publication of transparent inequality data and their ability to measure the evolution of inequality per country accurately is crucial.	

Source: Open-https://wid.world/news-article/2020-regional-updates/

The information is self-explanatory; however, the important point WID.WORLD (2021) raises is "Tackling inequality is a matter of political choice." Why? Because it is easy to manipulate to the marginalized people who have no time to raise a question and no energy to fight with the political elites.

Table 1.2 The specific case—The regional inequality.

Region	Situation
Africa	Inequality levels in the African region are extremely high: the average country-level top 10% income share equals half of the national income. South Africa is the most unequal country of the region: in 2019, the income share of top 10% households is estimated at 65%. Inequality levels seem to have changed very little, on average, over the last decades.
Asia	Three conclusions on the evolution of income inequality in Asia, for the period 1993 to 2019: The top 10% captured 49% of national income in 2019, showing a decrease in regional inequality since 1990, where the top 10% share amounted to 57%. While inequality has substantially risen in India and China, in the early 1990s and 2000s, it stabilized in China while it continued to rise in India. This suggests that national economic policies do matter, and lower inequality levels can accelerate both poverty reduction and overall economic growth rates. Within-country inequality has been rising significantly in major emerging economies since 1990, while regional inequality in Asia has been decreasing. This trend suggests a great economic convergence among Asian economies.
Europe	Europe is the least unequal region worldwide, with the top 10% capturing 35% of the average national income in 2019. Western European countries continue to lead on redistributive outcomes due to greater redistribution between the top 10% and the bottom 50%. However, regressive tendencies have emerged in both East and West over the last 10 years. Income inequality within countries has grown significantly since 1980, but less so since the Great Recession (2007–2009).
Russia	In Russia in 2019, the top 1% captured 20% of the national income, as much as half of the population. Russian incomes have polarized much more profoundly during the 1990s transformations than in Eastern Europe: the top 10% share has doubled, up to 46% in 2019. National income growth has stagnated since the late 2000s. In recent years, Eastern European national income per adult has overtaken Russia's adult national income.
Latin America	Latin America is among the most unequal regions, with the top 10% capturing 54% of the national income. Chile, Mexico, and Brazil are the three most unequal countries in the region, with the top 10% share capturing respectively 60%, 58%, and 57% of the average national income (2019).

Continued

Table 1.2 (*Continued*)

Region	Situation
	Data show a decline in inequality since 2000 in Ecuador, Argentina, and Uruguay, with the top 10% share capturing respectively 38%, 40%, and 42% of the national income (2019). Inequality levels are underestimated in official measures, as they are solely based on survey data. In WID.World, we complemented survey data with tax data and national accounts to provide a more accurate picture of the true level of inequality in Latin America.
Middle East 1990 and 2019	The Middle East is the most unequal region worldwide, with the top 10% capturing 56% of the average national income in 2019. Gulf countries are the most unequal countries in the region: 54% of national income accrues to the top 10%. Inequality levels have remained unchanged over the last three decades. This is driven by the extreme levels of within-country inequality.
North America and Oceania	In terms of pre-tax, top 10% income shares in 2019, Australia (35%) and New Zealand (37%) remain significantly more equal than Canada (43%) and the United States (45%). The United States shows a significant rise in the concentration of incomes, unseen in other rich nations: the top 10% captured 45% of the national income, close to half of it (2019). In all four countries, we see declining wages as a share of national income. This decline does not reflect an increase in self-employment. Instead, corporate profits have surged. Without any change in the concentration of capital income, this increase in corporate operating surplus already does much of the work to explain the observed increases in inequality.

Source: open-source data and texts (https://wid.world/news-article/2020-region al-updates/November10,2020;PreparedandWrittenBy~WID.WORLD)

We may assume that political barriers may exist because of the limited resources that have to be reallocated to support strategic decisions in response to the challenges. The political institutions are socially inertial and sustained, even though the dynamic of the economy, business, and social institutions is rapid. Political power can be a barrier and a driver of the change in the movement toward equality.

Among the possible actions toward equality, there can be a redistribution of wealth through the progressive tax system.

Another action may be focused on the development of the ladder of opportunities or "social elevator" that unites all initiatives in empowering people and the community in lifelong learning, retraining, and acquiring new skills. The combination of public and private universities that aim to provide equal access to education may offer a solution to social inequality.

The best cases are France, Turkey, and post-Soviet countries, where education is offered both in public and private sectors. The various scholarships are established to support talented people in their intentions to gain the degree.

The empowerment initiatives need to be established at the local community level to encourage the population to upgrade their level of life and income. These are concise but powerful actions that can be implemented at any level of governance but need sufficient support from society within the election process. The long process of decision-making is another barrier to rapid political changes in response to crises. Nevertheless, the network of international funds and organizations provides their help, guidance, and resources, which can help the process of policymaking and policy changing.

We agree with the Wid.World statement that "Tackling inequality is a matter of political choice." The people in political power thrive to remain in power no matter what, and finance is the main pillar to save them in power. Therefore, they normally do not bother with the inequality issues because they are not a victim of it, and it also may not support to dominate the marginalized people. There is a similar view even in the United Nations (2018) report, where most of the scholars' state that the "Main barrier to tackling inequalities is lack of political will."

Here, we present few representative views without mentioning the name of the scholars. The detailed report with names and affiliations can be found at the link:(https://www.un.org/development/desa/en/news/social/main-barrier-to-tackling-inequalities-is-lack-of-political-will.html).

Table 1.3 Representative views—why inequality is increasing.

Why have so many countries had difficulty reducing inequality?	"The main barrier to tackling inequalities is lack of political will. Over the last few decades, a mass of evidence has been produced documenting levels and trends in economic inequality and some other key inequalities such as health and education, and many research programs and projects, conferences and seminars have explored the main drivers and policy options............ The main problem is that reducing inequality will create winners and losers, and in terms of tackling inequality, the losers include the rich; a group with a strong voice, who have influence and are powerful. Working in their own interest, an economically rational thing to do, this highly influential group have the power to block government initiatives that could potentially leave them and their families worse off.

Continued

Table 1.3 (*Continued*)

	To tackle inequality, it will be necessary to convince this group that reducing inequality is in their interest (not necessarily their financial interest) or to find ways of getting policies through despite their opposition."
How does gender inequality impact the rest of the SDGs?	"To achieve gender equality and empower all women and girls is of intrinsic value, but also of instrumental importance as it pushes forward the wider SDG agenda. Women and children remain over-represented amongst the poorest populations, are often more food insecure, may face stigmatization and discrimination that results in a range of multidimensional deprivations........... Structural constraints often contribute to these inequalities. Social exclusion and harmful gender norms can block vulnerable women's mobility and agency and circumvent existing laws that lack the power of implementation or monitoring. Structural and individual deprivations may also be compounded. Inequalities concerning poverty and gender—for example, those affecting chronically poor women and girls- are especially severe. Yet, research and policymaking in these fields typically address issues of poverty or gender, but rarely its intersection. A focus on intersecting inequalities can address SDGs in a range of its dimensions."
What must be done to reduce global inequalities? Where are we on the road to achieve SDG 5 and 10?	"Economic inequality is largely driven by the unequal ownership of capital, which can be either privately or publicly owned, and by the policy. Since 1980, very large transfers of public to private wealth occurred in nearly all countries, whether rich or emerging. While national wealth has substantially increased, public wealth is now negative or close to zero in rich countries. Arguably this limits the ability of governments to tackle inequality; certainly, it has important implications for wealth inequality among individuals. The future of global inequality depends on the pace of economic growth, particularly in emerging economies, and on trends of inequality within countries. No one knows which of these forces will dominate and whether current trends are sustainable"

Source: adopted from United Nations 2018—online html web link: https://www.un.org/development/desa/en/news/social/main-barrier-to-tackling-inequalities-is-lack-of-political-will.html September 21, 2018, New York (downloaded on March 29, 2021)

The above notes of the eminent scholars are very powerful and self-explanatory. Agreeing with those views, we would add that if the government works as a service-oriented institution (not self-power oriented), the challenges of inequality could be transformed into the opportunity to create a society based on economic, social, and political equity. The service-oriented inclusive government can encourage creating an inclusive society, where the citizens of the nations can enjoy their lives according to their capacity. In so far, there are rare examples where governments are motivated to govern for the people. The evaluation of the governments' role to empower its citizens can be a topic of new research. In this book, we have presented the existing challenges of inequality, how it is widening, and how it can be reduced.

There have been several research papers that unveiled the nature of inequality, why inequality matters, how inequalities are institutionalized; how inequalities or their effects might be reduced (World Bank 2008—Institutional Pathways to Equity: Addressing Inequality Traps, by Bebbington, Anthony J.; Dani, Anis A.; de Haan, Arjan; Walton, Michael. (2008). The World Bank (2008) provides a good scenario, which helps to understand the challenges of inequality.

In Table 1.1, the column "how inequalities might be reduced" needs additional attention. The points raised and recommended by Sabates Wheeler; Moncrieffe; de Haan; Lucero; Barrientos; Ross and Boix to minimize inequality are still very valid and applicable. The report itself was prepared by World Bank, which is one of the major economic stakeholders in financing the development in developing world; however, why the policy recommendations were not applied or even applied why there is no effects? Who is creating the obstacles in policy making and implementation?

As cited above, "The main barrier to tackling inequalities is lack of political will. .... Economic inequality is largely driven by the unequal ownership of capital, which can be either privately or public owned, and by policy.. .. ... The future of global inequality depends on the pace of economic growth, particularly in emerging economies, and on trends of inequality within countries. No one knows which of these forces will dominate and whether current trends are sustainable" (United Nations 2018), and the points raised by Sabates Wheeler; Moncrieffe; de Haan; Lucero; Barrientos; Ross and Boix (2008) show how inequality can be reduced and who create the obstacles. There are thousands of gray and academic literature, where authors try to provide the pathways to reduce inequality.

For example, Nick Galasso & Marjorie Wood, 2021, list that

- Stop Illicit Outflows
- In developing countries, inadequate resourcing for health, education, sanitation, and investment in the poorest citizens drives extreme inequality.

Table 1.4 Inequality traps.

Nature of Inequality	Why Inequality Matters	How Inequalities are Institutionalized	How Inequalities might be Reduced
Inequalities among social groups—in particular, gender groups—that lead such groups to have differential access to productive assets and to social and political institutions	Inequalities in asset distribution can impede rates of agricultural growth. The combination of asset inequality, market failure, and unequal access to resources and institutions not only reproduces patterns of inequality but also can cause persistent poverty. Gendered and ascribed forms of inequality lead to unequal access to resources and also underlie unequal access to political institutions.	Through local relationships of power among land users, in particular through gendered power relations; through ideologies that sustain and naturalize unequal power relationships; through political institutions that exclude the voice of the disadvantaged	Policies can be implemented to deal with market failures (thus improving access to complementary factors of production) and to change the asset distribution. The inefficiency of extreme inequality cannot be cured simply by remedying market failures; other types of public action are required.
Inequalities among ethnic and racial groups based on their ethnic identities and markers	Inequalities that are inherent in the ways in which different groups are categorized lead to systematic imposition of disadvantages on some groups. These inequalities also lead those who are adversely categorized to internalize the same sense of low worth and to act in ways that perpetuate their own inequalities.	Through the exercise of power; through ethnic relationships and colonial and postcolonial institutions and policies that apportioned labels, values, and prejudices to particular ethnic and racial groups	Addressing inequality is a deeply psychological challenge. It also implies addressing power relations through education, human rights,

Table 1.4 (*Continued*)

Nature of Inequality	Why Inequality Matters	How Inequalities are Institutionalized	How Inequalities might be Reduced
		and that formalized particular relations of ethnic inequality; through processes of internalization among the poor	and reduction of discrimination. Merely redistributing assets or changing rules governing formal institutions will not be enough. Policy makers should rigorously probe the classifications and labels they adopt, should question assumptions of communities and groups, and should investigate the differing experiences of inequality and poverty that classifications tend to mask and even enforce. Without this disaggregated approach, policies may profit some and exclude others.

Continued

Table 1.4 (*Continued*)

Nature of Inequality	Why Inequality Matters	How Inequalities are Institutionalized	How Inequalities might be Reduced
Regional and spatial inequalities within a country	Regional inequalities constrain the growth potential of disadvantaged regions and contribute to the reproduction of their poverty. Spatial inequalities have particularly negative effects for the poorest of most excluded social groups.	Through other social inequalities in which regional inequalities are grounded and that are particularly resistant to changes; through state and other institutions that the social inequalities have structured and whose workings therefore contribute to the reproduction of these inequalities	
Through other social inequalities in which regional inequalities are grounded and that are particularly resistant to changes; through state and other institutions that the social inequalities have structured and whose workings therefore contribute to the reproduction of these inequalities	Inequalities that are inherent in the ways in which different groups are categorized lead to systematic exclusion of some groups and thus to unequal access to land, natural resources, services, assets, and political institutions. These same inequalities lead to uneven distribution of wealth and of economic opportunities.	Through interethnic relationships and the everyday practices of racism Through public and other agencies, because identity-based inequalities are built into the ways in which these agencies operate, so that the actions of these agencies go a long way in institutionalizing inequalities in broader sets of relationships	inequalities can be addressed through the progressive social and political organization of ethnic and race-based groups, through the quality of leadership in these organizations, and through organizational strategies

Table 1.4 (*Continued*)

Nature of Inequality	Why Inequality Matters	How Inequalities are Institutionalized	How Inequalities might be Reduced
			that are well crafted in relation to the existing political opportunity structure. Such mobilization and protest have reduced political inequalities but have not yet reduced economic inequalities.
Inequalities among age groups, in particular those inequalities that lead the elderly to have reduced access to income and opportunity	Age-based inequalities contribute to reduced access of old people to services and economic opportunities and thus contribute directly to their poverty proneness and vulnerability. The poverty status of the elderly can affect the poverty status and asset portfolios of the households of which they are a part	Through categories associated with age and through political institutions in which the elderly and poor lack voice	Noncontributory pension programs can address effects of inequality experienced by the poor and elderly. Social mobilization precedes emergence of such noncontributory programs. Changes in public opinion in favor of such programs

Continued

Table 1.4 (*Continued*)

Nature of Inequality	Why Inequality Matters	How Inequalities are Institutionalized	How Inequalities might be Reduced
			provide important support for the emergence of these programs.
Inequalities between extractive industry companies and social groups in areas affected by mining that lead some social groups to enjoy reduced access to the fruits of mining; geographic, ethnic, and racial inequalities among social groups within countries	Inequalities of power can lead to uneven access to information about mining and the revenues generated by it. These inequalities can deepen tendencies toward social and political violence. In extreme instances, they can elicit, or strengthen the cause of, militarized and separatist movements.	Through the cultures and practices of national political and mineral development institutions, which cause these institutions to fail to address or foresee such problems; through company cultures, which institutionalize practices that sustain inequalities in access to information and resources; through community-extractive industry conflicts, which are also embedded in broader social inequalities that can further institutionalize local inequalities	Well-designed programs of information provision, participatory monitoring of mining and revenue transfer programs, and periodic adaptation of program design can reduce inequalities and especially reduce the potential for them to lead to violence. Fiscal transfers to offset some of the economic inequalities fostered by extractive industries are insufficient.

Table 1.4 (*Continued*)

Nature of Inequality	Why Inequality Matters	How Inequalities are Institutionalized	How Inequalities might be Reduced
Inequalities in access to land, economic opportunity, and political participation	Inequalities in access to land and other assets can fuel civil unrest. Inequalities in political participation can limit the ability of a country to participate in multilateral institutions that would otherwise facilitate economic growth.	Through political and social institutions that are backed up and sustained by practices of authoritarian and military rule and that sustain inequalities in political participation and economic opportunity	International demands are needed for reduced (political) inequality as a precondition for membership of multilateral organizations. Policy technocrats with some room to maneuver can introduce design changes in institutions. Organized middle-class demands for increased access to economic opportunity can elicit change.

Source: World Bank 2008—Institutional Pathways to Equity: Addressing Inequality Traps, by Bebbington, Anthony J.; Dani, Anis A.; de Haan, Arjan; Walton, Michael. (2008:28-31). Chapters by: Sabates Wheeler; Moncrieffe; de Haan; Lucero; Barrientos; Ross and Boix- in chronological order.

- Progressive Income Tax
- After falling for much of the 20th century, inequality is worsening in rich countries today.
- A Global Wealth Tax

- Enforce a Living Wage
- Workers' Right to Organize
- The right of workers to organize has always been a cornerstone of more equal societies and should be prioritized and protected wherever this basic right is violated. Extreme inequality requires the disempowerment of workers
- Despite Article 23 of the Universal Declaration of Human Rights—which declares the right to organize as a fundamental human right—workers worldwide
- Stop Other Labor Abuses
- Open and Democratic Trade Policy
- A New Economics
- Economists are often imagined as stuffy academics who value arcane economic theory above humanitarian values.

Source: Nick Galasso & Marjorie Wood, 2021, Eight Ways To Reduce Global Inequality, Inequality.org is a project of the Institute for Policy Studies https://inequality.org/great-divide/8-ways-reduce-global-inequality/

Similarly, the Challenging Inequalities: Pathways to a Just World (summary of ISSC, IDS and UNESCO 2016) states that:

- Unchecked inequality could jeopardize the sustainability of economies, societies, and communities.
- Inequalities should not just be understood and tackled in terms of income and wealth: they are economic, political, social, cultural, environmental, spatial and knowledge based.
- The links and intersections between inequalities need to be better understood to create fairer societies.
- A step change toward a research agenda that is interdisciplinary, multiscale, and globally inclusive is needed to inform pathways toward greater equality. In short, too many countries are investing too little in researching the long-term impact of inequality on the sustainability of their economies, societies, and communities. Unless we address this urgently, inequalities will make the cross-cutting ambition of the Sustainable Development Goals (SDGs) to "leave no one behind" by 2030 an empty slogan. (ISSC, IDS and UNESCO 2016:back cover page)

ISSC, IDS and UNESCO (2016), World Social Science Report (2016), Challenging Inequalities: Pathways to a Just World, UNESCO Publishing, Paris. en.unesco.org/wssr2016

There is no question of validity of such recommendation; however, who is responsible for implementing these recommendations—the obvious answer is the government. The politicians who are in power. It seems that the political system, unknowingly reluctant to create the socially, economically, and politically inclusive society. But why it is happening, what is the fear, is still unknown. As noted earlier, and also in the chapters, the world is facing a global crisis due to Covid-19 and its impacts will remain more than a decade. In this scenario, it seems that the inequality will still be widening. How to tackle the Covid-19 crisis and how or in what way inequality could be minimized?

These are hard questions—and we have no readymade answers. That is why we said Inequality—unbeatable challenges; however, nothing is unbeatable, every challenge can be turned into opportunity.

We should add that another driving force and barrier to the positive change is economic and social resistance to the idea of cooperation between institutions, governments, even companies—competitors. The cooperation at the institutional level may enable the implementation of the long-run scaled-up initiatives such as projects aimed at building the assets for the households (i.e., subsidies), tax redistribution, educational incentives, inclusion policy, and so on.

The policy interventions toward equalizing power should be aimed at giving a voice to disadvantaged groups in the process of adaptive decision-making.

Policy transformation may indeed take years of effort both by activists, governments, policymakers, donors, and other stakeholders, yet not bringing the desirable state of equality. However, new technologies give additional resources to people who may take control of economic, social, and political processes via data processing, sharing economy, and collective actions to equalize the access to goods and services.

As noted in the Social Inequality a Global Challenges (Bhandari ad Shvindina 2021); it is important to restate that, when we talk about inequality, the focus remains on economic inequality; however, the major inequality persists due to discrimination to gender, age, origin, ethnicity, disability, sexual orientation, class, and religion within the society. It is true that primarily the world is divided between "haves" and "have nots"; and social strata are created to create the comfort zones to them who dominate the social, economic, and political systems. The rules and regulations, social norms and values are also created and implemented always by the social elite, therefore, the marginalized groups have no or very minimal stake in the society, and they are always victims. There is a need for complete social, economic, and political reform, so that the designed division can be minimized. Until or unless the marginalized or disadvantaged group is empowered, the inequality issue cannot be solved or even minimized (Bhandari and Shvindina, 2021). As UN notes, there is a clear need to pursue inclusive, equitable, and sustainable growth, ensuring a balance among economic, social, and environmental dimensions of sustainable development (United Nations 2020a website). "The major goal of any society is sustained harmony in the society; however, it rarely exists in human sociopolitical history. The divisive element of inequality—is oppression to weak according to strata which can be economic, social, political, cultural or other forms—grounded and influenced by societal circumstances. Theoretically, it is hard to fully pinpoint the root cause of inequality—why oppressed are being oppressed, and oppressors have been maintaining the domination" (Bhandari, 2017, Bhandari, 2018; Bhandari and Shvindina, 2019).

> *"We must all realize that inequality reduction does not occur by decree; neither does it automatically arise through economic growth, nor through policies that equalize incomes downward via blind taxing and spending. Inequality reduction involves a collaborative effort that must motivate all concerned parties, one that constitutes a genuine political and social innovation, and one that often runs counter to prevailing political and economic forces" (Genevey et.al. 2013:6).*

In our opinion, the major role belongs to those who create the knowledge—therefore, it is necessary to develop a new epistemology, which basically makes people sensitive to other. That will change the motive from "I" to "we"—the collectivism—so that, when we prepare policy, no one remains behind (United Nations 2018). "In the 2030 Agenda, Governments envisage 'a world of equal opportunity permitting the full realization of human potential and contributing to shared prosperity' (A/RES/70/1, para. 8). A growing literature has attempted to quantify the broad concept of equality of opportunity for policy purposes. The basic proposition of this literature is that inequality results from two sets of factors: those that are in some way assumed to be under an individual's control, such as effort or personal responsibility, and those that are not. A person's circumstances, such as place of birth, parental socioeconomic status and other attributes highlighted in the 2030 Agenda, including gender, ethnicity and race, and disability status, are beyond one's control. These and other circumstances affect access to education, health, income, and other resources as well as participation in social and political life" (United Nations 2020:35).

"The impact of inequality—Slower economic growth and poverty reduction, Limited upward mobility and Captured political processes, mistrust of institutions and growing unrest" (United Nations 2020)

What to do in reducing inequality within countries: "Promote equal access to opportunities; Institute a macroeconomic policy environment conducive to reducing inequality; Tackle prejudice and discrimination and promote the participation of disadvantaged groups in economic, social and political life" (United Nations 2020:14-15)

With the agreement of the above statements and texts, this edited book follows theoretical frameworks, raises the questions, pinpoints problems, provides recommendations, and fulfills the knowledge gap of complex social discourse in reducing the global, regional, national, and local inequality. This book tries to provide some of the basic answers of why inequality is growing in general and especially how Covid-19 is playing to widen inequality. "It is a well-known fact that the world is already facing various challenges: political, social, religious, and other inequalities, aging, aids, atomic energy, children, climate change, economic colonization, democracy fights, poverty, food insecurity, gender inequality, lack of access to healthcare, human rights, questions on international law and justice, migration, challenges on peace and security, population growth, refugees challenges, scarcity of water, misuse of technology, drugs, growing individualistic approach among youth, deviant behavior to involve in temporary pleasures, corruption, wars, violation of social norms and values and so on.

In addition to that, inequality is growing, and the discriminations on the basis of race, caste, color, country of origin, religion, migratory status, gender, and sexual orientations are still in practice. The economic inequality is widening and expanding, and the Covid-19 is playing a favorable role in increasing it. At the same time, climate change induces challenges and plays an important role in widening inequality, where the poor are the main victims.

Women are more vulnerable to the effects of climate change *than men—primarily as they constitute the majority of the world's poor and are more dependent for their livelihood on natural resources that are threatened by climate change. Furthermore, they face social, economic, and political barriers that limit their coping capacity. Women and men in rural areas in developing countries are especially vulnerable when they are highly dependent on local natural resources for their livelihood. Those charged with the responsibility to secure water,*

Continued

> *food and fuel for cooking and heating face the greatest challenges. Secondly, when coupled with unequal access to resources and to decision-making processes, limited mobility places women in rural areas in a position where they are disproportionately affected by climate change. It is thus important to identify gender-sensitive strategies to respond to the environmental and humanitarian crises caused by climate change. It is important to remember, however, that women are not only vulnerable to climate change, but they are also effective actors or agents of change in relation to both mitigation and adaptation. Women often have a strong body of knowledge and expertise that can be used in climate change mitigation, disaster reduction and adaptation strategies. Furthermore, women's responsibilities in households and communities, as stewards of natural and household resources, positions them well to contribute to livelihood strategies adapted to changing environmental realities. (52nd session of the Commission on the Status of Women (2008)- https://www.un.org/womenwatch/feature/climate_change/factsheet.html)*

As shown in various reports of UN (https://www.un.org/womenwatch), climate change also plays a triggering role in widening the inequality adding more burdens to women and girls.

Table 1.5 Climate change—role in rising inequality (Women).

Area	Evidence
Women, Gender Equality and Climate Change	Detrimental effects of climate change can be felt in the short-term through natural hazards, such as landslides, floods, and hurricanes; and in the long-term, through more gradual degradation of the environment. The adverse effects of these events are already felt in many areas, including in relation to, inter alia, agriculture and food security; biodiversity and ecosystems; water resources; human health; human settlements and migration patterns; and energy, transport, and industry (whereas main victims are always women and children).
Food Security	Women face loss of income as well as harvests—often their sole sources of food and income. Related increases in food prices make food more inaccessible to poor people, in particular to women and girls whose health has been found to decline more than male health in times of food shortages.
Impact of Biodiversity Loss	In many parts of the world, deforestation has meant that wood—the most widely used solid fuel—is located further away from the places where people live. In poor communities in most developing countries, women and girls are responsible for collecting traditional fuels, a physically draining task that can take from 2 to 20 or more hours per week. As a result, women have less time to fulfil their domestic responsibilities, earn money, engage in politics or other public activities, learn to read

Table 1.6 Climate change—role in rising inequality (Women).

Area	Evidence
	or acquire other skills, or simply rest. Girls are sometimes kept home from school to help gather fuel, perpetuating the cycle of disempowerment.
Water Resources	Climate change has significant impacts on fresh water sources, affecting the availability of water used for domestic and productive tasks. The consequences of the increased frequency in floods and droughts are far reaching, particularly for vulnerable groups, including women who are responsible for water management at the household level.........Given the changing climate, inadequate access to water and poor water quality does not only affect women, their responsibilities as primary givers, and the health of their families,' it also impacts agricultural production and the care of livestock; and increases the overall amount of labor that is expended to collect, store, protect, and distribute water.
Health	Climate change scenarios include increased morbidity and mortality due to heat waves, floods, storms, fires, and droughts.........floods—increasing consistently with climate change—may also increase the prevalence of water-related diseases, especially water- and vector-borne diseases, which affect millions of poor people each year. In addition, an increase in prevalence of diseases will likely aggravate women's caregiving of family and community members who are ill.
Human Settlements and Migration Patterns	Climate change adds a new complexity to the areas of human mobility and settlement by exacerbating environmental degradation.........the migratory consequences of environmental factors result in higher death rates for women in least developed countries, as a direct link to their socioeconomic status, to behavioral restrictions, and poor access to information.
Women's Human Rights	Global warming and extreme weather conditions may have calamitous human rights consequences for millions of people. Global warming is one of the leading causes and greatest contributors to world hunger, malnutrition, exposure to disease, and declining access to water. Moreover, it poses limitations to adequate housing, spurring the loss of livelihoods as a result of permanent displacement. Climate change affects the economic and social rights of countless individuals; this includes their rights to food, health, and shelter. As climate change will inevitably continue to affect humanity, a key UN priority is safeguarding the human rights of people whose lives are most adversely affected.

Continued

Table 1.5 (*Continued*)

Area	Evidence
Women, Gender Equality, and Energy	Linkages between energy supplies, gender roles, and climate change are strongest in countries with low availability of basic electricity and modern fuels, as well as high dependence on biomass fuels for cooking, heating, and lighting—and close to 2 billion people in the developing world use traditional biomass fuels as their primary source of energy. In these countries, cultural traditions make women responsible for gathering fuel and providing food, even when this involves long hours performing heavy physical labor or traveling longer distances. With the onslaught of aggravated environmental changes, women are likely to continue spending long (perhaps even longer) hours fetching firewood, drawing water, working the land, and grinding cereal crops.
Technology is Never Gender-Neutral	Technology is never gender-neutral and when coupled with the negative effects of the changing climate, it is even less gender-sensitive. In many developing countries, the access of girls and women to information and communication technology is constrained by social and cultural bias, inadequate technological infrastructure in rural areas, women's lower educations levels (especially in the fields of science and technology), the fear of or lack of interest in technology, and women's lack of disposable income to purchase technology services.
Vulnerability of Women to Disasters	The vulnerability of women to disasters is increased for a number of reasons. Post-disaster, women are usually at higher risk of being placed in unsafe, overcrowded shelters, due to lack of assets, such as savings, property, or land. In the context of cyclones, floods, and other disasters that require mobility, cultural constraints on women's movements may hinder their timely escape, access to shelter or access to health care.

Source: open—UN: https://www.un.org/womenwatch (2020). These texts are adopted from https://www.un.org/womenwatch/feature/climate_change/factsheet.html (April 5, 2021), the additional information of gender and climate change is listed in website reference.

Table 1.2, Climate Change—role in rising inequality (Women), is self-explanatory, which gives a clear picture of how women are especially victimized due to climate change and how climate change is playing a crucial role in widening the inequality among men and women.

> "The effects of climate change are experienced to varying degrees across and within countries due to differences in exposure, susceptibility and coping capacities. If left unaddressed, climate change will lead to increased inequality both within and among countries

> and could leave a substantial part of the world further behind; Developing countries, particularly small island developing States, face disproportionate risks from an altered climate, while high-income countries are generally less vulnerable and more resilient; Within countries, people living in poverty and other vulnerable groups—including smallholder farmers, indigenous peoples and rural coastal populations—are more exposed to climate change and incur greater losses from it, while having fewer resources with which to cope and recover; Climate change can generate a vicious cycle of increasing poverty and vulnerability, worsening inequality and the already precarious situation of many disadvantaged groups; Just as the effects of climate change are distributed unevenly, so too are the policies designed to counter them. As countries take climate action, there will be trade-offs to consider between the positive and negative effects of mitigation and adaptation measures and distributional impacts......" (United Nations 2020:82).

What to do in reducing inequality within countries: "Promote equal access to opportunities; Institute a macroeconomic policy environment conducive to reducing inequality; Tackle prejudice and discrimination and promote the participation of disadvantaged groups in economic, social and political life" (United Nations 2020:14-15)

As seen in the scenario above, there are various forms of inequalities supported by the power-centric mentality of governance. Climate change has been another factor of rising inequality, especially in the developing world. Additionally, in recent years Covid-19 is playing a role in widening inequality. On the basis of this situation, this edited book presents some unexplored issues of economic inequality, including case studies of various countries. Inequality is a chronic and divisive factor of society. Inequality exists as an integral attribute of human development. Communities, nations, and systems are not evolving at the same speed and rate and thus require different resources in different amounts. We may have no doubt that inequality is an unsolved problem, but now we need to find out whether it is unbeatable? There is still lack of knowledge around how inequality has been grounded throughout human civilization, why society is stratified and classified economically, politically, socially, and religiously, and why there is discrimination due to gender, sexual orientation, country of origin, language differences, immigration status, caste, race, and ethnicity? This book addresses these issues in a holistic way as well as including case studies of various countries. It tries to find out why inequality has been unbeatable and what would be the best policies to overcome this challenge.

In the process of the evolution of humankind, the unachievable goals were always stated as the targets. The era of incremental changes is over; the rapid transitions in the environment, markets, demography, perceptions, human behavior, and other abovementioned trends, especially exogenous shock of Covid-19 shifted our perception from the unachievable from the first sight to unavoidable transformations toward equalization.

The long-run strategies and short-term ploys should be implemented at different levels—countries, regions, and local communities, at meta level in cooperation between countries and institutions, in diverse arenas—political, economic, educational, social, and environmental to overcome the multi-faceted crises of inequality.

Repeated again, there is a need for factual knowledge and practices which can bridge the gap in the divided society. Awareness needs to come from individual to individuals, society to societies, and nation to nations. To overcome the inequality problems, the concept of "Vasudhaiva Kutumbakam—The entire world is our home, and all living beings are our relatives" and "Live and let other live—the harmony within, community, nation and global" is needed. There is a need for trust, love, and respect for each other.

Acknowledgment and Funding

Funding for some of the research reported here was gratefully provided by the Ministry of Education and Science of Ukraine Reforming the lifelong learning system in Ukraine for the prevention of the labor emigration: a coopetition model of institutional partnership (No0120U102001).

References

Aashe (2018), Stars Technical Manual. (The Association for the Advancement of Sustainability in Higher Education). Version 2.1. Available online: https://stars.aashe.org/pages/about/technical-manual.html

Adomßent, M., Fischer, D., Godemann, J., Herzig, C., Otte, I., Rieckmann, M., Timm, J. (2014), Emerging Areas in Research on Higher Education for Sustainable Development – Management Education, Sustainable Consumption and Perspectives from Central and Eastern Europe, Volume 62, 1 January 2014, p. 1–7. http://dx.doi.org/10.1016/j.jclepro.2013.09.045

Adomßent, M., Godemann, J., Leicht, A., Busch, A. (Eds.), 2006. Higher Education for Sustainability: New Challenges from a Global Perspective. VAS –Verlag für Akademische Schriften, Frankfurt am Main.

Alghamdi, N.; den Heijer, A.; de Jonge, H. (2017), Assessment Tools' Indicators for Sustainability in Universities: An Analytical Overview. Int. J. Sustain. High. Educ., 18, 84–115

Andrew A., S. Cattan, M. Costa Dias, C. Farquharson, L. Kraftman, S. Krutikova, A. Phimister and A. Sevilla. (2020). How are mothers and fathers balancing work and family under lockdown? Institute for Fiscal Studies. https://www.ifs.org.uk/publications/14860

Arima A (2009), A plea for more education for sustainable development. Sustain Sci 4(1):3–5

Assa J. and C. Calderon. (2020). Privatization and Pandemic: A cross-country analysis of COVID-19 rates and health-care financing structures. UNDP/HDRO. https://www.researchgate.net/profile/Jacob_Assa2/publication/341766609_Privatization_and_Pandemic_A_Cross-Country_Analysis_of_COVID-19_Rates_and_Health-Care_Financing_Structures/links/5ed29f994585152945 1c5df9/Privatization-and-Pandemic-A-Cross-Country-Analysis-of-COVID-19-Rates-and-Health-Care-Financing-Structures.pdf

Barbier, E. B., (1987), "The concept of sustainable economic development," Environmental Conservation, Vol. 14, No. 2 (1987), pp. 101–110.

Barker, Chris (2004), The SAGE Dictionary of Cultural Studies, SAGE Publications, London / Thousand Oaks / New Delhi https://zodml.org/sites/default/files/%5BDr_Chris_Barker%5D_The_SAGE_Dictionary_of_Cultural__0.pdf

Barnosky, Anthony, and others (2012). Approaching a state shift in Earth's biosphere. Nature, vol. 486, No. 7401 (7 June), pp. 52–58.

Barrera-Algarín E, F. Estepa-Maestre, J. Sarasola-Sánchez-Serrano, and A. Vallejo-Andrada. (2020). COVID-19, neoliberalism and health systems in 30 European countries: relationship to deceases. https://europepmc.org/article/med/33111713

Barth M, Godemann J (2006), Study program sustainability—a way to impart competencies for handling sustainability? In: Adomßent M, Godemann J, Leicht A, Busch A (eds) Higher education for sustainability: new challenges from a global perspective. VAS, Frankfurt, pp 198–207

Barth, M., Rieckmann, M., Sanusi, Z.A. (Eds.), 2011. Higher Education for Sustainable Development. Looking Back and Moving Forward. VSA – Verlag für Akademische Schriften, Bad Homburg.

Baumert, Kevin A., Timothy Herzog, and Jonathan Pershing (2005). Navigating the Numbers: Greenhouse Gas Data and International Climate Policy. Washington, D.C.: World Resources Institute.

Beall, Jo, Basudeb Guha-Khasnobis and Ravi Kanbur, eds. (2012). Urbanization and Development in Asia. Multidimensional Perspectives. New York: Oxford University Press.

Bebbington, Anthony J.; Dani, Anis A.; de Haan, Arjan; Walton, Michael. (2008), Institutional Pathways to Equity : Addressing Inequality Traps. New Frontiers of Social Policy. World Bank, Washington, DC. © World Bank. https://openknowledge.worldbank.org/handle/10986/6411 License: CC BY 3.0 IGO." https://openknowledge.worldbank.org/bitstream/handle/10986/6411/439640PUB0Box310only109780821370131.pdf?sequence=1&isAllowed=y

Berg, Andrew, and Jonathan Ostry (2011). Inequality and unsustainable growth: two sides of the same coin? IMF Staff Discussion Note. SDN/11/08. Washington, D.C.: International Monetary Fund. 8 April.

Bertelsmann Stiftung and Sustainable Development Solutions Network (2018), SDG Index and Dashboards Report 2018-Global Responsibilities, Implementing the Goals, G20 and Large Countries Edition. www.pica-publishing.com, http://www.sdgindex.org/assets/files/2018/00%20SDGS%202018%20G20%20EDITION%20WEB%20V7%20180718.pdf

Berzosa, A.; Bernaldo, M.O.; Fernández-Sanchez, G. (2017), Sustainability Assessment Tools for Higher Education: An Empirical Comparative Analysis. J. Clean. Prod. 161, 812–820

Bhandari Medani P. and Shvindina Hanna (2019), The Problems and consequences of Sustainable Development Goals, in Bhandari, Medani P. and Shvindina Hanna (edits) Reducing Inequalities Towards Sustainable Development Goals: Multilevel Approach, River Publishers, Denmark / the Netherlands- ISBN: Print: 978-87-7022-126-9 E-book: 978-87-7022-125-2

Bhandari, Medani P (2017), Climate change science: a historical outline. Adv Agr Environ Sci. 1(1) 1–8: 00002. http://ologyjournals.com/aaeoa/aaeoa_00002.pdf

Bhandari, Medani P. (2018), Green Web-II: Standards and Perspectives from the IUCN, Published, sold and distributed by: River Publishers, Denmark / the Netherlands ISBN: 978-87-70220-12-5 (Hardback) 978-87-70220-11-8 (eBook),

Bhandari, Medani P. (2019), Inequalities with reference to Sustainable Development Goals in Bhandari, Medani P. and Shvindina Hanna (edits)

Reducing Inequalities Towards Sustainable Development Goals: Multi-level Approach, River Publishers, Denmark / the Netherlands- ISBN: Print: 978-87-7022-126-9 E-book: 978-87-7022-125-2

Bhandari, Medani P. (2020), Second thoughts, In the COVID-19 Regime, What Role Does Intellectual Society Play?, The Society of Transnational Academic Researchers (STAR Scholars Network), USA, Bulletin 20/2, https://starscholars.org/in-the-covid-19-regime-what-role-does-intellectual-society-play/

Bhandari, Medani P. (2020a), The Phobia Corona (COVID 19) - What Can We Do, Scientific Journal of Bielsko-Biala School of Finance and Law, ASEJ 2020, 24 (1): 1–3, GICID: 01.3001.0014.0769, https://asej.eu/resources/html/article/details?id=202946

Bhandari, Medani P. (2020b), The Phobia Corona (COVID 19) - What Can We Do, Scientific Journal of Bielsko-Biala School of Finance and Law, ASEJ 2020, 24 (1): 1–3, GICID: 01.3001.0014.0769, https://asej.eu/resources/html/article/details?id=202946

Bhandari, Medani. P. (2020c). In the Covid-19 Regime – What Role Intellectual Society Can Play. International Journal of Science Annals, 3(2), 5–7. doi:10.26697/ijsa.2020.2.1 https://ijsa.culturehealth.org/en/arhiv https://ekrpoch.culturehealth.org/handle/lib/71

Bhattacharyya, Subhes C. (2012), Energy access programs and sustainable development: A critical review and analysis, Energy for Sustainable Development, Volume 16, Issue 3, September 2012, Pages 260–271

Blau, Francine D., and Lawrence M. Kahn. (2017). "The Gender Wage Gap: Extent, Trends, and Explanations." Journal of Economic Literature, 55(3): 789–865.

Brockmann, Hilke, and Jan Delhey (2010), "The Dynamics of Happiness". Social Indicators Research 97, no.1 (2010): 387–405.

Brussevich,M, E. Dabla-Norris and S. Khalid. (2020). Who will Bear the Brunt of Lockdown Policies? Evidence from Tele-workability Measures Across Countries. IMF Working Paper. https://www.imf.org/en/Publications/WP/Issues/2020/06/12/Who-will-Bear-the-Brunt-of-Lockdown-Policies-Evidence-from-Tele-workability-Measures-Across-49479

Caeiro, S.; Jabbour, C.; Leal Filho, W. (2013), Sustainability Assessment Tools in Higher Education Institutions Mapping Trends and Good Practices around the World; Springer: Cham, Gemany, p. 432.

Callanan, Laura and Anders Ferguson (2015), A New Pilar of Sustainability, Philantopic-Creativity, Foundation Center, New York, https://pndblog.typepad.com/pndblog/2015/10/creativity-a-new-pillar-of-sustainability.html

Carson, Rachel (1962), Silent Spring, A Mariner Book, Houghton M1fflin Company, Boston, New York

Cheng, V. (2018) Views on Creativity, Environmental Sustainability and Their Integrated Development. Creative Education, 9, 719–743. doi: 10.4236/ce.2018.95054.

Chhabria, Sheetal (2016), Inequality in an Era of Convergence: Using Global, Histories to Challenge Globalization Discourse, World History Connected Vol. 13, Issue 2.

Christiansen C.O., Jensen S.L.B. (2019) Histories of Global Inequality: Introduction. In: Christiansen C., Jensen S. (eds) Histories of Global Inequality. Palgrave Macmillan, Cham. https://doi.org/10.1007/978-3-030-19163-4_1

Clark, W. C., and R. E. Munn (Eds.) (1986), Sustainable Development of the Biosphere, Cambridge: Cambridge University Press

Colyvas, Jeannette A. and Walter W.Powell (2006), Roads to Institutionalization: The Remaking of Boundaries between Public and Private Science, Research in Organizational Behavior, Volume 27, Pages 305–353

Crossan, M. & Bedrow, I. (2003), "Organizational learning and strategic renewal". Strategic Management Journal, 24, 1087–1105.

Daly, H. E (2007), Ecological Economics and Sustainable Development, Selected Essays of Herman Daly, Advances in Ecological Economics, MPG Books Ltd, Bodmin, Cornwall http://library.uniteddiversity.coop/Measuring_Progress_and_Eco_Footprinting/Ecological_Economics_and_Sustainable_Development-Selected_Essays_of_Herman_Daly.pdf

Desa (2013), World Economic and Social Survey 2013, Sustainable Development Challenges, Department of Economic and Social Affairs, The Department of Economic and Social Affairs of the United Nations Secretariat, NY https://sustainabledevelopment.un.org/content/documents/2843WESS2013.pdf

E. Seery and A. Caistor Arendar. (2014). Even It Up: Time to end extreme inequality. Oxfam. https://policy-practice.oxfam.org.uk/publications/even-it-up-time-to-end-extreme-inequality-333012

Esteban Ortiz-Ospina (2018). - "Economic inequality by gender". Published online at OurWorldInData.org. Retrieved from: 'https://ourworldindata.org/economic-inequality-by-gender' [Online Resource]

Esty, K., R. Griffin, and M. Schorr-Hirsh. (1995), Workplace diversity. A manager's guide to solving problems and turning diversity into a competitive advantage. Avon, MA: Adams Media Corporation.

Fischer, D.; Jenssen, S.; Tappeser, V. Getting (2015) an Empirical Hold of Thesustainable University: A Comparative Analysis of Evaluation Frameworks across 12 Contemporary Sustainability Assessment Tools. Assess. Eval. High. Educ. 40, 785–800

Fobes (2019), America's Wealth Inequality Is At Roaring Twenties Levels, Contributor-Jesse Colombo, Forbes (https://www.forbes.com/sites/jessecolombo/2019/02/28/americas-wealth-inequality-is-at-roaring-twenties-levels/#62f244642a9c).

Freistein, K., and. Mahlert, B. (2015), The Role of Inequality in the Sustainable Development Goals, Conference Paper, University of Duisburg-EssenSee discussions, stats, and author profiles for this publication at: https://www.researchgate.net/publication/301675130

Furceri D., P. Loungani and J.D. Ostry. (2020). How Pandemics Leave the Poor Even Farther Behind. IMF Blog. https://blogs.imf.org/2020/05/11/how-pandemics-leave-the-poor-even-farther-behind/

Galbraith, James K. (2012). Inequality and Instability: The Study of the World Economy Just before the Great Crisis. Oxford: Oxford University Press

Girard, Luigi Fusco (2010), Sustainability, creativity, resilience: toward new development strategies of port areas through evaluation processes, Int. J. Sustainable Development, Vol. 13, Nos. 1/2, 2010 161

Gneiting U., N. Lusiani and I. Tamir. (2020). Power, Profits and the Pandemic: From corporate extraction for the few to an economy that works for all. Oxfam International. https://www.oxfam.org/en/research/power-profits-and-pandemic

Goldin, C., & Rouse, C. (2000). Orchestrating impartiality: The impact of" blind" auditions on female musicians. *American Economic Review*, 90(4), 715–741

Gould E. and V. Wilson. (2020). Black workers face two of the most lethal preexisting conditions for coronavirus—racism and economic inequality. Economic Policy Institute. https://www.epi.org/publication/black-workers-covid/

Guidetti, Rehbein (2014), Theoretical Approaches to Inequality in Economics and Sociology, Transcience, Vol. 5, Issue 1 ISSN 2191-1150 https://www2.hu-berlin.de/transcience/Vol5_No1_2014_1_15.pdf

Håvard Mokleiv Nygård (2017), Achieving the sustainable development agenda: The governance – conflict nexus, International Area Studies Review, Vol. 20(1) 3–18

Holvino, E., Ferdman, B. M., & Merrill-Sands, D. (2004), Creating and sustaining diversity and inclusion in organizations: Strategies and approaches. In M. S. Stockdale & F. J. Crosby (Eds.), The psychology and management of workplace diversity (pp. 245–276). Malden, Blackwell Publishing. http://reports.weforum.org/global-gender-gap-report-2020/dataexplorer/ http://www.ilo.org/ilostat/ https://wir2018.wid.world/files/download/wir2018-full-report-english.pdf

IISD, (2005), Indicators. Proposals for a way forward. Prepared L. Pinter, P. Hardi & P. Bartelmus. International Institute for Sustainable Development Sustainable Development, Canada Retrieved January 8, 2015, from https://www.iisd.org/pdf/2005/measure_indicators_sd_way_forward.pdf.

IISD, (2013), The Future of Sustainable Development: Rethinking sustainable development after Rio+20 and implications for UNEP. International Institute for Sustainable Development Retrieved November 5, 2015, from http://www.iisd.org/pdf/2013/future_rethinking_sd.pdf

ILO (2020). Gender Statistics, International Labor Organization (ILO), 1990–2016

Institute of Employment Rights. (2020). Covid-19 has exacerbated wealth gap, research finds https://www.ier.org.uk/news/covid-19-has-exacerbated-wealth-gap-research-finds/;

International Labor Organization. (2020). ILO Monitor: COVID-19 and the world of work. Sixth edition Updated estimates and analysis. https://www.ilo.org/wcmsp5/groups/public/---dgreports/---dcomm/documents/briefingnote/wcms_755910.pdf. Due to the pandemic, working hours are estimated to have declined by 17.3% in the second quarter of 2020 (compared with the fourth quarter of 2019), which is equivalent to 495m full-time jobs. Working hour losses in the third and fourth quarters eased slightly but the jobs deficit at the end of 2020 remains significant.

International Labor Organization. (2020). Social protection responses to the COVID-19 pandemic in developing countries: Strengthening resilience by building universal social protection. https://www.ilo.org/wcmsp5/groups/public/---ed_protect/---soc_sec/documents/publication/wcms_744612.pdf

International Labour Organization. (2020). COVID-19 crisis and the informal economy: Immediate responses and policy challenges. https://www.ilo.org/wcmsp5/groups/public/---ed_protect/---protrav/---travail/documents/briefingnote/wcms_743623.pdf

IUCN (1980), World Conservation Strategy: Living Resource Conservation for Sustainable Development. Retrieved November 7, 2015, from https://portals.iucn.org/library/efiles/documents/WCS-004.pdf.

IUCN, UNDP & WWF, (1991), Caring for the Earth. A Strategy for Sustainable Living. International Union for Conservation of Nature and Natural Resources, United Nations Environmental Program & World Wildlife Fund Retrieved November 8, 2015, from https://portals.iucn.org/library/efiles/documents/CFE-003.pdf

IUCN, UNDP & WWF, (1991), Caring for the Earth. International Union for Conservation of Nature and Natural Resources, United Nations Environmental Program & World Wildlife Fund

Jackson, Tim (2009). Prosperity without Growth: Economics for a Finite planet. Abingdon, United Kingdom: Earthscan.

Jan-Peter Voß, Jens Newig, Britta Kastens, Jochen Monstadt† & Benjamin Nö Lting (2007), Steering for Sustainable Development: a Typology of Problems and Strategies with respect to Ambivalence, Uncertainty and Distributed Power, Journal of Environmental Policy & Planning Vol. 9, Nos. 3-4, September–December 2007, 193–212 https://www.researchgate.net/profile/Jochen_Monstadt/publication/233049753_Steering_for_Sustainable_Development_A_Typology_of_Problems_and_Strategies_with_Respect_to_Ambivalence_Uncertainty_and_Distributed_Power/links/577ff29608ae5f367d370a97/Steering-for-Sustainable-Development-A-Typology-of-Problems-and-Strategies-with-Respect-to-Ambivalence-Uncertainty-and-Distributed-Power.pdf

Jayachandran S. (2015), The Roots of Gender Inequality in Developing Countries, Annu. Rev. Econ. 2015.7:63–88. https://faculty.wcas.northwestern.edu/~sjv340/roots_of_gender_inequality.pdfDownloadedfromwww.annualreviews.org

Jonathan M. Harris (2000), Basic Principles of Sustainable Development, GLOBAL Development And Environment Institute, Working Paper 00–04, Global Development and Environment Institute, Tufts University, https://tind-customer-agecon.s3.amazonaws.com/11dc38b4-a3e2-44d0-b8c8-3265a796a4cf?response-content-disposition=inline%3B%20filename%2A%3DUTF-8%27%27wp000004.pdf&response-content-type=application%2Fpdf&AWSAccessKeyId=AKIAXL7W7Q3XHXDVDQYS&Expires=1560578358&Signature=T%2BpMgFFZjQvmVL8EHsy74Ds%2FKAM%3D

Julia Dehm, (2018), "Highlighting Inequalities in the Histories of Human Rights: Contestations Over Justice, Needs and Rights in the 1970s" *https://doi.org/10.1017/S0922156518000456 (published online 19 September 2018).*

K. Crenshaw. (1989). Demarginalizing the Intersection of Race and Sex: A Black Feminist Critique of Antidiscrimination Doctrine, Feminist Theory and Antiracist Politics. African Journal of International and Comparative Law,139–167. https://chicagounbound.uchicago.edu/cgi/viewcontent.cgi?article=1052&context=uclf,

Karlsson-Vinkhuyzen, Sylvia; Arthur L Dahl and Asa Persson (2018), The emerging accountability regimes for the Sustainable Development Goals and policy integration: Friend or foe? Environment and Planning C: Politics and Space, Vol. 36(8) 1371–1390

Khanal L. , B.K. Paudel and B.K. Acharya. (2020). Community vulnerability to epidemics in Nepal: A high-resolution spatial assessment amidst COVID-19 pandemic. https://www.medrxiv.org/content/medrxiv/early/2020/07/02/2020.07.01.20144113.full.pdf

Klarin, Tomislav (2018), The Concept of Sustainable Development: From its Beginning to the Contemporary Issues, Zagreb International Review of Economics & Business, Vol. 21, No. 1, pp. 67–94, DOI: https://doi.org/10.2478/zireb-2018-0005 https://content.sciendo.com/view/journals/zireb/21/1/article-p67.xml

Lakner C., D. G. Mahler, M. Negre, and E. B. Prydz. (2020). How Much Does Reducing Inequality Matter for Global Poverty? Global Poverty Monitoring Technical Note 13 (June), World Bank. http://documents1.worldbank.org/curated/en/328651559243659214/pdf/How-Much-Does-Reducing-Inequality-Matter-for-Global-Poverty.pdf

Lang, D.J., Wiek, A., Bergmann, M., Stauffacher, M., Martens, P., Moll, P., Swilling, M., Thomas, C.J., (2012), Transdisciplinary research in sustainability science: practice, principles, and challenges. Sustainability Science 7 (1), 25–43.

Lélé, Sharachchandra M. (1991), Sustainable development: A critical review. World Development, Vol 19, No 6, 607–621 https://edisciplinas.usp.br/pluginfile.php/209043/mod_resource/content/1/Texto_1_lele.pdf

LGBT Foundation. (2020.) Why LGBT People are Disproportionately Impacted by COVID-19. https://lgbt.foundation/coronavirus/why-lgbt-people-are-disproportionately-impacted-by-coronavirus

Mair, Simon, Aled Jones, Jonathan Ward, Ian Christie, Angela Druckman, and Fergus Lyon (2017), A Critical Review of the Role of Indicators in Implementing the Sustainable Development Goals in the Handbook of Sustainability Science in Leal, Walter (Edit.) https://www.researchgate.net/publication/313444041_A_Critical_Review_of_the_Role_of_Indicators_in_Implementing_the_Sustainable_Development_Goals

Marin, C., Dorobantu, R., Codreanu, D. & Mihaela, R. (2012), The Fruit of Collaboration between Local Government and Private Partners in the Sustainable Development Community Case Study: County Valcea. Economy Transdisciplinarity Cognition, 2, 93–98. In Duran, C.D.,

Marshall, Alfred (1961) (originally 1920), Principles of Economics, 9th edn, New York: Macmillan.

Marx, Karl (1953ff), Marx-Engels-Werke (MEW). Berlin: Dietz

Meadows, D.H. (1998), Indicators and Information Systems for Sustainable Development. A report to the Balaton Group 1998. The Sustainability Institute.

Meadows, D.H., Meadows, D.L., Randers, J. & Behrens III, W.W. (1972), The Limits of Growth. A report for the Club of Rome's project on the predicament of mankind. Retrieved September 20, 2015, from http://collections.dartmouth.edu/published-derivatives/meadows/pdf/meadows_ltg-001.pdf.

Meakin. (2020). U.K. Low-Income Households Turn to Debt as Rich Save in Lockdown. Bloomberg. https://www.bloombergquint.com/global-economics/u-k-low-income-households-turn-to-debt-as-rich-save-in-lockdown;

Neumark, D., Bank, R. J., & Van Nort, K. D. (1996). Sex discrimination in restaurant hiring: An audit study. The Quarterly Journal of Economics, 111(3), 915–941.

Nick Galasso & Marjorie Wood, (2021), Eight Ways To Reduce Global Inequality, Inequality.org is a project of the Institute for Policy Studies https://inequality.org/great-divide/8-ways-reduce-global-inequality/

Nicolau, Melanie and Rudi W Pretorius (2016), University of South Africa (UNISA): Geography at Africa's largest open distance learning institution, (in book), The Origin and Growth of Geography as a Discipline at South African Universities, Gustav Visser, Ronnie Donaldson, and Cecil Seethal, eds. Stellenbosch, South Africa: Sun Press

Norgaard, R. B., (1988), "Sustainable development: A coevolutionary view," Futures, Vol. 20, No. 6 pp. 606–620.

North, Douglass C. (1981), Structure and Change in Economic History. New York: W.W. Norton & Co.

Olivetti, C., & Petrongolo, B. (2008). Unequal pay or unequal employment? A cross-country analysis of gender gaps. Journal of Labor Economics, 26(4), 621–654.

Oxfam (2021), Even It Up Time to End, Extreme Inequality, OXFAM

Oxfam (2021), The Inequality Virus, bringing together a world torn apart by coronavirus through a fair, just and sustainable economy Embargoed Until January 25, 2021 00:01 GMT, OXFAM

Oxfam. (2020). The hunger virus: how COVID-19 is fueling hunger in a hungry world. Media briefing. Oxfam International: Oxford. https://www.oxfam.org/en/research/hunger-virus-how-covid-19-fuelling-hunger-hungry-world.

Pap/Rac, (1999), Carrying capacity assessment for tourism development, Priority Actions Program, in framework of Regional Activity Centre Mediterranean Action Plan Coastal Area Management Program (CAMP) Fuka-Matrouh – Egypt, Split: Regional Activity Centre

Purvis, Ben, Yong Mao and Darren Robinson (2018), Three pillars of sustainability: in search of conceptual origins, Sustainability Science, Springer,23https://doi.org/10.1007/s11625-018-0627-5r15%20-low%20res%2020100615%20-.pdf

Robertson T., E.D. Carter, V.B. Chou, A.R. Stegmuller, B.D. Jackson, Y. Tam, T. Sawadogo-Lewis and N. Walker. (2020). Early estimates of the indirect effects of the COVID-19 pandemic on maternal and child mortality in low-income and middle-income countries: A modelling study. The Lancet Global Health, Vol. 8, Issue 7. https://doi.org/10.1016/S2214-109X(20)30229-1

Rockström, J., Steffen, W., Noone, K., Persson, Å, Chapin, S., Lambin, E., Lenton, T., Scheffer, M., Folke, C., Schellnhuber, H., Nykvist, B., de Wit, C., Hughes, T., van der Leeuw, S., Rodhe, H., Sörlin, S., Snyder, P., Costanza, R., Svedin, U., Falkenmark, M., Karlberg, L., Corell, R., Fabry, V., Hansen, J.,Walker, B., Liverman, D., Richardson, K., Crutzen,P., Foley, J., (2009), A safe operating space for humanity. Nature 461, 472–475.

Sayed, A.; Asmuss, M. (2013), Benchmarking Tools for Assessing and Tracking Sustainability in Higher Educational Institutions. Int. J. Sustain. High. Educ. 14, 449–465

Scott, W. R. (2003). "Institutional carriers: reviewing models of transporting ideas over time and space and considering their consequences". Industrial and Corporate Change, 12, 879–894.

Soares, Maria Clara Couto, Mario Scerri and Rasigan Maharajh -edits (2014), Inequality and Development Challenges, Routledge, https://prd-idrc.azureedge.net/sites/default/files/openebooks/032-9/

ST/ESA/372, United Nations publication, Sales No. E.20.IV.1, ISBN 978-92-1-130392-6, eISBN 978-92-1-004367-0 https://www.un.org/development/desa/dspd/wp-content/uploads/sites/22/2020/01/World-Social-Report-2020-FullReport.pdf

Sterling, S. (2010), Learning for resilience, or the resilient learner? Towards a necessary reconciliation in a paradigm of sustainable education. Environmental Education Research, 16, 511–528. DOI: 10.1080/13504622.2010.505427.

Sumner A., E. Ortiz-Juarez and C. Hoy. (2020). Precarity and the pandemic: COVID-19 and poverty incidence, intensity, and severity in developing countries. WIDER Working Paper 2020/77. https://www.wider.unu.edu/sites/default/files/Publications/Working-paper/PDF/wp2020-77.pdf

The Independent. (2020). Coronavirus crisis "exacerbating inequalities as richer build savings faster". https://www.independent.co.uk/news/uk/home-news/coronavirus-inequality-savings-banks-ifs-b1395750.html;

Tomislav Klarin (2018), The Concept of Sustainable Development: From its Beginning to the Contemporary Issues, Zagreb International Review of Economics & Business, Vol. 21, No. 1, pp. 67–94, DOI: https://doi.org/10.2478/zireb-2018-0005 https://content.sciendo.com/view/journals/zireb/21/1/article-p67.xml

UN Women. (2020). From Insights to Action: Gender equality in the wake of COVID-19. https://www.unwomen.org/-/media/headquarters/attachments/sections/library/publications/2020/gender-equality-in-the-wake-of-covid-19-en.pdf?la=en&vs=5142

UN, United Nations (1972), Report of the United Nations Conference on the Human Environment. Stockholm. Retrieved September 20, 2015, from http://www.un-documents.net/aconf48-14r1.pdf.

UN, United Nations (1997), Earth Summit: Resolution adopted by the General Assembly at its nineteenth special session. Retrieved November 4, 2015, from http://www.un.org/esa/earthsummit/index.html.

UN, United Nations (2002), Report of the World Summit on Sustainable Development, Johannesburg; Rio +10. Retrieved November 4, 2015, from http://www.unmillenniumproject.org/documents/131302_wssd_report_reissued.pdf.

UN, United Nations (2010), The Millennium Development Goals Report. Retrieved September 20, 2015, from http://www.un.org/millenniumgoals/pdf/MDG%20Report%202010%20En%20

UN, United Nations (2012). Resolution "The future we want". Retrieved November 5, 2015, from http://daccess-dds-ny.un.org/doc/UNDOC/GEN/N11/476/10/PDF/N1147610.pdf?.

UN, United Nations (2015). SDGS, UN, New York, Retrieved September 21, 2015, from http://www.un.org/en/index.html.

UN, United Nations (2015b), 70 years, 70 documents. Retrieved September 21, 2015, from http://research.un.org/en/UN70/about.

UN, United Nations (2015c), Resolution "Transforming our world: the 2030 Agenda for Sustainable Development. Retrieved November 5, 2015, from http://www.un.org/ga/search/view_doc.asp?symbol=A/RES/70/1&Lang=E.

UN, United Nations (2015d), The Millennium Development Goals Report 2015. Retrieved November 5, 2015, from http://www.un.org/millenniumgoals/2015_MDG_Report/pdf/MDG%20

UN. (2020). Tackling the Inequality Pandemic: A New Social Contract for a New Era. UN Secretary-General's Lecture for Nelson Mandela's International Day. https://www.un.org/sg/en/content/sg/statement/2020-07-18/secretary-generals-nelson-mandela-lecture-%E2%80%9Ctackling-the-inequality-pandemic-new-social-contract-for-new-era%E2%80%9D-delivered

UNDESA-DSD –(2002). United Nations Department of Economic and Social Affairs Division for Sustainable Development, 2002. Plan of Implementation of the World Summit on Sustainable Development: The Johannesburg Conference. New York. UNESCO – United Nations Educational, Scientific and Cultural Organization, 2005. International Implementation Scheme. United Nations Decade of Education for Sustainable Development (2005–2014), Paris.

UNDP (2020). UN Human Development Report, United Nations Development Program http://hdr.undp.org/en/data# http://data.worldbank.org/data-catalog/world-development-indicators

UNDP. (2020). Gender Gaps in the Care Economy during the COVID-19 Pandemic in Turkey. https://www.tr.undp.org/content/turkey/en/home/library/corporatereports/COVID-gender-survey-report.html

UNEP (2010), Background paper for XVII Meeting of the Forum of Ministers of Environment of Latin America and the Caribbean, Panamá City, Panamá, 26 -30 April 2010, UNEP/LAC-IG.XVII/4, UNEP, Nairobi, Kenya http://www.unep.org/greeneconomy/AboutGEI/WhatisGEI/tabid/29784/Default.aspx.

UNESCO (2013), UNESCO's Medium-The Contribution of Creativity to Sustainable Development Term Strategy for 2014–2021, http://www.unesco.org/new/fileadmin/MULTIMEDIA/HQ/CLT/images/CreativityFinalENG.pdf

UNESCO, UNICEF and the World Bank. (2020). What Have We Learnt? Overview of findings from a survey of ministries of education on national responses to COVID-19. https://data.unicef.org/resources/national-education-responses-to-covid19/

UNICEF. (2016). Girls spend 160 million more hours than boys doing household chores everyday [sic]. https://www.unicef.org/press-releases/girls-spend-160-million-more-hours-boys-doing-household-chores-everyday

United Nations (2015), Transforming our world: the 2030 agenda for sustainable development. New York (NY): United Nations; 2015 (https://sustainabledevelopment.un.org/post2015/transformingourworld,accessed5October2015).

United Nations (2008), 52nd session of the Commission on the Status of Women (2008) "Gender perspectives on climate change," Issue's paper for interactive expert panel on Emerging issues, trends and new approaches to issues affecting the situation of women or equality between women and men. http://www.un.org/womenwatch/daw/csw/csw52/issuespapers/ Gender%20and%20climate%20change%20paper%20final.pdf

United Nations (2002), 46th Session of the Commission on the Status of Women (2002) "Agreed Conclusions,". Report on the forty-sixth session of Commission on the Status of Women. Official Records, 2002 supplement No. 7 (E/2002/27-E/CN.6/2002/13). p12. Economic and Social Council, United Nations. http://daccessdds.un.org/doc/UNDOC/GEN/N02/397/04/PDF/N0239704.pdf?OpenElement

United Nations (2020). – Gender Statistics, United Nations (Minimum Set of Gender Indicators, as agreed by the United Nations Statistical Commission in its 44th Session in 2013) https://genderstats.un.org

United Nations (2020). Inequality in A Rapidly Changing World, World Social Report 2020, UN Department of Economic and Social Affairs, New York

United Nations (2020a), Inequality – Bridging the Divide- Shaping our future together- United Nations 75–2020 and Beyond, United Nations, New York https://www.un.org/en/un75/inequality-bridging-divide

United Nations General Assembly. (1987), Report of the world commission on environment and development: Our common future. Oslo, Norway: United Nations General Assembly, Development and International Co-operation: Environment.

Vander-Merwe, I. & Van-der-Merwe, J. (1999), Sustainable development at the local level: An introduction to local agenda 21. Pretoria: Department of environmental affairs

Vos, Robert O. (2007), Perspective Defining sustainability: a conceptual orientation, Journal of Chemical Technology and Biotechnology, 1 82: 334–339 (2007)

Wals, A., (2009), United Nations Decade of Education for Sustainable Development (DESD, 2005–2014): Review of Contexts and Structures for Education for Sustainable Development Learning for a sustainable world 2009. Paris.

WB, The World Bank (2015), World Development Indicators. Retrieved September 2, 2015, from http://data.worldbank.org/data-catalog/world-development-indicators.

WCED (1987), Our Common Future World Commission on Environment and Development New York: Oxford University Press

Wickramage. K. et al. (2018). Missing: Where are the migrants in pandemic influenza preparedness plans? Health and Human Rights Journal, 2018 Jun; 20(1): 251–258. https://www.ncbi.nlm.nih.gov/pmc/articles/PMC6039731/

WID (2018), World Inequality Report 2018, The Paris School of Economics, Inequality Lab, WID.world,

Williams O. (2020). How Wealth Managers Helped Millionaire Clients Grow Richer During Lockdown. Forbes, 3 June 2020. https://www.forbes.com/sites/oliverwilliams1/2020/06/03/how-wealth-managers-helped-millionaire-clients-grow-richer-during-lockdown/#49d27bda6951

Wiseman, Erica (2007), The Institutionalization of Organizational Learning: A Noninstitutional Perspective, Proceedings of OLKC 2007 – "Learning Fusion", UK https://warwick.ac.uk/fac/soc/wbs/conf/olkc/archive/olkc2/papers/wiseman.pdf

World Bank (2020). – World Development Indicators, World Bank

World Bank (2020). Gender Statistics- World Bank – https://datacatalog.worldbank.org/dataset/gender-statistics

World Bank. (2011). Gender Equality and Development, World Bank http://siteresources.worldbank.org/INTWDR2012/Resources/7778105-1299699968583/7786210-1315936222006/Complete-Report.pdf

World Bank. (2020). Poverty and Shared Prosperity 2020: Reversals of Fortune. Chapter 2. https://www.worldbank.org/en/publication/poverty-and-shared-prosperity

World Bank. (2020). Poverty and Shared Prosperity 2020: Reversals of Fortune. https://www.worldbank.org/en/publication/poverty-and-shared-prosperity.

World Economic Forum (2020). – Global Gender Gap Report, World Economic Forum

World Economic Outlook (2007), Globalization and Inequality, World Economic Forum, DC

World Food Program. (2020). World Food Programme to assist largest number of hungry people ever, as coronavirus devastates poor nations. https://www.wfp.org/news/world-food-programme-assist-largest-number-hungry-people-ever-coronavirus-devastates-poor

Wright, E. O., (1994), Interrogating Inequalities. New York: Verso

Wu,SOS, Jianguo (Jingle) (2012), Sustainability Indicators Sustainability Measures: Local-Level SDIs494/598–http://leml.asu.edu/Wu-SIs2015F/LECTURES+READINGS/Topic_08-Pyramid%20Method/Lecture-The%20Pyramid.pdf

Yarime, M.; Tanaka, Y. (2012), The Issues and Methodologies in Sustainability Assessment Tools for Higher Education Institutions—A Review of Recent Trends and Future Challenges. J. Educ. Sustain. Dev. 6, 63–77.

Websites Resources

Gender and Sustainable Development in Drylands: An Analysis of Field Experiences
ftp://ftp.fao.org/docrep/fao/005/j0086e/j0086e00.pdf

Gender and Desertification: Expanding roles for women to restore drylands
http://www.ifad.org/pub/gender/desert/gender_desert.pdf

Gender and Desertification: Making ends meet in drylands
http://www.ifad.org/pub/gender/desert/gender_desert_leaf.pdf

Gender and Equity Issues in Liquid Biofuels Production. Minimizing the risks to maximize the opportunities
ftp://ftp.fao.org/docrep/fao/010/ai503e/ai503e00.pdf

Rural Households and Sustainability: Integrating environmental and gender concerns into home economics curricula
http://www.fao.org/DOCREP/V5406e/V5406e00.htm

Energy and Gender: In rural sustainable development, Rome, 2006
ftp://ftp.fao.org/docrep/fao/010/ai021e/ai021e00.pdf

2004. Gender Perspectives on the Conventions on Biodiversity, Climate Change and Desertification. Rome, FAO.
http://www.fao.org/sd/dim_pe1/pe1_041002_en.htm

Environment and Natural Resource Management: IFAD' s Growing Commitment
http://www.ifad.org/pub/enviorn/EnvironENG.pdf

Adivasi Women: Engaging with Climate Change
http://www.unifem.org/materials/item_detail.php?ProductID=149

Gender and Water – Securing Water for Improved Rural Livelihoods: The multiple-uses system approach
http://www.ifad.org/gender/thematic/water/gender_water.pdf
2008. Climate change, water and food security. High Level Conference on World Food Security - Background Paper HLC/08/BAK/2. FAO
ftp://ftp.fao.org/docrep/fao/meeting/013/ai783e.pdf
Women, Health and the Environment
Our Planet (Volume 15, No 12)
http://www.unep.org/ourplanet/imgversn/152/images/Our_Planet_15.2.pdf
Migration, Climate Change and the Environment (Policy Brief)
http://www.iom.int/jahia/webdav/shared/shared/mainsite/policy_and_researc h/policy_documents/policy_brief.pdf
Climate Change, Migration and Human Rights, Address by Ms. Kyung-wha Kang Deputy High Commissioner for Human Rights Office of the United Nations Conference on Climate Change and Migration: Addressing Vulnerabilities and Harnessing Opportunities, 19 February 2008, Geneva
http://www.unhchr.ch/huricane/huricane.nsf/view01/
BA5B630BFFAD7FC1C12573F600386398?opendocument
State of the World Population 2009- Facing a changing world: women, population and climate
http://www.unfpa.org/swp/2009/en/
Human Rights and Climate Change Resolution 7/23, March 2008
http://www2.ohchr.org/english/issues/climatechange/docs/Resolution_7_23.pdf
Resource Guide on Gender and Climate Change
http://content.undp.org/go/cms-service/download/asset/?asset_id=1854911
Training Manual on Gender and Climate Change
http://www.reliefweb.int/rw/lib.nsf/db900sid/ASAZ-7SNCA9/\protect\T 1\textdollarfile/UNDP_Mar2009.pdf
Energy and Gender for Sustainable Development
http://www.energyandenvironment.undp.org/undp/index.cfm?module=Library &page=Document&DocumentID=5108
Resource Guide on Gender and Climate Change
http://content.undp.org/go/cms-service/download/asset/?asset_id=1854911
Training Manual on Gender and Climate Change
http://www.reliefweb.int/rw/lib.nsf/db900sid/ASAZ-7SNCA9/\protect\T 1\textdollarfile/UNDP_Mar2009.pdf?
Gender Perspectives: Integrating Disaster Risk Reduction into Climate Change Adaptation – Good Practices and Lessons Learned

http://www.unisdr.org/eng/about_isdr/isdr-publications/17-Gender_Perspectives_Integrating_DRR_CC_Good%20Practices.pdf

WHO (2008) "Gender inequities in environmental health"? 25th Session of the European Environment and Health Committee. (EUR/5067874/151). http://www.euro.who.int/Document/EEHC/25th_EEHC_Milan_edoc15.pdf

WHO (2008) "Gender inequities in environmental health"? 25th Session of the European Environment and Health Committee. (EUR/5067874/151). http://www.euro.who.int/Document/EEHC/25th_EEHC_Milan_edoc15.pdf

Update on IDA Contribution to COVID-19 Pandemic Response (2021). Washington, D.C. : World Bank Group.
http://documents.worldbank.org/curated/en/673321588557820754/Update-on-IDA-Contribution-to-COVID-19-Pandemic-Response

2

Social Inclusion, Sustainable Development, and ASTA-JA in Nepal

Durga D. Poudel

The Founder of Asta-Ja Framework, Professor, University of Louisiana at Lafayette, Lafayette, Louisiana, USA
E-mail: durga.poudel@louisiana.edu

Abstract

This chapter unveils the major challenges of social inequalities, economic inequality, gender inequality, healthcare inequality, education inequality, employment inequality, and spatial inequality and argues that until or unless society adopts the measures of equity, social justice, and social inclusion, the goals of sustainable development cannot be achieved. This chapter elaborates Asta-Ja Framework, the interconnectedness of eight resources in Nepali letter *Ja—Jal* (water), *Jamin* (land), *Jungle* (forest), *Jadibuti* (medicinal and aromatic plants), *Janashakti* (manpower), *Janawar* (animal), *Jarajuri* (crop plants), and *Jalabayu* (climate) and argues that proper management, development, and utilization of these resources are necessary to attain sustainability. Chapter shows that inequitable access to natural resources such as land, water, rangelands, and forests coupled with entrenched social inequality in the society, relatively a large population is marginalized, resource-deprived, and is under poverty. Due to pervasive discriminations, social exclusions and poverty, the economic growth, community resiliency, and socio-economic transformation of the nation have become very challenging.

Keywords: Inequalities, social inclusion, ownerships, equity, Asta-Ja, sustainability

2.1 Introduction

Equity, social justice, and sustainable development of a society are some of the very challenging issues worldwide. Even though societies across the globe have tried to address these very complex issues of equitable access to natural resources, alleviating inequalities in various forms and realizing economic growth and environmental quality for a long time, the outcomes are not satisfactory. Discontent in the societies with respect to equitable access to resources and addressing the issue of inequalities are manifested in various forms including resentments, lawsuits, protests, riots, and even wars. International efforts on addressing these issues by requiring governments granting human rights and imposing bans on any form of discriminations are commendable. Despite this global drive on the alleviation of inequalities and promoting social inclusions in development, success on mainstreaming marginalized communities in economic growth and socio-economic transformation is far from satisfactory. Therefore, it is important to assess the issue of equity, social inclusion, and economic development from the grassroot levels to national level and come up with practical and sustainable strategies in systematically addressing the issue of equity, social justice, and sustainable development.

Nepal is an agricultural country where almost 68% of its nearly 30 million people depend on agriculture for livelihoods. Nepalese agriculture contributes about 33% to national GDP. Of the total surface area of 147,181 km^2, 28.7% include agricultural land. The National Sample Census of Agriculture 2011/12 showed a total of 3,715,555 households engaged in crop production (having > 0.01 ha land) and 115,538 households engaged in the production of livestock only (having < 0.01 ha land) (NPC, 2013). The Terai, Hills, and Mountain regions contain, respectively, 70%, 26%, and 4% of total agricultural land with corresponding 50.3%, 43.0%, and 6.7% of total population. Top 5% farmer households control more than 37% of agricultural land while bottom 47% of farmer households control only 15% of agricultural land. Despite several efforts on land reform, 25% of Nepalese population is landless or near landless (Leitner Center, 2011). In an agrarian society, the ownership of land is the single most important factor deciding the livelihood, self-esteem, family income and nutrition, education, and community resiliency of the tillers. Nepalese agrarian society is seriously suffering primarily due to: (1) a very small landholding size for most tillers, while a large amount of agricultural land is in the hands of just 5% of the farmer households, (2) a large number of landless populations cultivating government lands without land titles, and (3) massive fragmentation of land holdings. It is important to address all these land ownership issues appropriately for agricultural and socio-economic development of the country.

The issue of ownership rights and access is important not only in land resource but also in other critical resources such as forests, water, pasturelands, and minerals and mines for economic growth and fast-paced socio-economic transformation of the society. This becomes even more important in smallholder mixed-farming system like in Nepal. Smallholder mixed-farming system is characterized by an integrated production of crops, raising livestock, and utilizing pasturelands and forest resources for grazing, forest products, and for other necessities and water resource for irrigation (Poudel, 2015). Typically, a smallholder mixed farmer in Nepal produces rice, maize, vegetables, fruits, and raises animals mainly buffalo, cattle, goats, sheep, and chickens. Smallholder mixed-farming system is both the challenge and opportunity for agricultural development in Nepal. It is a challenge in the sense that landholding size, labor constraint, and willingness and capability of farmers in managing smallholder mixed-farming system profitably in the backdrop of outmigration, climate change impacts, land use conversions, and resource degradation create challenging situation in Nepal due to its geographical condition (to the other countries which have challenges of geographical and social diversity). On the other hand, smallholder mixed-farming system is well suited for organic production (Poudel, 2018). Diversified agro-ecoregions with favorable growing conditions throughout the year give competitive advantage for Nepal in production of organic fruits and vegetables, dairies, spices, and other products. In fact, over 90% of agricultural production especially in the hills and high mountains remain to be organic. As sustainable management and utilization of natural resources such as land, water, forest, and pasturelands is a pre-requisite to profitable agricultural production, ownerships and users' rights of these critical resources become very important for the success of smallholder mixed-farming system.

As Hardin, 1968 demonstrated in his famous essay "The Tragedy of The Commons" that a complete freedom in the use of public resources by its stakeholders ultimately results in the ruins of the resources itself, a clear definition on the users' rights is critical for natural resources sustainability. A country's constitutions, laws and regulations, and policies define the ownership rights on natural resources. Any disproportionate access to critical natural resources results in serious economic and social inequalities including health, education, and gender discrimination and poverty in the society. In the subsequent sections, I will discuss the issues of access to land, water, forests, and other natural resources in Nepal, the issue of economic and social inequality, and the issue of social inclusion and sustainable development. Then, I will present Asta-Ja Framework for sustainable conservation, development, and utilization of natural resources for economic growth and socio-economic transformation. Finally, I will present policy recommendations for addressing the issue of equity, social justice, and sustainable community development.

2.2 Inequitable Access

2.2.1 Land resource

Inequitable access to land resource has been a major problem in Nepalese society. Land ownership is strongly associated with social prestige, farm income guarantee, food security, family nutrition, and health at the farm level. Leitner Center, 2011 presents an excellent account of land ownership rights in Nepal. With the national unification campaign that was initiated and led by the Great King Prithvi Narayan Shah, all lands were owned by the State, but the tillers could cultivate the lands and offer produce to the Chief as Kut. The tillers did not have the land rights. This system of land rights was called Raikar system. Then, land rights were given to the members of ruling families and their functionaries under the Birta system. Land rights were given mostly to Royal and Rana families, high-level government official, and individuals with high connection to the rulers. The owner of Raiker grant not only owned the land but also peasants who worked in the land. Under the Raikar system, land rights were given as Birta land to upper class people, Guthi land to religious organizations, and also lands were given to individual employees for good performance. Almost one-third of total cultivated land was Birta land in 1950. Rana regime families and their intermediaries owned most of the land. Absentee landlordism was very common. Another category of land ownership was Kipat system, where land ownership rights were given to community groups. After the fall of Rana regime, a new constitution in 1951 incorporated land ownership rights to the common people. Traditional Raikar system, Jamindari system, and Birta system were subsequently abolished. King Mahendra took the land issue very seriously. A land commission was established in 1956. Land Act was enacted in 1964. Efforts were made to transfer land ownership rights to the tillers. At present, almost 28% of Nepal's land is privately owned or in leasehold. Policy of Joint land Ownership (JLO) 2011 is also available in which husbands and wives can register their lands for joint ownership. Despite several attempts in tackling land ownership issue, 25% of Nepalese population is still landless or near landless (Leitner Center, 2011). In 2019, the Government of Nepal made the Eighth amendment to Land Act and has made provisions of giving land rightsof certain parcel of such lands to those landless people who are using government lands for last 10 years. Over a million hectares of agricultural land is reportedly left fallow in recent years primarily due to outmigration of youths for foreign employment.

2.2.2 Water resource

Access to water resources is generally associated with the location of the settlements, especially in the hills and mountain regions. Settlements located on the summits or hill slopes have less access to water resource compared to those settlements, which are located in the valley floors or nearby rivers or streams. Communities located further away from the drinking water sources in the summits or hill slopes spend much more time in fetching drinking water than those located nearby the drinking water sources. Communities located nearby water sources have access to irrigation water whereas, communities located at a distance from sources mostly lack irrigation water supply. Similarly, financial status of individuals is also equally responsible for inequitable access to water resource. For example, in the case of utilizing groundwater, some can afford to install pumps for getting drinking water or for irrigation but some cannot. Individuals who are using irrigation water traditionally continue enjoying user rights even today. WaterAid Nepal, 2005 nicely summarizes Nepali water laws, regulations, and policies including licensing water use and several other water-related topics. The Essential Commodity Protection Act 1955 protects the drinking water as an essential commodity and the Muluki Ain 1963 sets the priority of water uses. These acts established rights of individuals, group of individuals, or the community to divert water from sources like streams, rivers, or ground water. Diversion for irrigation must not have adverse downstream impacts with regard to government irrigation schemes or hydropower plants. Whoever built Kula (small canal) has the authority to use its water first. No one is allowed to restrict those people who are using it since from the beginning. The Nepal Water Supply Corporation Act 1989 established Nepal Water Supply Corporation as an autonomous government-controlled corporation responsible for supplying drinking water. However, this system was changed to privatized system after the political change of 1990. The Water Resource Act 1992, the Umbrella

Act, gives ownership of water to the State and people have user-rights so that the resources are utilized for creating national assets and contribute to revenue. Sectoral prioritization under this act includes Drinking water, Irrigation, Agricultural Uses (Livestock), Hydropower, Cottage Industries, Industries and Mining, Navigation, and Recreation Use. Under this act, Water User Associations were formed, and licenses were issued. The Electricity Act 1992 established a system of licensing for hydropower production. The Water Resource Regulation 1993 established the rights and obligations of Water User Associations and license holders. Similarly, the Drinking Water Regulation 1998 authorizes the formation of the Drinking Water User Associations and deals with the licensing of drinking water use. The Irrigation Regulation 2000 deals with the Irrigation Water User Associations. Thus, various laws and regulations developed after 1990 have privatized water use and hydropower generation in Nepal. Nepal Constitution 2015 Part 3, Article 35, Clause 4, protects right to drinking water in Nepal. International treaties and agreements have also affected access to water resources in Nepal. A more in-depth analysis in this topic is necessary.

2.2.3 Rangelands

Rangelands consist of nearly 12% of total land of Nepal. Rangelands are comprised of grasslands, pasturelands, grazing lands in riparian areas and in the forest, shrub-lands, and other grazing lands (Pande, 2009). Ninety-six percent of total rangelands lie in the mid-hills, the mountains, and the high Himalayan regions and only 4% of the total rangelands lie in the Terai region. Livestock-based farming system and rangelands provide livelihoods for the communities in high altitudes. Of the total 1,701,671 ha of rangelands, the Himalayas, hills, and the Terai regions contain 1,082,235 ha, 545,335 ha, and 74,101 ha, respectively. In the mountains and the high Himalayan regions, local communities had historically owned the rangelands. Rangelands were nationalized following the promulgation of Grazing Lands Act in 1974. In recent years, rangelands user's rights are also transferred to Forest User Groups. However, elites are still controlling some rangelands in different parts of the country and, obviously, there is a great deal of confusion in relation to their ownership rights. The lack of ownership rights coupled with increased number of livestock, overgrazing of rangelands, and climate change impacts have resulted in severe degradation of rangelands.

2.2.4 Forests

Forestlands in Nepal consist of 6,584,878 hectares (which is 44.74% of the total land of Nepal) out of the country's total area of 147,181 km^2 (Rastriya Samachar Samiti, 2016). The Forest Act 1957, nationalized all forests with limited access to local communities. After the first Community Forest was established in Thokarpa Village in Sindhupalchowk district in 1973, the National Forest Plan 1976 put forwarded the policy of Community Forestry in Nepal (DoFSC, 2020). Subsequent Forestry Master Plan 1989 approved the transfer of user's rights of forests in the hills to local communities. Forest Act 1993 defined the national forest into five categories: Government Managed Forest, Protected Forest, Community Forest, Leasehold Forest, and Religious Forest. While Government Managed Forests include national forest managed by the government, the Protected Forests include national forest declared by the Government of Nepal as a protected forest because of its special environmental, scientific, or cultural significance. Community Forest is defined as the national forest handed over to a user group for the protection, development, and utilization of forest resource in the community. Similarly, Leasehold Forest is defined as national forest handed over to an institution or community to support forest-based industry by producing raw materials, producing and selling forest products, operating tourism industry, agroforestry, or to operate farm of insects, butterflies, and wildlife. Religious Forest is a national forest handed over to a religious body, group or community who is interested in the conservation, development, and utilization of the national forest of a religious place or its surroundings. Private Forests include the forests that are grown in private lands. Thirty-four percent (i.e., 2,237,670 hectare) of the total forestlands in Nepal is under Community Forestry (DoFCFD, 2018). Community Forests are supplying fodder and forages, fuelwoods, leaf litters, timber, and other commercial products to the communities (Poudel, 2015). There are 22,266 Community Forest User's Groups across the country in which 2,907,871 households are associated. Distribution of forest cover across the physiographic regions include 32% in high mountains, 38% in mid-hills, 23% in Chure region, and 7% in the Terai. Almost half of the country's population live in the Terai region. Since there was no Community Forestry program approved for the Terai region, the 2057 Forest Policy made a provision of Collaborative Forests and Chakla Forests (i.e., Block Forest). This provision requires involvement of the Government and the local communities in forest management. Communities can get fodder and forages, logs, and other products from the forests. There are 30 Collaborative Forests in 12 districts and 6 Chakla Forests (i.e., Block Forest) in 6 districts in the Terai region. The differences in the modalities of forest management between the hills, mountains, and the Terai region clearly show an example of inequitable access to forest resource based on spatial (geographic) differences in Nepal.

2.2.5 Mines and Minerals

Nepal is enriched with many minerals and mines. Some of the commonly cited minerals found in Nepal include limestone, iron, copper, slate, marble, lead, nickel, pyrite, and gold. The Phulchoki iron mine alone is estimated to contain over 10 million tons of iron. There are 31 copper mines identified in Nepal. The Mines and Minerals Act 1985 establishes state's ownership rights on mines and minerals. It states, "All minerals lying or discovered on the surface or underground in any land belonging to an individual or the government within Nepal shall be the property of the Government of Nepal." The Government of Nepal has exclusive power in carrying out the mining operations. This act also has provisions of licensing mining operations to private individuals or companies who must follow necessary measures for environmental protection. The 2015 Constitution vests ownerships of minerals and mines to federal, provincial, and local governments. In this context, local or provincial governments that lack minerals and mines will have disadvantage over those that have mines and minerals in terms of revenue generation.

2.3 Economic and Social Inequalities

2.3.1 Economic Inequality

Although poverty rate in Nepal has dropped to 21% from 42% in past 20 years, the gap between the richest and the poorest has reportedly widened extensively in recent years. Available literatures suggest that 10% richest Nepalese are earning three times more than the poorest 40% and the 10% richest own 26% more wealth than the 40% poorest. Economic inequality makes the society very divisive and discriminatory and it impedes the overall progress. Economic inequality results in social conflicts, political unrest, and adversely affects individual's access to healthcare, education, and personal wellbeing. Economic inequality occurs due to differences in income-generating activities, employment opportunities, geographic regions, ancestral asset inheritance, and business opportunities. While cash flow in

Nepalese society has increased primarily due to remittances, sharp increase on imports of food items and other goods has heavily strained Nepalese economy, and the divide between haves and have nots is further widening.

2.3.2 Social Inequality

Nepalese society has a massive social inequality issue. Social status in Nepalese society comes from many complex factors that include: caste system and hierarchy within the caste; the ancestry of an individual; relationships due to marriages; location of residence such as city versus rural area; how much agricultural land (or irrigated land) a family owns; whether a family owns a piece of land for housing in city areas or by the roadside; whether a family has a house in a city or a town; structure of the house (e.g., thatched roof vs. galvanized iron roof houses or concrete vs. mud and stone houses); job status of a person; and the level of education. The caste system, which was legalized by the Muluki Ain (Civil Code) of 1854, was abolished in 1963. However, caste-based discrimination is still widespread. There are serious social restrictions on marriages, worships, gatherings, and festivities between and within caste system. Talents, skills, integrity, and qualities of individuals especially from lower caste in the hierarchy are often undermined and not respected. Caste system segregates people in the society. People in the lower caste are looked down and are often considered inferior by their births. Children from lower caste groups are considered less intelligent and less capable of education. Social inequality affects the accessibility to health facilities, employment and business opportunities, and access to the state's resources. Social discrimination leaves very deep scars on people's personal growth, feelings, and psychology, which may affect the future generations. Therefore, it is necessary to fight against social discrimination and inequality jointly by the people who fall in different social strata in the society.

2.3.3 Gender Inequality

Nepal ranks 115th out of 160 countries in Gender Inequality and experts say that it will take more than 200 years to fix this issue. Literacy rates differ by at least 17% between males and females. Female population suffers from widespread violence, rapes, and harassments. Females, especially in rural areas, are considered as household chores servers; the caregivers of children, sickly people and elderly people, and animals; food makers, and dish and house cleaners. Females in rural areas are also tied-up with many agricultural activities such as rice planting and growing vegetables. They have much lower access to decision-making process in their household matters, money management, and investment. However, guaranteeing 33% of the total seats in the parliament for females and provisions of males and females in the highest positions like President, Vice President, and similar provisions on other governmental structures by the 2015 Constitution of Nepal are commendable. The Government of Nepal is committed in addressing gender inequality since its First-Five-Year Plan of 1956 and since then it has continued its gender equality mainstreaming strategy through subsequent Five-Year Periodic Plans (ADB, 2010).

2.3.4 Healthcare Inequality

Lack of hospital facilities, especially in rural areas, have resulted in severe healthcare inequality. Many rural districts lack basic facilities on maternity and children health, adult healthcare, and emergency responses. In rural areas, people have to walk for days just to get some basic health services. The divide between urban and rural population on the access to healthcare is widening. Private hospitals are providing services largely to rich people. Irrespective of its location, the low-income population is lacking access to healthcare. National healthcare system must be strengthened in order to provide access to healthcare facilities for all to lead healthy lives. Health equity should be guaranteed by providing free services, medicine supplies, vaccination, and therapies that are specific to ethnic groups, marginalized population, or age groups who cannot afford them.

2.3.5 Education Inequality

Nepal's education system is largely theory-based, severely lacking the practical aspect. It is important to deliver practical and problem-solving education in order to develop skilled human resources. A practical education system relies on activity-based, interactive, experimental, and multidisciplinary learning. It helps in developing student's critical thinking ability. A practical education system is flexible enough in accommodating student's needs and interests. With the emergence of private schools coupled with dwindling educational quality of public schools, the educational inequality is increasing rapidly in the country. Private schools are the choices of the people who can afford them. Many primary schools in rural areas are at the verge of collapse due to low enrollments. Many of these schools are running with a small number of students whose parents cannot afford to admit their children into private schools. Students from poor family backgrounds are left behind in schools, which lack good teachers and educational materials. Those students who attend private schools receive relatively better education and eventually end up in better colleges and future jobs.

2.3.6 Employment Inequality

Employment inequality due to gender, caste, and origin is very high and it is detrimental to the society. Various reports suggest that only 22% of working age females (11.53 million as opposed to 9.2 million males) are working in formal sector of employment. While governmental jobs, clean works, and high paid jobs tend to go to higher caste people, especially from urban areas, low paid and difficult jobs tend to go to the lower caste people and people from rural areas. Survey studies have reported quite contrasting wage differences between the genders, with about 30% less wages to females compared to males.

2.3.7 Spatial Inequality

Spatial inequality exists because of the spatiality or geographic differences in the society. For example, people born in the cities naturally have better access to schools, health, and other opportunities, whereas, people born in remote districts have to struggle for access to education, health facilities, and employment opportunities. Similarly, people born in a village located on the hilltop will have trouble fetching drinking water for their households throughout the lifetime or carrying fodder and forages for livestock from far distance daily. In contrast, people living near the sources of drinking water or nearby fodder and forage sources will have more time available for other households' chores and will be less stressed. The productivity of farm will also differ depending on access to natural resources.

2.4 Social Inclusion and Sustainable Development

Like any other society in the world, Nepalese society is also suffering from many forms of inequalities and discriminations. Because of inequitable access to natural resources such as land, water, rangelands, and forests coupled with entrenched social inequality in the society, relatively a large population is marginalized, resource-deprived, and is under poverty. Due to pervasive discriminations, social exclusions and poverty, the economic growth, community resiliency, and socio-economic transformation of the nation have become highly challenging. It is important that all marginalized people and the communities get the fruits of development. However, due to the lack of income generating opportunities and, in many cases, lack of access to education and health services, many marginalized communities are not fully able to benefit from developmental initiatives. Social inclusion is necessary from early stage of economic development in order to make it possible in the society that everybody enjoys the fruits of development. United Nation's global Sustainable Development Goals (SDGs) declared at the UN's New York Convention on September 25–27, 2015 (UN, 2015) captured very well the issues of equitable access to resources, social inclusions, guarantee on human rights, end of any form of inequalities in the world, environmental and ecological sustainability, sustainable management of natural resources, eradication of poverty and hunger from the world, and resilient societies. The 17 SDGs can be stated in their brief forms as follows:

[Goal 1]: *End poverty in all its forms everywhere*
[Goal 2]: *End hunger, achieve food security and improved nutrition and promote sustainable Agriculture*
[Goal 3]: *Ensure healthy lives and promote well-being for all at all ages*
[Goal 4]: *Ensure inclusive and equitable quality education and promote lifelong learning opportunities for all*
[Goal 5]: *Achieve gender equality and empower all women and girls*
[Goal 6]: *Ensure availability and sustainable management of water and sanitation for all*
[Goal 7]: *Ensure access to affordable, reliable, sustainable, and modern energy for all*
[Goal 8]: *Promote sustained, inclusive, and sustainable economic growth, full and productive employment, and decent work for all*
[Goal 9]: *Build resilient infrastructure, promote inclusive and sustainable industrialization and foster innovation*
[Goal 10]: *Reduce inequality within and among countries*
[Goal 11]: *Make cities and human settlements inclusive, safe, resilient, and sustainable*
[Goal 12]: *Ensure sustainable consumption and production patterns*
[Goal 13]: *Take urgent action to combat climate change and its impacts*

[Goal 14]: *Conserve and sustainably use the oceans, seas, and marine resources for sustainable development*

[Goal 15]: *Protect, restore, and promote sustainable use of terrestrial ecosystems, sustainably manage forests, combat desertification, and halt and reverse land degradation and halt biodiversity loss*

[Goal 16]: *Promote peaceful and inclusive societies for sustainable development, provide access to justice for all and build effective, accountable, and inclusive institutions at all levels*

[Goal 17]: *Strengthen the means of implementation and revitalize the global partnership for sustainable development*

Many multilateral development aid agencies have started incorporating gender equality and social inclusion element in their developmental programs (ADB, 2010). This is a highly praiseworthy initiative. The Government of Nepal is suggested to incorporate social inclusion component in its all-development project to expedite the process of alleviation of social and economic inequalities. A district-level Social Inclusion Authority (SIA) could be established for coordination and implementation of these activities. As shown in Figure 2.1, a developmental project whether it is highway construction, irrigation, drinking water project, or any other project that comes from governmental funding should allocate certain amount for social inclusion

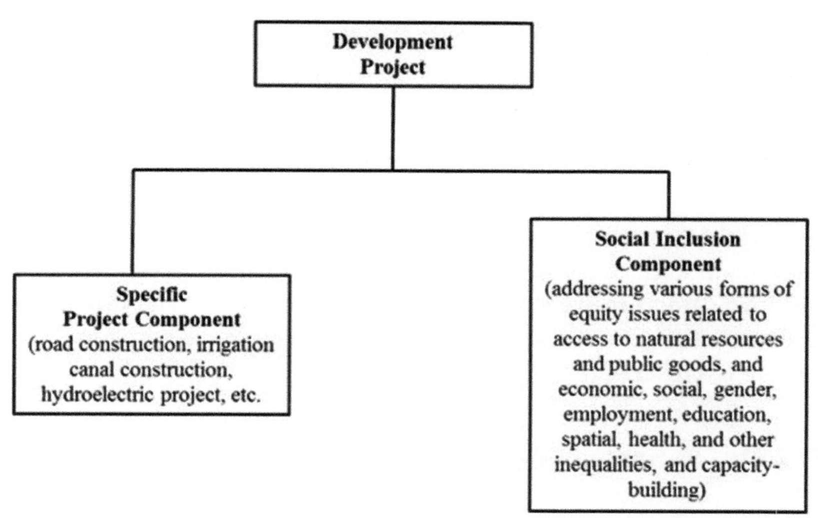

Figure 2.1 Incorporation of social inclusion component in developmental projects.

component in its budget. These funds can be used for many different activities such as housing, education, health support, relief funds, community projects, capacity-building, etc. for the target communities.

In my earlier publication, Poudel, 2012, I have discussed at length on community capacity-building process with reference to sustainable conservation, development, and utilization of Asta-Ja in Nepal, and have illustrated community capacity-building as a continuous cycle. The basic steps that go through the cycle include: (1) education, outreach, and community awareness, (2) motivation and problem solving, (3) community organizations and partnerships development, (4) community capacity assessment for problem solving, (5) policies, laws, regulations, and guidelines, (6) community capacity-building projects development, (7) implementation of community capacity-building projects, and (8) monitoring, evaluation, and reporting.

Climate change has added new dimension to the issue of equitable access to natural resources and even widened income inequality. As many drinking water sources are dried up in recent years especially in the mid-hill regions of Nepal (Poudel and Duex, 2017, Poudel et al., 2020), local communities are spending more and more time on fetching drinking water from far distance for their families and water supply for their livestock. Similarly, due to changes on rainfall patterns and drying of or decline in the flow of irrigation water sources, crop production is also affected adversely impacting on the income generation capacity of local communities (Poudel, 2015). Climate change impacts have been observed in various forms including frequent flooding, incidence of new diseases and pests, changes on rainfall patterns, droughts, declining agricultural productivity, and land degradation (Bhandari, 2019a; Hali et al., 2019a, Hali et al., 2019b, Hali et al., 2019c). A developmental project must be very comprehensive and must account for environmental, natural resources, socioeconomic, public health, climate change, social justice, and other pertinent issues in the society. The ultimate goal of any developmental initiative is sustainable community development at the local, regional, and national level. For this, a theoretically grounded grassroots-based comprehensive natural resources conservation, development, and utilization and human resource planning and development framework is necessary, which I am going to present in the subsequent section.

2.5 Asta-Ja Framework

Envisioning an equitable, socially just, environmentally sound, and sustainable society, a decade ago, I published a groundbreaking framework of Asta-Ja meaning eight *Ja* in Nepali letter: *Jal* (water), *Jamin* (land), *Jungle* (forest), *Jadibuti* (medicinal and aromatic plants), *Janshakti* (manpower), *Janawar* (animal), *Jarajuri* (crop plants), and *Jalabayu* (climate) as a comprehensive resources planning and development approach for fast-paced economic growth and socio-economic transformation in Nepal (Poudel, 2008). I have published several journal articles on this topic after publishing this first groundbreaking article on Asta-Ja in 2008. Some of the major articles specific to Asta-Ja Framework are Asta-Ja environmental and natural resources policy framework (Poudel, 2009), Asta-Ja strategic framework (Poudel, 2011), Asta-Ja management capacity-building framework (Poudel, 2012), and management of Asta-Ja system (Poudel, 2016). Furthermore, I have published numerous articles applying Asta-Ja Framework in the context of Nepal such as Restructuring National Planning Commission focusing on Asta-Ja and Nepal Vision 2040 (Poudel, 2018), Asta-Ja for Grassroots-based Economic Development of Nepal (Poudel, 2019), Asta-Ja Crusade for a Fast-paced Agro-Jadibuti Industrialization of Nepal (Poudel, 2019), Agricultural and Natural Resources Development and Management Strategy in Nepal (Poudel, 2019), and Management of Cooperatives Focusing on Asta-Ja and Globalization (Poudel, 2018). These publications have enormously enriched the Asta-Ja Framework by covering its theoretical, scientific, and operational dimensions, and competitive advantages.

Asta-Ja Framework suggests "*Jalabayu*" as the driving force for all other elements (*Jal, Jamin, Jungle, Jadibuti, Janashakti, Janawar, and Jarajuri*) and requires full consideration of all eight elements while utilizing Asta-Ja resources for economic development. The *Jalabayu* (climate) serves as the most critical and central driving factor in Asta-Ja system. Any changes on weather or climatic conditions will influence all other *Jas*. Simply put, if a farmer in a smallholder mixed-farming system likes to be successful in organic agriculture, he/she should emphasize sustainable management and development of Asta-Ja resources so that a great deal of synergy would develop within Asta-Ja system resulting in a higher level of farm productivity and environmental quality. Asta-Ja Framework emphasizes community capacity-building, self-reliant national, regional, and local level planning and development of environmental and natural resources for socio-economic transformation of the nation. Asta-Ja Framework is a holistic, system-based, scientific, collaborative, interdisciplinary, participatory, self-reliant, and grassroots-based environmental and natural resources planning and management approach for conservation, development, and sustainable utilization of the eight resources within the Asta-Ja system. Asta-Ja is the backbone of

Nepal's economy. Therefore, the best governance of Asta-Ja is the ultimate goal of a government.

The eight principles of Asta-Ja serve as the guidelines for implementation of Asta-Ja Framework at grassroots level. These eight principles include: (1) community awareness, (2) community capacity-building, (3) policy decision making, (4) interrelationships and linkages, (5) comprehensive assessment, (6) sustainable technologies and practices, (7) institutions, trade and governance, and (8) sustainable community development and socio-economic transformation. It starts functioning from community (principle 1) and ends with the community (principle 8). Detailed elaboration of these principles can be found in (Poudel, 2008, 2009, 2011 and 2012). The principle of community awareness plays the central pivotal role in the Asta-Ja Framework. Grassroots communities are the ultimate change agents and the beneficiaries. Communities participating in the process of sustainable development and utilization of Asta-Ja resources, income generation, and socio-economic transformation must be sufficiently aware, engaged and educated on Asta-Ja system and its development. The principle of community capacity-building emphasizes environmental and natural resources community capacity-building for conservation, utilization, and sustainable development of natural resources for socio-economic transformation of Nepal. The principle of policy decision making underscores effective policy measures for sustainable development, conservation and utilization of natural resources for socio-economic transformation. To develop effective policy measures, it emphasizes multisector, participatory, and holistic approach in problem assessment and analysis and the identification of alternative solutions (Poudel, 2009). The principle of interrelationships and linkages stresses that Asta-Ja resources are intricately linked to each other and among themselves and require a deep understanding of these interrelationships and linkages for their sustainable development and utilization. Similarly, the principle of comprehensive assessment emphasizes detailed assessment of Asta-Ja resources for their sustainable development and utilization and socio-economic transformation of the country. The principle of sustainable technologies and practices emphasizes research and development in technologies and practices and innovations for sustainable conservation, utilization, and development of Asta-Ja resources. The principle of institutions, trade and governance focuses on institutional strengthening, governance, and domestic and foreign trades. It emphasizes handling diverse Asta-Ja–related concerns including ownerships, decision making, resource sharing, customs and duties, trade barriers and restrictions, and international relations.

Finally, the overarching principle of sustainable community development and socio-economic transformation emphasizes implementation of integrated developmental initiatives across the nation targeting specific communities for sustainable utilization and development of Asta-Ja resources, income generation, and socio-economic transformation. Asta-Ja Framework envisions communities to be free from poverty, hunger and malnutrition, have basic infrastructures for quality education, health, and services, have employment opportunities, have peace and security, have excellent environmental quality, and are resilient.

A close look on the SDGs and the eight principles of Asta-Ja Framework reveals that there is a great deal of synergy between them. While Goal 1 through 8, 12, 13, 15, and 16 directly relate to Asta-Ja Framework's principle 8 (i.e., sustainable community development and socio-economic transformation) and principle 6 (i.e., sustainable technologies and practices), Goal 9 and 11 relate to principle 2 (i.e., community capacity-building) and Goals 10 and 17 relate to principle 7 (i.e., institutions, trade, and governance). According to Bhandari, 2019b, the SDGs' goals and agendas require high level of engagement of stakeholders and public–private partnerships in achieving success. Since the concept of Asta-Ja Framework is very holistic and is well-established at grassroots level in Nepalese society (Poudel, 2016), it is prudent for the Government of Nepal in adopting Asta-Ja Framework as its developmental platform and working closely with UN agencies on sustainable development of Nepal. The Asta-Ja Framework emphasizes grassroots community capacity-building for sustainable conservation and management of environmental and natural resources and socio-economic transformation both at the macro and micro levels in Nepal.

Asta-Ja Framework serves as a unifying framework for environmental and natural resources planning and management. Through this framework, all governmental and nongovernmental agencies, private businesses, community organizations, academia, international aid agencies, and other stakeholders who are concerned with the Asta-Ja resources, can come together in natural resources planning and management. Asta-Ja Framework helps us in comprehending and understanding our critical natural and human resources system in a more effective way and serves as an invaluable platform for all stakeholders engaged in Asta-Ja system to work together efficiently in the assessment, understanding, conservation, and sustainable utilization of Asta-Ja resources for sustainable development. Through Asta-Ja Framework, international programs and initiatives such as Climate Actions, SDGs, and Sendai Disaster Risk Reduction can be effectively linked to grassroots and strengthen global efforts on effectively tackling issues such as natural resource management, sustainable development, food and energy, income generation, poverty eradication, natural disaster management, community resiliency, environmental quality, and global climate change.

2.6 Summary and Recommendations

Nepal is facing multidimensional socio-economic developmental challenges including the development of infrastructures such as roads, irrigation system, hydroelectricity, hospitals, educational facilities, and industries, in the meantime overcoming serious issues of equity, inequalities, and social justice. Nepal Constitution 2015, Part 3, Fundamental Rights and Duties, Article 43, Right to social security, guarantees social security to indigent citizens, incapacitated and helpless citizens, citizens with disabilities, helpless single women, citizens who cannot help themselves, and children. Similarly, Article 24, Right against untouchability and discrimination, deals with the subject of untouchability and discrimination in any private and public places on grounds of his or her origin, caste, tribe, community, occupation, or physical condition. Article 33, Right to employment, guarantees that every citizen shall have the right to employment. Other rights guaranteed by the 2015 Constitution of Nepal include Article 31, Right to education; Article 29, Right against exploitation; Article 35, Right relating to health; and Article 38, Rights of women. These are certainly commendable initiatives. However, proper implementation of these provisions with appropriate policies, practices, and strategies in addressing various inequalities is a quite challenging task. The Government of Nepal should give the highest priorities in solving these various forms of inequalities and discriminations. Accordingly, I would like to make the following recommendations:

1) The Government of Nepal is suggested to include Social Inclusion Component (SIC) in its developmental projects targeting the marginalized and poverty-stricken population in their capacity-building in order to realize the full benefits of developmental projects while addressing the critical issues of equity, social justice, and marginalization. It is important that comprehensive policies, strategies, and practices are developed and implemented, bringing together various community-based organizations, leaders, community groups, governmental agencies, marginalized population, and other stakeholders while addressing these issues.

2) Asta-Ja Framework is a theoretically grounded natural and human resources planning and development approach for sustainable community development. The eight principles of the Asta-Ja Framework provide guidelines for its implementation. Since the Asta-Ja Framework is well accepted in Nepalese society and there is a great deal of synergy between the UN's 2030 SDGs and the Asta-Ja Framework, it is suggested that the Government of Nepal adopt Asta-Ja Framework as its national framework for economic development and alleviation of all forms of inequalities in the society.
3) The ownership and user's rights issues in relation to land, water, rangelands, and forests are very complex and challenging. Promulgation of Nepal Constitution 2015 and the federalization of governance and administration have opened up a new dimension on the governance of natural resources in Nepal. There is a need for developing a large number of new laws, regulations, and policies in relation to natural resource ownerships and users' rights. In addition, evaluation of existing laws, policies, and regulations on natural resource ownerships and rights and making necessary improvements are suggested.

Disclaimer: Considering the utmost relevancy of this article on Nepal's socio-economic transformation, this author anticipates publishing additional similar articles in the future. Some of the contents, especially in sections 2.2 and 2.3 of this chapter, were published in Telegraphnepal.com as part of this author's weekly columns on the Governance of Natural Resources in Nepal.

References

ADB (Asian Development Bank), (2010), Overview of gender equality and social inclusion in Nepal, Asian Development Bank, Mandaluyong City, Philippines. Available at: https://www.adb.org/sites/default/files/institutional-document/32237/cga-nep-2010.pdf

Bhandari, M.P. (2019a), "BashudaivaKutumbakkam"- The entire world is our home and all living beings are our relatives. Why we need to worry about climate change, with reference to pollution problems in the major cities of India, Nepal, Bangladesh and Pakistan. *Adv Agr Environ Sci.* (2019);2(1): 8–35. DOI: 10.30881/aaeoa.00019

Bhandari, M.P. (2019b), Live and let other live- the harmony with nature /living beings-in reference to sustainable development (SD)- is contemporary world's economic and social phenomena is favorable for the sustainability of the planet in reference to India, Nepal, Bangladesh, and Pakistan? *Adv Agr Environ Sci.*(2019);2(2): 37−57. DOI: 10.30881/aaeoa.00020

DoFSC (Department of Forest and Soil Conservation), (2020), Acts and Regulations, Available at: http://www.dofsc.gov.np/page/acts-regulations/en

DoFCFD (Department of Forest Community Forest Division), (2018), Community Forestry Bulletin, No. 18, 2017/2018, Available at: http://www.dofsc.gov.np/search/en?query=Community+Forestry

Hali C., Katie Eddings, George Bailey, Andrew Braun, Aubrey Mann, Victoria Gomez, Holly Heafner, William Faulk, Luke Immel, Allison Hingdon, Brandon Stelly, Brittany N. Broussard, Layken Willis, Timothy C. Martin, Thomas J. Mizelle, Avery J. Baker, Timothy Duex, Durga D. Poudel. (2019a), Enriching college students through study abroad: A case of Nepal Field Experience Part 1, ASEJ 23(4):24–29. DOI:10.5604/01.3001.0013.6832 (online)

Hali C., Katie Eddings, George Bailey, Andrew Braun, Aubrey Mann, Victoria Gomez, Holly Heafner, William Faulk, Luke Immel, Allison Hingdon, Brandon Stelly, Brittany N. Broussard, Layken Willis, Timothy C. Martin, Thomas J. Mizelle, Avery J. Baker, Timothy Duex, Durga D. Poudel. (2019b), Enriching college students through study abroad: A case of Nepal Field Experience Part 2, ASEJ 23(4):30–37. DOI:10.5604/01.3001.0013.6850 (online)

Hali C., Katie Eddings, George Bailey, Andrew Braun, Aubrey Mann, Victoria Gomez, Holly Heafner, William Faulk, Luke Immel, Allison Hingdon, Brandon Stelly, Brittany N. Broussard, Layken Willis, Timothy C. Martin, Thomas J. Mizelle, Avery J. Baker, Timothy Duex, Durga D. Poudel. (2019c), Enriching college students through study abroad: A case of Nepal Field Experience Part 3, ASEJ 23(4):38–44. DOI:10.5604/01.3001.0013.6852 (online)

Hardin, G. (1968), The Tragedy of the Commons, Science 162 (3859), 1243–1248, DOI: 10.1126/science.162.3859.1243

Leitner Center, (2011), Land is Life, Land is Power: Landlessness, Exclusion, and Deprivation in Nepal, Leitner Center for International Law and Justice at Fordham Law School, New York City, New York, USA. Available at: http://www.leitnercenter.org/files/Crowley%20Program/2011%20Leitner%20Nepal%20Report%20v2.pdf

NPC (National Planning Commission). (2013), National Sample Census of Agriculture Nepal 2011/12, Government of Nepal, National Planning Commission Secretariat, Central Bureau of Statistics, Kathmandu, Nepal.

Pande, R.S. (2009), Status of rangeland resources and strategies for improvements in Nepal, CABI Wallingford, UK, CAB Reviews, Available at: https://www.cabi.org/cabreviews/review/20093276266, DOI 10.1079/PAVSNNR20094047

Poudel, D.D. (2008), Management of Eight 'Ja' for Economic Development in Nepal. *Journal of Comparative International Management*, 11(1): 1–13.

Poudel, D.D. (2009), The *Asta-Ja* Environmental and Natural Resources Policy Framework (Asta-Ja ENRPF) for Sustainable Development in Nepal. *Journal of Comparative International Management,* 12(2): 49–71.

Poudel, D.D. (2011), A strategic framework for environmental and sustainable development in Nepal. *International Journal of Environment and Sustainable Development*, 10(1):48–61.

Poudel, D.D. (2012), The Asta-Ja Management Capacity-building Framework for Sustainable Development in Nepal, *International Journal of Sustainable Development,* Vol. 15, No. 4, pp. 334–352.

Poudel, D.D. (2015), Factors associated with farm-level variation, and farmers' perception and climate change adaptation in smallholder mixed-farming livestock production system in Nepal, *Int. J. Environmental and Sustainable Development* 14(3): 231–257.

Poudel, D.D. (2016), Management of Asta-Ja System, *Journal of Comparative International Management,* 19(2):19–40.

Poudel, D.D. and T.W. Duex. (2017), Vanishing Springs in Nepalese Mountains: An Assessment of Water Sources, Farmer's Perceptions, and Climate Change Adaptation, *Mountain Research and Development*, 37(1):35–46. DOI: http://dx.doi.org/10.1659/MRD-JOURNAL-D-16-00039.1

Poudel, D.D. (2018), Management of Cooperatives Focusing on Asta-Ja and Globalization, *Journal of Comparative International Management*, 21(1):77–84.

Poudel, D.D. (2018), Restructuring National Planning Commission Focusing on Asta-Ja and Nepal Vision 2040, *Asian Profile*, 46(2):151–167.

Poudel, D.D. (2019), Agricultural and Natural Resources Development Strategy in Nepal, *Asian Profile*, 47(1): 1–17.

Poudel, D.D. (2019), Asta-Ja for Grassroots-based Economic Development of Nepal, Telegraphnepal, Published on January 14, 2019. Available at http://telegraphnepal.com/asta-ja-for-grassroots-based-economic-development-of-nepal/

Poudel, D.D. (2019), Asta-Ja Crusade for a fast-paced agro-jadibuti industrialization of Nepal, Telegraphnepal, Published on January 20, 2019.

Available at http://telegraphnepal.com/asta-ja-crusade-for-a-fast-paced-agro-jadibuti-industrialization-of-nepal/

Poudel, D.D., T.W. Duex, and R. Poudel. (2020), Drinking water security in the mid-hill region of Nepal, ASEJ - Scientific Journal of Bielsko-Biala School of Finance and Law, 24(1):44–48. DOI: 10.5604/01.3001.0014.1351 (online)

Rastriya Samachar Samiti, (2016), Forest cover has increased in Nepal of late, The Himalayan, Published on May 13, 2016. Available at: https://thehimalayantimes.com/nepal/forest-cover-increased-nepal-late

UN (United Nations). (2015), Transforming our World: The 2030 Agenda for Sustainable Development, A/RES/70/1, United Nations, Available at: https://sustainabledevelopment.un.org/content/documents/21252030%20Agenda%20for%20Sustainable%20Development%20web.pdf,

WaterAid Nepal, (2005), Water Laws in Nepal: Laws Relating to Drinking Water, Sanitation, Irrigation, Hydropower and Water Pollution, WaterAid Nepal, Lalitpur, Nepal, Available at: https://washmatters.wateraid.org/publications/all/nepal

3

Forecasting Inequality—The Innovation Implementation Benefits for Countries with Different Economic Development Levels

Anna Rosohata[1] **and Liubov Syhyda**[2]

[1]PhD in Economics, Senior Lecturer, Department of Marketing, Sumy State University, Sumy, Ukraine
[2]PhD in Economics, Senior Lecturer, Department of Marketing, Sumy State University, Sumy, Ukraine

Abstract

Enterprises and companies without industrial workers, high-speed trains and subways without drivers, robotic cleaners and automated buildings, multifunctional gadgets and 3D printers, e-marketing, freelancing, and sharing economy—these and many other factors are the results of modern innovation. These are the basis for exploring the benefits for countries which stepped on the path of transformational technological changes, the path of Industry 4.0.

There are countries with different levels of economic development in the current context we can see that industrialization, computerization, and robotization cause inequality between the competitiveness among leading and developing countries. And it needs to be investigated in detail.

This research is devoted to investigating the inequalities in the distribution of benefits as a result of the radical and improving innovation implementation in countries with different levels of economic development; to create a forecast for further distribution of benefits; to develop ways to overcome identified inequalities; to determine the main prospects for the economic development of developing societies (in particular, Ukraine) in terms of Industry 4.0.

The methodical approach in this study is based on an analytic and descriptive approach. Also, the graphical method of trend analysis and a new trend forecasting method are used.

The database of the research is formed by studies of leading world scientists, official information of technological innovations' influence in some countries, statistical information of international organizations, such as the World Bank.

The dataset for analysis was taken from the Report of IMD World Digital Competitiveness Ranking 2019. Based on the data for 2015–2019, pessimistic, normal, and optimistic forecasts for the development positions of some leading and developing countries in terms of knowledge, technology, and future-readiness were generated. And the gap between them in terms of technological innovation impact was estimated.

As a result of the research, digital and innovation competitiveness ranking trends for 2020 and 2021, were developed. They helped to define the peculiarities of the further development of countries with different levels of economic development and to determine the synergistic impact of such indicators as knowledge and future-readiness.

So, the research found that the main factor of inequality is industrialization, computerization, and robotization in economically developed societies. Firstly, they have an expensive labor force, which is more profitable to be replaced with computers and robotics. Secondly, they have more opportunities for innovations' development and are the centers of robotics production. And thirdly, they can provide mechanisms to neutralize the negative impact of industrialization, computerization, and robotization on the unemployment rate. But there are also a few paradoxes. For example, under special arrangements, some developing countries also have high rates of technological innovation implementation and thus gain certain benefits. They balance global inequality in this way.

To counterbalance the inequalities in the benefits distribution in the countries, some measures have been proposed. All mentioned measures will help to balance the capacity of innovative development and its economic and social effect for countries with different levels of economic development.

3.1 Introduction

The problem of inequality is burning in the modern world. Nowadays, we see inequality in different spheres of human lives, in activities and society development. The problem of inequality in countries is especially up to time.

One of the reasons for this inequality is uneven implementation and the use of innovative technologies and uneven profit (Saher et al., 2018).

The modern innovative process has a complex multifaceted nature. Periods of the qualitative transition of the system from one stage of development to another have several characteristics. The use of different models of the innovative process largely depends on the macro- and microeconomic conditions of business activity of certain economic objects, participants in the innovative process. As sources of innovation at this stage can be used research sources of new knowledge; market needs; existing knowledge; acquired knowledge, etc. Some economic systems independently form demand (potential needs) for future markets.

The cyclical development of economic systems and its wave-like nature of innovative processes are relevant for the countries' economic development (Bhandari, 2020; Bhandari and Shvindina, 2019). Modernity shows that all countries develop differently and have their characteristics (Liulov et al., 2019). Anyway, there are certain formalized factors by which they can be characterized and investigated further considering the causes and consequences of benefit-sharing, as well as the features of innovative development.

3.2 Methods

To study the consequences of the innovations' implementation and their impact on the countries' economies, we used the analytical method to define the peculiarities of previous innovations replacing with new ones in the process of economic system development. Also, we grouped and analyzed quantitative indicators at the level of GDP and GDP per capita to identify and rank countries as developed and developing.

The Human Development Index Ranking of the analyzed countries was studied using the grouping method. Therefore, an approach involving the comparison of the Human Development Index values and the Global Innovation Index scores was proposed. To study the innovative activity of analyzed countries further, the World Digital Competitiveness was investigated. According to the main indicators, the forecast values of further businesses and innovative activities of these countries were developed, namely: for knowledge, technology, and future-readiness digital and innovation competitive ranking trends for 2020 and 2021 were made via Excel Trendline. They helped to define the peculiarities of further development of countries with different levels of economic development and to determine the synergistic impact of such indicators as knowledge and future-readiness.

The authors used the deductive method to form the main factors that are caused by the uneven implementation of innovations in countries with different levels of economic development. Combining the elements of SWOT-analysis and the principles of the modern method for innovative development forecasting (trend watching) (Rosokhata, 2013, 2014), the trends of strengths and weaknesses of this process were grouped, and alternative opportunities and threats were formalized. Applying analytical advanced methods, the ways of inequalities elimination in benefits distribution between countries with different levels of economic development were proposed.

3.3 Results

The role of innovation sources differs significantly for different companies, industries, countries and depends on the stages of their life cycles. Turning to the theory of technological systems' development and in terms of systems development life cycles, there are certain steps of transition between stages of development and they are cyclical (Figure 3.1).

Points AB, BC, CD, presented in Figure 3.1 show the stages of transition from one innovation system to another following the qualitative characteristics. This occurs at a time when life cycle of the previous innovation is at the

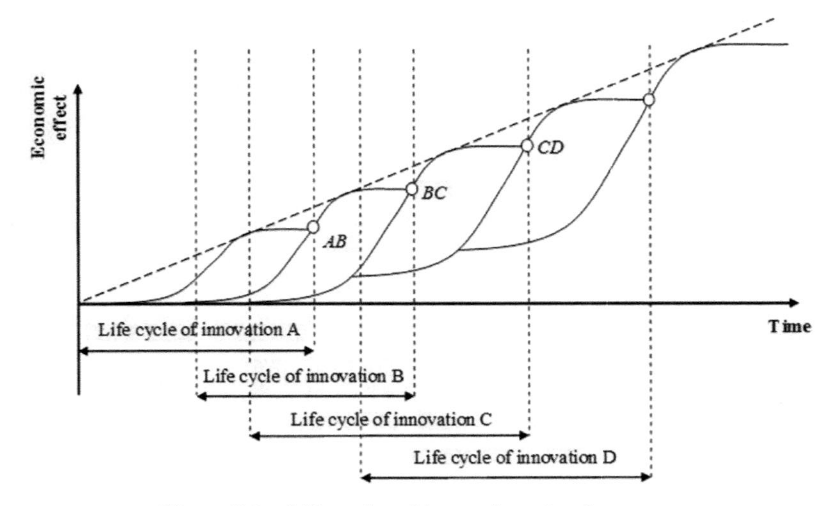

Figure 3.1 Life cycles of innovations development.

stages of market maturity, and the next innovation structure is at the stage of introduction and growth. Thus, accordingly, there is a change from one innovation structure to another. Additionally, if we consider the general development of the economic system of innovations' birth and implementation, it has a certain decline in development at the transition points of innovation structures (AB, BC, CD). Anyway, after the decline, there is a qualitatively new step to further development.

This wave-like approach can characterize the development of any economic system. So, it is important to consider forecasting options for economic systems' development.

Thus, periods of change and the emergence of innovations in the market are always defined by previous stabilization or reduction of previous innovations sales. At the same time, when the innovation structure is completely formed, the system and the country's economy can increase efficiency rapidly and reach a qualitatively new level of development.

On the one hand, the innovative process can be considered as a process of transforming inputs (resources) into outputs (products, technologies). Also, it is extremely important to clarify that a necessary condition for the innovative process is the use of available resources in other ways (Cebula & Pimonenko, 2015).

On the other hand, the innovative process is a process of achieving the potential of economic systems, countries, industries, and enterprises. So, the results received by economic systems from the innovations' implementation can differ significantly (Lyulyov et al., 2015).

As an aim of the research, we consider the countries, so, it should be said that there are different approaches to the countries' division on developed and developing. For this purpose, various metrics are used. Thus, in classical economic literature, it is customary to classify countries depending on their development by GDP, GNP, level of education, life expectancy, etc. (Filipenko et al., 2000).

The following common characteristics are helping to categorize countries as developed (Top 25, 2019):

1. High industrialization;
2. Birth and death rates are stable (e.g., the infant mortality rate is not higher than 10 per 1000 live births; life expectancy is mostly greater than 70 years; many averages 80);
3. More women working, particularly in high-ranking executive positions;

4. Use of a disproportionate amount of the world's resources, such as oil; and
5. Higher levels of debt.

Different countries also offer their classification features and relevant groups. For example, for economic analysis, the UN divides countries into (Country classification, 2014):

- developed countries (market economies);
- countries with economies in transition (formerly socialist countries or countries with centralized planning); and
- developing countries.

According to (World Economic Situation, 2019; Felipe et al., 2010; Perspectives, 2018), the most important and the most used metrics for countries' economic development levels determination and their division into developing and developed ones are the total gross domestic product (GDP) and the gross domestic product (GDP) per capita. These are useful metrics for countries' categorizing (Vasylieva et al., 2019). In Table 3.1, the GDP and GDP per capita of some countries are given.

According to Table 3.1, Canada, the USA, Japan, Great Britain, Republic of Korea, and Germany have a high level of GDP and respectively high level of GDP per capita (not less than 30.00 US$). At the same time, the level of GDP in China is 13.608.151.86 million US$ that is much more than that

Table 3.1 Level of GDP and GDP per capita of analyzed countries in 2018.

Country	GDP (millions of US$)	GDP per capita (US$)
Canada	1.713.341.70	46.233.0
USA	20.544.343.46	62.794.6
Japan	4.971.323.08	39.290.0
Great Britain	2.855.296.73	42.943.9
Republic of Korea	1.619.423.70	31.362.8
Germany	3.947.620.16	47.603.0
Switzerland	705.140.35	82.796.5
India	2.718.732.23	2.010.0
Ukraine	130.832.37	3.095.2
Thailand	504.992.76	7.273.6
Mexico	1.220.699.48	9.673.4
Poland	585.663.81	15.420.9
China	13.608.151.86	9.770.8
South Africa	368.288.94	6.374.0

Source: compiled based on (GDP per capita, n.d.)

of Canada, Japan, Great Britain, Republic of Korea, Germany, and close to the GDP of the USA. But, GDP per capita is only 9.770.8 US$ (almost 6.5 times less compared to the GDP per capita of the USA). The common situation is with Mexico, whose GDP (1.220.699.48 million US$) is close to the GDP of the Republic of Korea (1.619.423.70 million US$) and Canada (1.713.341.70 million US$); however, their GDP per capita differ much—9.673.4 US$, 31.362.8 US$, and 46.233.0 US$ for Mexico, Republic of Korea, and Canada respectively. Ukraine has the smallest GDP and GDP per capita among analyzed countries.

Consequently, the GDP cannot be taken as the key metric for countries' division by the level of economic development. The more informative metric is GDP per capita, as it shows part of GDP on each person in a country.

As mentioned in (Top 25, 2019), countries with developed economies have GDP per capita of at least 12.000 US$, although some economists believe 25.000 US$ is a more realistic measurement threshold.

Thus, according to the metric of GDP per capita, Canada, the USA, Japan, Great Britain, Republic of Korea, Germany, and Switzerland are developed countries as their GDPs per capita is higher than 25.000 US$. Other analyzed countries (India, Ukraine, Thailand, Mexico, Poland, and South Africa) can be classified as developing. Poland's GDP per capita is 15.420.9, that is more than 12.000 US$, but less than 25.000 US$. So, according to the metric of GDP per capita, Poland is on its way to be a developed country.

However, using only the GDP per capita metric is not enough to divide countries into groups. Also, according to World Development Report (World Development, 2016), two more important metrics must be considered in defining countries' economic development inequality, which determines inequality of countries' development is their innovative development and innovative potential. The second metric is human capital. It was chosen as for now human capital formation (a population's education, particularly secondary education, and health status) plays a significant role in a country's economic development (Economic Growth, 2008).

Thus, to identify countries' inequality, we propose to use two more metrics—the Human Development Index (HDI) and the Global Innovation Index (GII). The GII and the HDI were chosen to characterize not only economic aspects of countries' growth but additionally their social (people and their capabilities) and innovative aspects.

The value of the HDI is 1. This index includes three key dimensions of human development: a long and healthy life, access to knowledge, and a decent standard of living. And four indicators are used to count it: (1)

life expectancy at birth, (2) expected years of schooling, (3) mean years of schooling, and (4) gross national income per capita (Beyond income, 2019; Human Development Index, 2019).

The GII relies on two sub-indices—the Innovation Input Sub-Index and the Innovation Output Sub-Index. The Innovation Input Sub-Index combines five indicators: (1) Institutions, (2) Human capital and research, (3) Infrastructure, (4) Market sophistication, and (5) Business sophistication. And the Innovation Output Sub-Index consists of two output pillars: (6) Knowledge and technology outputs and (7) Creative outputs. The maximum value of the GII is 100 (History of the Global, 2019).

The HDI and the GII of the analyzed countries in 2018 are given in Table 3.2.

To compare the positions of analyzed countries and divide them into two groups—developed and developing countries, we plot them on the matrix (Figure 3.2).

As we can see from Figure 3.2, the Global Innovation Index and the Human Development Index are good factors to cluster analyzed countries into developed and developing groups. According to Figure 3.2, there are two clusters. The first cluster is formed by the countries with high GII and HDI (also, these countries have high GDP and GDP per capita). The other cluster is formed by countries with the low and medium score in GII and medium score in HDI (also, they have low GDP and GDP per capita). Poland's HDI

Table 3.2 The Human Development Index and the Global Innovation Index of the analyzed countries in 2018.

Country	Global Innovation Index	Human Development Index
Canada	53.88	0.922
USA	61.73	0.920
Japan	54.68	0.915
Great Britain	61.30	0.920
Republic of Korea	56.55	0.906
Germany	58.19	0.939
Switzerland	67.24	0.946
India	36.58	0.647
Ukraine	37.40	0.750
Thailand	38.63	0.765
Mexico	36.06	0.767
Poland	41.31	0.872
China	54.82	0.758
South Africa	34.04	0.705

Source: Compiled based on (2019 Human Development, 2019; Dutta et al., 2019)

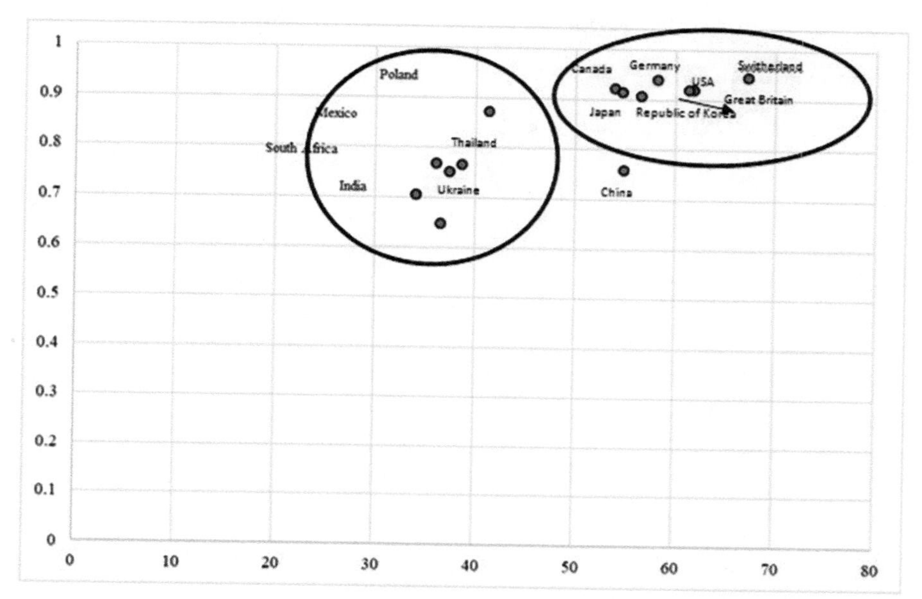

Figure 3.2 Comparison of the Human Development Index values and the Global Innovation Index scores.

is high too and is close to HDI of Republic of Korea and Japan; however, the country still must work over its GII. China has a middle position between developing and developed countries. China has a high score of HDI, but its GII score is lower than that of countries from the first cluster.

Using the data received from Table 3.1 and Figure 3.2, we can combine all analyzed countries into two groups: (1) developed countries (this group includes countries with GDP per capita more than 25,000 US$ and with high level of GII and HDI)—Canada, the USA, Japan, Great Britain, Republic of Korea, Germany, and Switzerland; (2) developing countries (this group includes countries with GDP per capita less than 25,000 US$ and with low or medium level of GII and HDI)—India, Ukraine, Thailand, Mexico, Poland, China, and South Africa.

Moreover, according to World Bank reviews (World Development, 2016) a lot of world development reports are based on information and different indicators about digital dividends. So, an important aspect of the identification of inequality between developing and developed countries is the level of Digital Competitiveness (IMD World digital, 2019). The World Digital Competitiveness consists of three factors: (1) Knowledge

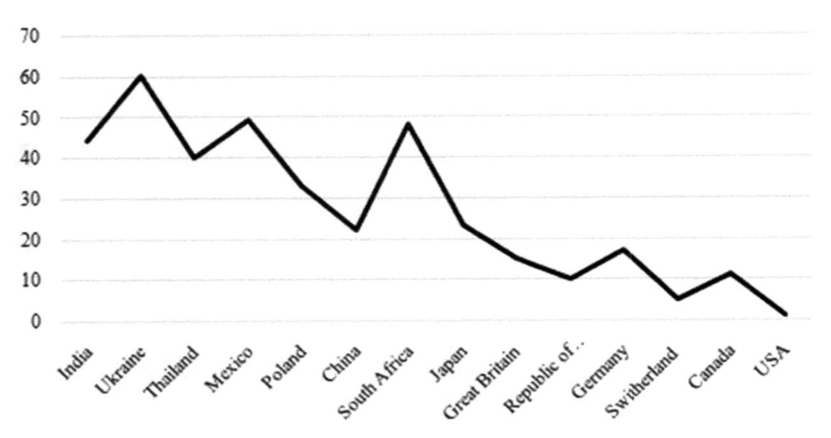

Figure 3.3 The World Digital Competitiveness Index of analyzed countries in 2019.
Source: Compiled based on (IMD World digital, 2019).

(Talent, Training and Education, and Scientific Concentration), (2) Technology (Regulatory Framework, Capital, and Technological Framework), and (3) Future Readiness (Adaptive Attitudes, Business Agility, and IT Integration) (IMD World digital, 2019). Firstly, we explore the index of the World Digital Competitiveness within analyzed countries in 2019 (Figure 3.3).

The investigation of the overall index of the World Digital Competitiveness shows that there is a gap between developed and developing countries. Developed countries take first ranks in Digital Competitiveness (the first rank belongs to the USA, rest of analyzed developed countries are in Top-25).

Developing countries have lower ranks and Ukraine has the worst position among analyzed countries. Thus, in 2019 Digital Competitiveness of Ukraine is almost three times lower than that of Japan. But at the same time, China has a high rank in the World Digital Competitiveness Index.

In more detail we examine the overall index of Digital Competitiveness and its indicators over the 5 years from 2015 to 2019 for three countries—the USA (as a leader of 2019), Ukraine, and Poland (it has middle position in index ranking).

To understand the problems and the influence of some indicators, we need to investigate in further detail each of three factors (Knowledge, Technology, and Future Readiness) for Ukraine, Poland, and the USA (Figures 3.4, 3.5 and 3.6).

As we can see from the Figures above, the “Knowledge” and “Technology” indicators of the World Digital Competitiveness Index for Ukraine are

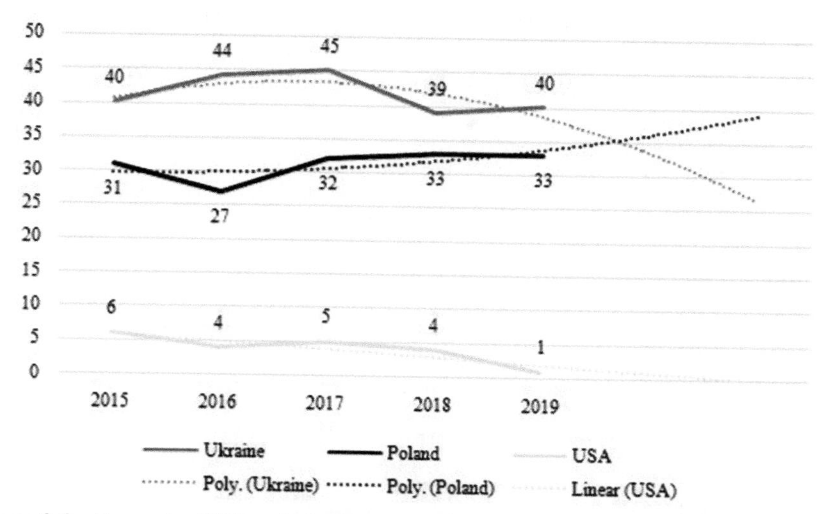

Figure 3.4 Dynamic of "Knowledge" indicator of the World Digital Competitiveness Index for Ukraine, Poland, and the USA in 2015–2019.
Source: Compiled based on (IMD World digital, 2019).

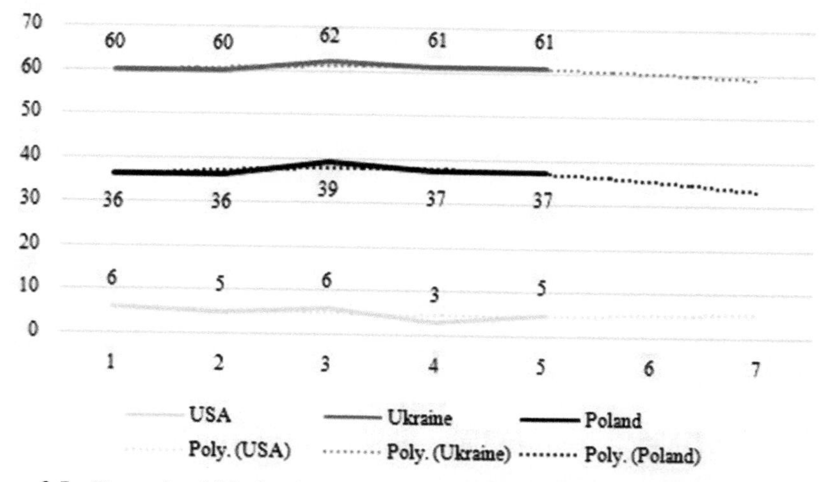

Figure 3.5 Dynamic of "Technology" indicator of the World Digital Competitiveness Index for Ukraine, Poland, and the USA in 2015–2019.
Source: Compiled based on (IMD World digital, 2019).

about to get better in the next 2 years. But "Future readiness" indicator tends to get worse.

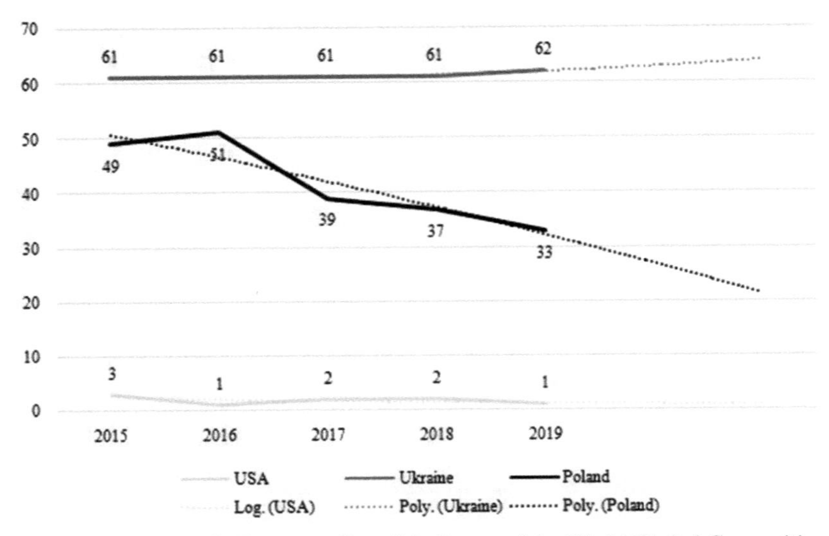

Figure 3.6 Dynamic of "Future readiness" indicator of the World Digital Competitiveness Index for Ukraine, Poland, and the USA in 2015–2019.
Source: Compiled based on (IMD World digital, 2019).

Poland is able to increase "Technology" and "Future readiness" indicators, but "Knowledge" indicator shows trend to get worse and reach 40 by 2021.

As for now, the USA already took first place in "Knowledge" and "Future readiness" indicators and has all chances to keep its place in the future.

Thus, according to the results of forecasting for 2020 and 2021, we can see that there will be gaps between developed and developing countries within all three indicators—knowledge, technology, and future-readiness.

A clear and formalized division of countries into developed and developing ones, and a quantitative description of forecasts for the future development show the directions of further research (OECD, 2016).

The reason to study the issue of inequalities in the process of distribution of benefits from the modern technologies and innovations introduction is that it is necessary to understand other social and economic processes in countries that may have an indirect impact. The rapid development in one area can have a significant impact on others (Ibragimov et al., 2019, Bilan et al., 2019). And this impact may not necessarily be positive. For example, the transition to automated production and the replacement of the human workforce by robotics can improve product quality, increase the speed of manufacturing, and as a result increase sales and affect GDP growth on the one hand.

On the other hand, it can provoke an increase in the unemployment rate, and consequently may harm other social and economic factors of the country's development (Citi & Oxford Martin School, 2016, 7). During crises, it can completely destabilize the economic situation (Tielietov et al., 2019).

Thus, because of the suspension of global cooperation during the quarantine period due to the COVID-19, the world's economic systems have undergone significant changes. In most cases these changes are negative. The economic collapse affected almost all spheres of human activity. Also, it revealed even more sharply the peculiarities of inequalities that currently exist in the global space.

Currently, developed countries having more opportunities to implement the latest technologies responded more quickly and efficiently to pandemic situations. Additionally, they created the conditions to eliminate shortcomings fast.

At the same time, inequality was observed in some countries. People with high-speed Internet access were able to successfully get through this crisis by transforming existing resources into a digital environment, creating new digital products, and using existing digital technologies to fulfill their tasks and responsibilities. Simultaneously, individuals with lower incomes, without quick access to the Internet and/or without the necessary equipment faced more significant problems and are in much worse conditions in this difficult period.

Taking into account the critical situation emerging in 2020 at the global level, it can be argued that the inequality between countries creates additional risks. Some of them are critical.

However, the tools of innovations' implementation at the world level, can have both advantages and disadvantages for different countries and provoke opportunities and threats in the future (World Bank Development Report, 2016). Based on SWOT-analysis, the obtained results of inequality research were grouped in Table 3.3.

Today, changes in globalization, economic conditions, scientific and technological progress, regional market specialization, socio-labor potential, as well as changes in geopolitical and geo-cultural processes are so rapid that situations occur which need to be reacted quickly and decisively.

Thus, the constant implementation of modern innovative technologies causes an inequality distribution of benefits between countries and as a result creates negative consequences and threats. In the future, these negative consequences and threats may intensify under certain conditions and lead to even greater exacerbation of inequality.

Table 3.3 Results of innovation implementation for countries with different levels of development.

Country	Advantages	Disadvantages	Opportunities for the Future	Threats for the Future
Developed countries	– receipt of hard currency (freely convertible currency) into the country; – investing to innovative industries by creating joint ventures with foreign founders; – increasing of knowledge-intensive and high-tech industries competitiveness; – freeing up time for the development of the creative population potential; – improving demographic situation in the country, slowing down the overall pace of population aging	– increasing the share of the unemployed population; – decrease in receipts to national GDP, decrease in tax revenue, due to nonpayment of taxes by those who have previously worked; – increasing the burden on local authorities and social institutions that take care of citizens' lives; – load on the social system of the country	– accelerating of technology exchange, increasing of business opportunities, and export facilities for producers; – increasing the country's intellectual potential, including rising competition in the labor market; – openness of the country; – following the globalization trends as a factor of improving the image in the international labor market	– reduction of production capacity within the country (reduction of production's automation) due to its reorientation to developing countries; – information leakage; – rise in crime and other negative phenomena
Developing countries	– increasing the country's innovative potential; – free training of new professional skills of the workforce for the exporting country, familiarity with advanced work organization, etc.;	– limitations of the social protection system; - deepening social inequalities through increased funding for families receiving money transfers from abroad;	– increasing demand for the social needs—the development of other social institutions, religions	– economic dependence on developed countries; – shortage of workers in certain specialties in the labor market;

Table 3.3 (*Continued*)

Country	Advantages	Disadvantages	Opportunities for the Future	Threats for the Future
	– reducing unemployment; – improving the well-being of poor working-class families; – increasing opportunities for international cooperation	– unwillingness to promptly level critical situations (such as the pandemic from COVID-19)	– increasing motivation for education or professional skills; – establishment of intercultural relations; – international-ization of the population;	– the loss of competitive-ness of countries, industries

Source: Own research.

3.4 Discussion

The research found that the main factors of inequality are industrialization, computerization, and robotization in economically developed societies. First, they have an expensive labor force, which is more profitable to be replaced with computers and robotics. Second, they have more opportunities for innovations development and are the centers of robotics production.

And third, they can provide mechanisms to neutralize the negative impact of industrialization, computerization, and robotization on the unemployment rate. But there are also a few paradoxes. For example, under special arrangements, some developing countries also have high rates of technological innovation implementation and thus gain certain benefits. They balance global inequality in this way.

Measures that can reduce inequality in the division of benefits and reduce the gap between the effects of innovative development are the following:

- investing in education;
- increasing the scientific and technical potential of the country;
- international cooperation and interaction in the field of knowledge exchange;
- attracting international investment to create high-tech industries within countries;
- increase in domestic demand;
- raising social standards;
- state support to meet the social needs of citizens; and

- the orientation of economic processes to support high-tech industries.

Thus, the main task of overcoming inequality between countries with different levels of economic development is to reduce the gaps between high- and low-skilled labor and, accordingly, high- and low-paid employment. By closing these gaps, there may be a change in employment demand. This will reduce the negative social effects within countries.

3.5 Conclusion

Positive and negative changes can occur at the international level. This can lead to situations when the active development of one country dramatically affects others. It seems that positive economic effects lead to impoverishment and poverty. This creates an imbalance and inequality.

Today, inequality can be presented in different aspects of human life, due to information accessibility, the ease of movement of people in the world, as well as the people's availability to cooperate through the rapid development of modern technology.

Using innovations and advanced technologies, different actions in the production and management process can be done faster and more productively. Mechanical work no longer requires human labor. The Internet provides access to information and global remote interaction between people. Thus, automation makes it possible to expand the horizons of cooperation, reduce working hours, and provide more opportunities for people's creative potential development.

However, in a market economy, access to the benefits of the new technologies' implementation, as well as access to any other resources, is not equal between countries and its representatives. In total, this has a significant impact on employment and average income. The real average wage increase is slower than the growth of productivity, and socio-economic inequality only deepens. Thus, considering the social effects of innovation and modern technologies, the main thesis is that this process does not reduce inequality between countries, groups of countries, and between people. On the contrary, it produces a hierarchy and creates new mechanisms for social divide, geographical economic imbalance, and tools for the exploitation of the human resources in a qualitatively new way. This becomes a negative factor in the processes' automation. Additionally, we see a further increase of gaps in the analyzed indicators "Technology," "Future readiness," and "Knowledge."

The logic of modern business processes orients on profit maximization and cost reduction. Based on it, developed countries often focus on production relocation to peripheral countries, due to cheap labor, and lower levels of labor protection. This process causes the reduction of automation in developed countries. The reason is that there is a replacement of production processes in the periphery. As a result, this provides a certain positive effect on developing countries. But it should be clearly understood that this process has a controlling position from the standpoint of developed countries. And in certain cases, it can be quickly modified.

References

Bhandari, Medani P. (2020) Second Edition- Green Web-II: Standards and Perspectives from the IUCN, Policy Development in Environment Conservation Domain with reference to India, Pakistan, Nepal, and Bangladesh, River Publishers, Denmark / the Netherlands. ISBN: 9788770221924 e-ISBN: 9788770221917

Bhandari, Medani P. (2020), Getting the Climate Science Facts Right: The Role of the IPCC, River Publishers, Denmark / the Netherlands- ISBN: 9788770221863 e-ISBN: 9788770221856

Bhandari, Medani P. and Shvindina Hanna (2019) Reducing Inequalities Towards Sustainable Development Goals: Multilevel Approach, River Publishers, Denmark / the Netherlands- ISBN: Print: 978-87-7022-126-9 E-book: 978-87-7022-125-2

Bilan, Y., Raišienė, A. G., Vasilyeva, T., Lyulyov, O., & Pimonenko, T. (2019). Public Governance efficiency and macroeconomic stability: examining convergence of social and political determinants. Public Policy and Administration, 18(2), 241–255.

Cebula, J., & Pimonenko, T. (2015). Comparison financing conditions of the development biogas sector in Poland and Ukraine. International Journal of Ecology and Development, 30(2), 20–30.

Citi GPS (2016). Global Perspectives and Solutions. Technology at work v2.0. Oxford Martin School. University of Oxford 2016

Country classification. (2014). World Economic Situation and Prospects 2014, 143–150. URL: https://www.un.org/en/development/desa/policy/wesp/wesp_current/2014wesp_country_classification.pdf, accessed on Feb. 15, 2020.

Dutta, S., Lanvin, Br., & Wunsch-Vincent, S. (2019). Global Innovation Index 2019 rankings. Creating Healthy Lives – The Future of Medical Innovation. 12th ed. Switzerland, Geneva, India, New Delhi, 451 p.

IIASA (2008). Economic Growth in Developing Countries: Education Proves Key. (August 2008). IIASA Policy Brief. 03. International Institute for Applied Systems Analysis. URL: https://iiasa.ac.at/web/home/resources/publications/IIASAPolicyBriefs/pb03-web.pdf,accessedonJan.10,2020.

Felipe, J., Kumar, U., & Abdon, A. (December 2010). How Rich Countries Became Rich and Why Poor Countries Remain Poor: It's the Economic Structure ... Duh! *Working Paper,* 644. URL: http://www.levyinstitute.org/pubs/wp_644.pdf, accessed on Jan. 10, 2020.

Filipenko, A.S., Rohach, O.I, & Shnyrkov, O.I. (2000). World economy. Kyiv: Lybid.

World Bank (2020). GDP per capita (current US$) / GDP (current US$). World Bank national accounts data, and OECD National Accounts data files. URL: https://data.worldbank.org/indicator/NY.GDP.PCAP.CD, accessed on Feb. 14, 2020.

History of the Global Innovation Index. (2019). About the Global Innovation Index. URL: https://www.globalinnovationindex.org/about-gii, accessed on Jan. 17, 2020.

Human development report (2019). Beyond income, beyond averages, beyond today: Inequalities in human development in the 21st century. (2019). *Technical notes.* URL: http://hdr.undp.org/sites/default/files/hdr2019_technical_notes.pdf, accessed on March 10, 2020.

Human Development Index Ranking. (2019). URL: http://hdr.undp.org/en/content/2019-human-development-index-ranking, accessed on Feb. 10, 2020.

Human Development Index. (2019). URL: http://hdr.undp.org/en/content/human-development-index-hdi, accessed on Jan. 17, 2020.

Ibragimov, Z., Vasylieva, T., & Lyulyov, O. (2019). The national economy competitiveness: effect of macroeconomic stability, renewable energy on economic growth. Economic and Social Development: Book of Proceedings, 877–886.

IMD World digital competitiveness ranking (2019). IMD World competitiveness center. 180 p.

Liulov O., Pimonenko T, Stoyanets N., Letunovska N. (2019). Sustainable Development of Agricultural Sector: Democratic Profile Impact Among Developing Countries. Research in World Economy. 2019. Vol. 10, No 4. P. 97–105.https://doi.org/10.5430/rwe.v10n4p97

Lyulyov, O., Chortok, Y., Pimonenko, T., & Borovik, O. (2015). Ecological and economic evaluation of transport system functioning according to the territory sustainable development. International Journal of Ecology and Development, 30(3), 1–10.

OECD (2016). "Automation and Independent Work in a Digital Society", Policy Brief on the Future of Work, OECD Publishing, Paris

Perspectives on Global Development (2018). Rethinking development strategies. Overview. OECD 2018. URL: https://www.oecd.org/dev/Overview_EN_web.pdf, accessed on Feb. 14, 2020.

Rosokhata, A. (2013). Generalized classification of methods for forecasting areas of innovation of industrial enterprises. "Economic space": Collection of scientific works, 80, 257–266.

Rosokhata, A. (2014). Trendwatching as a direction of forecasting innovative development for industrial enterprise on production of machines and equipment. Bulletin of NTU "KhPI", 33 (1076), 62–75.

Saher L.Yu., Melnyk Yu.M., Niño-Amézquita J. (2018). The problems of development of an effective management system of internal communications and ways to overcome them. Innovative Management: theoretical, methodical and applied grounds. 1st edition, Prague Institute for Qualification Enhancement: Prague, 83–96.

Tielietov O.S., Letunovska N., Provozin M. (2019). Social infrastructure of modern enterprises and territories. TOV: "Trytoriia".

Investopedia (2019). Top 25 Developed and Developing Countries. (Nov 21, 2019). Investopedia. URL: https://www.investopedia.com/updates/top-developing-countries/, accessed on March 10, 2020.

Vasylieva, T., Lyulyov, O., Bilan, Y., & Streimikiene, D. (2019). Sustainable economic development and greenhouse gas emissions: The dynamic impact of renewable energy consumption, GDP, and corruption. Energies, 12(17) doi:10.3390/en12173289

World Bank Development Report (2016). World Development Report 2016: Digital Dividends. (2016). World Bank. Washington, DC: World Bank. doi:10.1596/978-1-4648-0671-1.

United Nations (2019). World Economic Situation and Prospects 2019. (2019). United Nations. New York. URL: https://www.un.org/development/desa/dpad/wp-content/uploads/sites/45/WESP2019_BOOK-web.pdf, accessed on Jan. 10, 2020.

4

Impact of Income Inequality on Financial Development: A Cross-Country Analysis

Olha Kuzmenko, Anton Boyko and Victoria Bozhenko

Department of Economic Cybernetics at Sumy State University, Sumy, Ukraine

Abstract

This chapter deals with the impact of inequality with reference to income and analyze the problems of financial development.

4.1 Introduction

The sphere of financial relations is a key element in ensuring the macroeconomic stability of the state and supporting the welfare of the population. The main problem of recent decades, which is inherent in countries with different levels of financial and economic development, is the unequal income distribution, which leads to decreased well-being of citizens and increased social tension in society [1]. Today, in the countries that are members of the Organization for Economic Cooperation and Development (OECD), the income of 10% of wealth-holders exceeds the income of 10% of the poorest population by 9.6 times, while this ratio was 7:1 in the 1980s [2]. It is worth noting that the unequal distribution of income is not only a problem of countries with low economic development but also concerns highly developed countries of the world [3]. Over the past decades, there has been a gradual increase in the unequal distribution of financial resources between the poor and the rich. In particular, during 2000–2009, the Gini index in Singapore rose sharply by 15.9 percentage points compared to the previous decade, in Germany—by 9.5 percentage points, while in the United Kingdom, the USA, and Japan—by 3–4 percentage points (Figure 4.1). It is worth noting that France showed

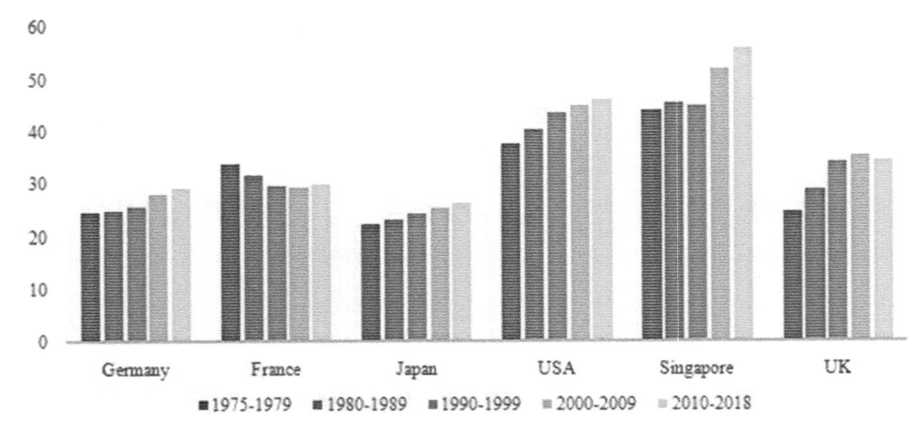

Figure 4.1 Gini index in some countries of the world for 1975–2018.
Source: World Bank Database.

a gradual decline in the Gini index as of 2018, namely by 16.3% compared to 1975, which indicates a more efficient and fair distribution of financial resources among households and a probable increase in the living standards of various segments of the population.

The income inequality of the population should be considered as a complex dynamic system, which is influenced by different factors and spheres of influence. The level of unequal income distribution in a particular country depends on various factors, of which the main ones include the intensification of globalization processes, rapid introduction of innovative technologies in various spheres of society, growing financialization of the economy, development of financial inclusion (gaining full access to key financial services in the country), as well as significant imbalances in the levels of social protection and security.

At the same time, despite the existence of an indisputable relationship between income inequality and various economic processes, the study of these patterns is of particular relevance in terms of identifying precisely the quantitative interdependencies between specific parameters of economic or social development, as well as income inequality.

Thus, Shin I. [4] hypothesizes the existence as a positive and negative impact of income inequality on economic growth, but at different stages of the latter. To achieve this goal, the scientist used the stochastic optimal growth model and concluded that fast-economic growth and low-income inequality could be achieved through a low-income tax at an early stage of economic development. At the same time, these two parameters cannot be reached simultaneously in a near-steady state.

Bénabou R. [5] considered the issue of determining the features of the relationship between income distribution and economic growth. The author used two unifying models and an empirical exercise to demonstrate the following results: the economy's growth rate is shown to be a decreasing function of interest groups' rent-seeking abilities, as well as of the gap between rich and poor; it is not income inequality per se that matters, however, but inequality in the relative distribution of earning and political power.

The relationship between the financial development of the country and income inequality was studied by Beck T., Demirgüç-Kunt A., and Levine R. [6]. As a result of their own calculations, scientists conclude that 40% of the long-run impact of financial development on the income growth of the poorest quintile is the result of reductions in income inequality.

It is worth highlighting the study by Reardon S. F. and Bischoff K. [7], which deals with quantifying the impact of growth in income inequality from 1970 to 2000 on patterns of income segregation. At the same time, income segregation was considered within three dimensions: the spatial segregation of poverty and affluence, race-specific patterns of income segregation, and the geographic scale of income distribution. The study of Tovmasyan and Minasyan [8] devoted to the analysis of the income distribution into account the gender factor deserves special attention.

Research on the relationship between income inequality and health has also been widely published in the scientific literature. Such groups of scientists as (1) Kawachi I., Kennedy B. P., Lochner K. and Prothrow-Stith D. [9], (2) Lynch J. W., Smith G. D., Kaplan G. A. and House J. S. [10], and (3) Pickett K. E. and Wilkinson, R. G. [11], despite studies of various aspects of public health, draw the same conclusions about the significant negative impact of income inequality on the health of the nation.

Researchers even pay attention to quantifying the level of per capita income at which pollutant volumes are transformed. Thus, Gene M. Grossman and Alan B. Krueger [12] define the reduced-form relationship between per capita income and various environmental indicators. Researchers found that the turning points for the different pollutants come before a country reaches a per capita income of $8,000.

Kuzmenko and Roienko [13] examined the technological and socio-economic drivers related to the Fourth Industrial Revolution on income inequality. One of the consequences of the rapid information and technology development is an increase in the share of self-employed population [14, 15].

Expanding the scope of research of the most successful experience of using mathematical tools to analyze the relationship of financial and economic patterns and socioeconomic processes, we note that Druhov O., Druhova V., and Pakhnenko O. [16] studied the relationship between financial innovation and the banking system using cluster analysis. The works by Bilan Y., Lyeonov S., Vasylieva T., and Samusevych Y. [17] are devoted to determining the impact of tax competence on entrepreneurship trends based on multifactor regression analysis. Bilan Y., Rubanov P., Vasylieva T., Lyeonov S., and Tiutiunyk I. determined the impact of the shadow economy on the investment market [18], industry 4.0 on financial services [19], macroeconomic stability on social progress [20], as well as corporate social responsibility on the banks' stability [21] using correlation-regression analysis. GMM model and Fishburne's method were used by Bilan Y., Raišienė A. G., Vasilyeva T., Lyulyov O., and Pimonenko T. [22] to establish patterns between public governance efficiency and macroeconomic stability. Kuzmenko O., Bozhenko V., Koibichuk V., and Kolotilina O [23, 24] used a more sophisticated econometric toolkit to define the relationship between different socioeconomic processes. Thus, bifurcation analysis [23] was used to establish the patterns of transformational changes in the economy, and relative normalization and Harrington's desirability function [24] were used to identify the impact of gender policy on the efficiency of a banking system. To determine the relevant factors of financial and economic development in the country Ordinary Least Square Method [25] is used.

Thus, it is fair to note that the study of quantitative characteristics of the relationship between the main parameters of the country's financial system and indicators of income inequality should be based on the use of a certain type of correlation-regression analysis, which is most widely used to determine these types of patterns despite its relative simplicity of calculation.

4.2 Methods

Canonical correlation analysis tools were used to determine the interconnections between the main parameters of the country's financial system development and indicators of the unequal income distribution. They, unlike other methods, allow evaluating the relationship between two sets of variables—two and more effective and factorial features at $p > q$, where p—the number of effective features; q—the number of factorial features. The general form of the canonical model for analyzing the relationship between lists of variables characterizing the financial state of a country's development and the unequal income distribution is as follows:

$$a_1y_1 + a_2y_2 + a_3y_3 + a_4y_4 + a_5y_5 + a_6y_6 + a_7y_7 + a_8y_8 \\ +a_9y_9 + a_{10}y_{10} = b_1x_1 + b_2x_2 \tag{4.1}$$

where y_1—domestic credit provided by financial sector (% of GDP);
y_2—foreign direct investment, net inflows (% of GDP);
y_3—market capitalization of listed domestic companies (current US$);
y_4—stocks traded, total value (current US$);
y_5—imports of goods and services (BoP, current US$);
y_6—exports of goods and services (current US$);
y_7—military expenditure (% of GDP);
y_8—government expenditure on education, total (% of GDP);
y_9—domestic general government health expenditure (% of GDP);
y_{10}—total reserves (includes gold, current US$);
x_1—Gini coefficient, equivalized household gross income;
x_2—share of top 10 per cent in gross income (*);
$a_1, a_2, a_3, a_4, a_5, a_6, a_7, a_8, a_9, a_{10}, b_1, b_2$—parameters.

Or in abbreviated form

$$\sum_{i=1}^{10} a_iy_i = \sum_{j=1}^{2} b_jx_j \tag{4.2}$$

Canonical variables have features of latent indicators that allow indirectly evaluating a process or phenomenon using linear equations.

The study involved the phased implementation of the following steps: the formation of the research information base, the determination of the canonical model, the construction of a correlation matrix between two groups of parameters, the verification of the statistical significance and adequacy of the constructed model, the determination of the latent roots of the canonical equation, the calculation of loading of canonical factors characterizing both the financial state of development and unequal income distribution in the country. The practical solution of the above tasks was carried out using the software product STATISTICA 10 ("Canonical analysis" module).

The result of the study should be the achievement of the following goals: assessing the tightness of the canonical correlation between the level of financial development of the country and the state of the income distribution, establishing the nature of the relationship between these processes, and determining the priority of the impact of financial development indicators on the state of the income distribution.

As part of the study, an information base of 44 periods was formed, but some data on some indicators were not available in certain periods. In this regard, the following methods were used: the mean of adjacent points in case of absence of one level of the time series inside the series; interpolation by the average growth coefficient method, if there are at least two levels of the time series inside the series or there are no levels at the beginning or end of the series.

4.3 Results

Within the framework of this study, it was decided to assess the relationship between income inequality of the population and the state of development of financial and economic relations by the example of some countries. Countries that have different approaches to government financial regulation, namely bank-centric (Germany, Sweden, France, Italy, Japan, etc.) and market (USA, UK, Singapore, Australia, Canada, etc.) models have been selected for the analysis. The peculiarity of the bank-centric model is the dominance of banking institutions in the implementation of monetary, loan and deposit, and investment operations. At the same time, the market model operates considering the advantages of the capital market with such participants as investment funds, stock exchange, hedge funds, etc. Annual data were selected for the study; the sample was 44 years (1975–2018).

To characterize the level of unequal income distribution, it is proposed to choose the Gini coefficient and the share of top 10 per cent in gross income, while the level of development of financial and economic relations in the country was decided to be considered within the framework of financial market development indicators (domestic credit provided by financial sector, market capitalization of listed domestic companies, and stocks traded), financial and economic indicators (foreign direct investment, imports of goods and services, exports of goods and services, and total reserves), as well as indicators of the distribution of public funds for socially important needs (military expenditure, government expenditure on education, and domestic general government health expenditure). The source of the statistical base was data from Worldbank, OECD, as well as studies by the Institute for New Economic Thinking at the Oxford Martin School.

The STATISTICA 10 software product was used to obtain the results of canonical correlation analysis, which reflect the degree of interconnection between the level of financial development of the country and the state of the income distribution, as well as the direction of influence between these processes. Figure 4.2 shows a fragment of the canonical correlation analysis for Germany, and generalized results for all the countries under consideration are presented in Table 4.1.

Analysis of Figure 4.2 and Table 4.1 indicates that there is a close relationship between the canonical variables characterizing the level of financial development and the state of the income distribution for the countries under consideration since the correlation coefficient in all cases is greater than 0.9. Analyzing the results of the canonical correlation analysis for Germany, we note that the total loss for the first group of indicators reflecting the financial and economic situation in Germany is 65.73%, while for the second group, which characterizes the income distribution, it is 86.34%. This means that

	Canonical Analysis Summary (Таблица Німеччина метод головних компонент2.sta) Canonical R: ,96831 Chi?(20)=116,24 p=0,0000	
N=44	Left Set	Right Set
No. of variables	10	2
Variance extracted	72,2591%	100,000%
Total redundancy	65,7307%	86,3658%
Variables: 1	Domestic credit provided by financial sector (% of GD	Gini coefficient, equivalised household gross incom
2	Foreign direct investment, net inflows (% of GDI	Share of top 10 per cent in gross income (*
3	Market capitalization of listed domestic companies (current US$)	
4	Stocks traded, total value (current US$	
5	Imports of goods and services (BoP, current US!	
6	Exports of goods and services (current US!	
7	Military expenditure (% of GDP	
8	Government expenditure on education, total (% of GDF	
9	Domestic general government health expenditure (% of GDF	
10	Total reserves (includes gold, current US$	

Figure 4.2 Results of canonical correlation analysis for Germany.

Table 4.1 Results of canonical correlation analysis for different countries of the world depending on the financial regulation model.

		Financial Development	Income Inequality	The Level of Accuracy, R2	*p*-value
Bank-based	Germany	65.7307	86.3658	0.9683	0.0000
	Sweden	59.0932	67.6744	0.9347	0.0000
	France	52.4488	81.0259	0.9471	0.0000
	Italy	61.1045	84.4019	0.9650	0.0000
	Japan	68.7098	94.1761	0.9789	0.0000
Market-based	USA	76.1286	92.228	0.9907	0.0000
	Singapore	64.9692	85.6033	0.9543	0.0000
	UK	61.2056	96.8225	0.9906	0.0000
	Australia	69.6198	86.8127	0.9797	0.0000
	Canada	66.2623	90.1304	0.9776	0.0000

65.7% of the variation in the values of the country's financial development is explained by changes in the income distribution, while the Gini index and the share of top 10 per cent in gross income in Germany, together determine 86.4% of the changes in the financial sector. A detailed analysis of Table 4.1 also demonstrates the following patterns: according to the development of the country's financial system, regardless of its state regulation model, it is possible to explain the changes in the income distribution system by an average of 64.5%; making adjustments to the mechanism of income distribution between households leads to changes in the development of financial relations of countries by approximately 86.5%. It is worth noting that a hypothesis that the model of financial regulation will determine the strength and direction of impact on the fair distribution of income in society has not been confirmed.

Correlation-canonical matrices are constructed to assess the tightness of the relationship between the parameters characterizing the income distribution and factors of the country's financial development. The results are summarized in Table 4.2. The significance of the canonical correlation coefficient is tested based on Bartlett's test.

Table 4.2 Results of correlation matrices between two groups of indicators for different countries of the world.

		y_1	y_2	y_3	y_4	y_5	y_6	y_7	y_8	y_9	y_{10}
Germany	x_1	0.738	0.280	0.875	0.706	0.928	0.935	-0.819	0.803	0.902	0.743
	x_2	0.583	0.319	0.818	0.756	0.856	0.860	-0.673	0.642	0.742	0.682
Sweden	x_1	0.514	0.340	0.559	0.546	0.596	0.614	-0.695	0.311	0.543	0.504
	x_2	0.700	0.373	0.816	0.814	0.790	0.803	-0.824	0.576	0.753	0.707
France	x_1	-0.703	-0.638	-0.635	-0.592	-0.640	-0.676	0.783	-0.809	-0.819	-0.484
	x_2	0.331	0.625	0.540	0.540	0.470	0.497	-0.663	0.479	0.529	0.232
Italy	x_1	0.460	0.014	0.103	0.304	0.138	0.174	-0.378	-0.797	0.141	0.154
	x_2	0.585	0.534	0.835	0.639	0.860	0.880	-0.800	-0.082	0.892	0.663
Japan	x_1	0.846	0.606	0.877	0.861	0.916	0.948	0.418	-0.745	0.912	0.889
	x_2	0.762	0.539	0.845	0.796	0.784	0.840	0.323	-0.821	0.741	0.750
USA	x_1	0.720	0.698	0.840	0.811	0.873	0.865	-0.677	0.772	0.789	0.730
	x_2	0.821	0.753	0.935	0.901	0.951	0.942	-0.719	0.780	0.886	0.784
Singapore	x_1	0.778	0.731	0.893	0.849	0.898	0.924	-0.734	-0.144	0.839	0.923
	x_2	0.769	0.663	0.776	0.725	0.815	0.834	-0.618	-0.126	0.796	0.836
UK	x_1	0.859	0.495	0.813	0.691	0.755	0.744	-0.881	-0.425	0.706	0.615
	x_2	0.855	0.510	0.917	0.774	0.845	0.834	-0.913	-0.252	0.795	0.666
Australia	x_1	0.626	0.281	0.585	0.631	0.538	0.527	-0.739	-0.089	0.604	0.493
	x_2	0.917	0.523	0.879	0.798	0.855	0.848	-0.781	-0.549	0.946	0.907
Canada	x_1	0.761	0.516	0.736	0.785	0.774	0.813	-0.880	-0.853	0.590	0.672
	x_2	0.817	0.552	0.830	0.833	0.856	0.882	-0.955	-0.797	0.713	0.756

An analysis of the data in Table 4.2 showed that the Gini index and the share of top 10 per cent in gross income as parameters of the equal income distribution are closely related to most financial indicators, namely: domestic credit provided by financial sector (y_1), market capitalization of listed domestic companies (y_3), imports of goods and services (y_3), exports of goods and services (y_6), and domestic general government health expenditure (y_9). At the same time, foreign direct investment, net inflows (y_2), and government expenditure on education, total (y_8) have a weak influence on the processes associated with the distribution of financial resources between different segments of the population. It is also worth noting that for some countries (Sweden, France, Italy, and Australia), it is advisable to adjust financial indicators that have a correlation coefficient of less than 0.7.

An important element of canonical analysis is determining the variation of variables in each of the indicator groups by calculating the canonical root. During the study, it was determined that for the countries under consideration it was advisable to consider only the first canonical root for both groups of indicators since it explains 60% or more of the variance in the set variables.

The next stage of the study is the determination of the values of canonical variables, presented in the form of canonical weights of each variable. The results of constructing canonical analysis models are presented in Table 4.3. An increase in the canonical weight of the variable leads to an increase in its share in the value of the canonical variable. Thus, canonical weights serve as a basis for determining the priority of the impact of indicators on the country's financial development and the equality of the income distribution.

4.4 Discussion

The analysis of Table 4.3 shows that foreign economic activity for most countries (Germany, Sweden, Italy, and Singapore) has a systemically important influence on the functioning of the financial system of countries, regardless of their state regulation model, since the canonical weights for the variables y_5 (import of goods and services) and y_6 (export goods and services) are greater than 1. The financial condition of the development of France is primarily determined by such factors as the volume of public expenditures on health (y_9) and the volume of the country's reserves (y_{10}). For Germany, France, Singapore and the UK, the equal distribution of the population is explained by the Gini index, while for Sweden, Italy, USA, Australia, and Canada—the share of top 10 per cent in gross income.

Table 4.3 Canonical analysis models characterizing the relationship between the level of financial development of the country and the state of income distribution.

Country		Model
Bank-based model	Germany	$-0.005y_1 - 0.063y_2 + 0.120y_3 + 0.092y_4 + 3.231y_5 - 4.431y_6 - 0.259y_7 + 0.028y_8 + 0.129y_9 + 0.151y_{10} = -0.771x_1 - 0.271x_2$
	Sweden	$0.199y_1 - 0.067y_2 + 0.022y_3 - 0.523y_4 + 4.242y_5 - 3.936y_6 + 0.495y_7 + 0.102y_8 - 0.356y_9 - 0.207y_{10} = -0.420x_1 - 0.743x_2$
	France	$0.915y_1 + 0.186y_2 - 0.021y_3 - 0.082y_4 - 1.074y_5 + 0.350y_6 + 0.122y_7 - 0.114y_8 + 1.494y_9 - 0.713y_{10} = -0.940x_1 + 0.119x_2$
	Italy	$-0.878y_1 - 0.004y_2 + 0.059y_3 - 0.089y_4 - 1.303y_5 + 1.828y_6 - 0.033y_7 + 0.029y_8 + 0.891y_9 + 0.234y_{10} = -0.299y_1 + 0.994y_2$
	Japan	$-0.442x_1 + 0.006x_2 - 0.098x_3 - 0.444x_4 - 0.130x_5 + 0.533x_6 - 0.110x_7 + 0.436x_8 + 0.399x_9 - 0.560x_{10} = -0.580x_1 - 0.450x_2$
Market-based model	USA	$0.013y_1 - 0.219y_2 - 0.079y_3 + 0.279y_4 - 1.191y_5 + 0.299y_6 + 0.071y_7 - 0.009y_8 - 0.117y_9 + 0.019y_{10} = 0.693x_1 - 1.653x_2$
	Singapore	$0.173y_1 - 0.150y_2 + 0.258y_3 - 0.124y_4 + 1.337y_5 - 3.164y_6 - 0.247y_7 - 0.300y_8 + 0.137y_9 + 0.301y = -1.067x_1 + 0.074x_2$
	UK	$-0.628y_1 + 0.015y_2 - 0.438y_3 + 0.255y_4 + 0.230y_5 - 0.133y_6 + 0.117y_7 + 0.353y_8 - 0.110y_9 - 0.024y_{10} = -0.716x_1 - 0.301x_2$
	Australia	$-0.482y_1 - 0.007y_2 - 0.607y_3 + 0.360y_4 + 0.312y_5 + 0.109y_6 + 0.249y_7 + 0.111y_8 - 0.341y_9 - 0.064y_{10} = -0.186x_1 - 0.890x_2$
	Canada	$-0.078y_1 - 0.016y_2 + 0.046y_3 - 0.437y_4 + 0.331y_5 + 0.124y_6 + 0.888y_7 - 0.064y_8 - 0.225y_9 - 0.017y_{10} = -0.110x_1 - 0.901x_2$

4.5 Conclusion

An empirical study using canonical analysis confirmed the hypothesis about a close relationship between the state of development of the country's financial system and income distribution. It has been established that the foreign economic activity has the main influence on the differentiation of income of the population for most countries of the world, regardless of the state financial regulation model. Based on the calculated canonical weights for further research, when assessing the dependence of the change in the effective parameter (equality of income distribution) on a number of factorial features (indicators of the country's financial development), it is advisable to exclude some parameters such as foreign direct investment, net inflows market capitalization of listed domestic companies, stocks traded, and government expenditure on education. Thus, intensive lending by financial institutions, the active use of stock market instruments to attract and place temporarily free financial resources, and the growth of the country's foreign trade turnover lead to an increase in the income gap between the poor and the rich.

Determining the prospects for further research on the quantitative relationship between income inequality and financial development, we note that the expansion of financial development is becoming relevant, especially in terms of the characteristics of retail banking [26, 27] and investment activities [28].

Given the territorial differences of the regions of Ukraine, it is advisable to take into account the financial decentralization in the future in the process of studying the impact of financial indicators on income inequality [29].

Special attention should be paid to the study of the impact of financial confidence of the population [30, 31] and their financial literacy [32] on income inequality, since these aspects increase the success of the investment activity of the population and as a result, their well-being.

References

[1] Gupta, R. (2017). Socioeconomic challenges and its inhabitable global illuminations. *SocioEconomic Challenges*, (1), 81–85. https://doi.org/10.21272/sec.2017.1-10

[2] In It Together: Why Less Inequality Benefits All (2015). OECD http://www.oecd.org/social/in-it-together-why-less-inequality-benefits-all-9789264235120-en.htm

[3] Nagy, Z. B., Kiss, L. B. (2018). The Examination of Appearance of Income Inequality in Scientific Databases with Content Analysis. *Business Ethics and Leadership, 2*(4), 35-45. http://doi.org/10.21272/bel.2(4).35-45.2018

[4] Shin, I. (2012). Income inequality and economic growth. *Economic Modelling, 29*(5), 2049–2057. https://doi.org/10.1016/j.econmod.2012.02.011

[5] Bénabou, R. (1996). Inequality and Growth. *NBER Macroeconomics Annual, 11*. https://doi.org/10.2307/j.ctvc77j93.14

[6] Beck, T., Demirgüç-Kunt, A., & Levine, R. (2007). Finance, inequality and the poor. *Journal of Economic Growth, 12*(1), 27–49. https://doi.org/10.1007/s10887-007-9010-6

[7] Reardon, S. F., & Bischoff, K. (2011). Income inequality and income segregation. *American Journal of Sociology*, *116*(4), 1092–1153. https://doi.org/10.1086/657114

[8] Tovmasyan, G., Minasyan, D. (2019). Gender Inequality Issues in the Workplace: Case Study of Armenia. *Business Ethics and Leadership, 3*(2), 6–17. http://doi.org/10.21272/bel.3(2).6-17.2019

[9] Kawachi, I., Kennedy, B. P., Lochner, K., & Prothrow-Stith, D. (1997). Social capital, income inequality, and mortality. *American Journal of Public Health*, *87*(9), 1491–1498. https://doi.org/10.2105/AJPH.87.9.1491

[10] Lynch, J. W., Smith, G. D., Kaplan, G. A., & House, J. S. (2000). Income inequality and mortality: Importance to health of individual income, psychosocial environment, or material conditions. *British Medical Journal*. https://doi.org/10.1136/bmj.320.7243.1200

[11] Pickett, K. E., & Wilkinson, R. G. (2015). Income inequality and health: A causal review. *Social Science and Medicine*. Elsevier Ltd. https://doi.org/10.1016/j.socscimed.2014.12.031

[12] Grossman G.M., Krueger A. B. (1994). Economic Growth and the Environment. Quarterly Journal of Economics, *110*, 353–378. https://www.nber.org/papers/w4634

[13] Kuzmenko, O., & Roienko, V. (2017). Nowcasting income inequality in the context of the Fourth Industrial Revolution. *SocioEconomic Challenges*, *1*(1), 5–12. https://doi.org/10.21272/sec.2017.1-01

[14] Santamaria, G. C., Villanueva Alvaro, J. J., Jimenez, J. M. (2018). Self-employment. The Case of Spain. *SocioEconomic Challenges, 2*(1), 35–39. DOI: 10.21272/sec.2(1).35-39.2018

[15] Balaraman, P. (2018). ICT and IT Initiatives in Public Governance – Benchmarking and Insights from Ethiopia. *Business Ethics and Leadership, 2*(1), 14–31. Doi: 10.21272/bel.2(1).14-31.2018

[16] Druhov, O., Druhova, V., Pakhnenko, O. (2019). The Influence of Financial Innovations on EU Countries Banking Systems Development. Marketing and Management of Innovations, 3, 167–177. http://doi.org/10.21272/mmi.2019.3-13

[17] Bilan, Y., Lyeonov, S., Vasylieva, T., & Samusevych, Y. (2018). Does tax competition for capital define entrepreneurship trends in Eastern Europe? *Online Journal Modelling the New Europe*, 27, 34–66. https://doi.org/10.24193/OJMNE.2018.27.02

[18] Bilan, Y., Vasylieva, T., Lyeonov, S., & Tiutiunyk, I. (2019). Shadow economy and its impact on demand at the investment market of the

country. *Entrepreneurial Business and Economics Review*, *7*(2), 27–43. https://doi.org/10.15678/EBER.2019.070202

[19] Bilan, Y., Rubanov, P., Vasylieva, T., & Lyeonov, S. (2019). The influence of industry 4.0 on financial services: Determinants of alternative finance development. *Polish Journal of Management Studies*, *19*(1), 70–93. https://doi.org/10.17512/pjms.2019.19.1.06

[20] Bilan, Y., Vasilyeva, T., Lyulyov, O., & Pimonenko, T. (2019). EU vector of Ukraine development: Linking between macroeconomic stability and social progress. *International Journal of Business and Society*, *20*(2), 433–450. http://www.ijbs.unimas.my/index.php/content-abstract/current-issue/588-eu-vector-of-ukraine-development-linking-between-macroeconomic-stability-and-social-progress

[21] Vasileva, T. A., & Lasukova, A. S. (2013). Empirical study on the correlation of corporate social responsibility with the banks efficiency and stability. *Corporate Ownership and Control*, *10*(4 A), 86–93. https://doi.org/10.22495/cocv10i4art7

[22] Bilan, Y., Raišienė, A. G., Vasilyeva, T., Lyulyov, O., & Pimonenko, T. (2019). Public Governance efficiency and macroeconomic stability: Examining convergence of social and political determinants. *Public Policy and Administration*, *18*(2), 241–255. https://doi.org/10.13165/VPA-19-18-2-05

[23] Vasilyeva, T., Kuzmenko, O., Bozhenko, V., & Kolotilina, O. (2019). Assessment of the dynamics of bifurcation transformations in the economy. In *CEUR Workshop Proceedings, 2422*, 134–146. https://doi.org/10.1051/shsconf/20196504006

[24] Kuzmenko, O. V., & Koibichuk, V. V. (2018). Econometric Modeling of the Influence of Relevant Indicators of Gender Policy on the Efficiency of a Banking System. *Cybernetics and Systems Analysis*, *54*(5), 687–695. https://doi.org/10.1007/s10559-018-0070-8

[25] Marcel, D. T. Am. (2019). The Determinant of Economic Growth Evidence from Benin: Time Series Analysis from 1970 to 2017. *Financial Markets, Institutions and Risks, 3*(1), 63–74. http://doi.org/10.21272/fmir.3(1).63-74.2019.

[26] Vasylieva T. A., Didenko, I. V. (2016) Innovations in marketing of deposit services. *Marketing and Management of Innovations*, 4. 56–64.

[27] Didenko, I. V., Kryvych, Y. M., & Buriak, A. V. (2018). Evaluation of deposit market competition: basis for bank marketing improvement. Marketing and Management of Innovations, 2, 129–141. http://doi.org/10.21272/mmi.2018.2-11

[28] Kolosok, S., Dementov, V., Korol, S., Panchenko, O. (2018). Public policy and international investment position in european integration of Ukraine. Journal of Applied Economic Sciences, *13*(8), 2375–2384.

[29] Vasylieva, T., Harust, Yu., Vynnychenko, N., & Vysochyna, A. (2018). Optimization of the financial decentralization level as an instrument for the country's innovative economic development regulation. *Marketing and Management of Innovations*, *4*, 382–391. http://doi.org/10.21272/mmi.2018.4-33

[30] Brychko, M., Polách, J., Kuzmenko, O., & Olejarz, T. (2019). Trust cycle of the finance sector and its determinants: The case of Ukraine. *Journal of International Studies*, *12*(4), 300–324. https://doi.org/10.14254/2071-8330.2019/12-4/20

[31] Bilan, Y., Brychko, M., Buriak, A., & Vasilyeva, T. (2019). Financial, business and trust cycles: The issues of synchronization. *Zbornik Radova Ekonomskog Fakultet Au Rijeci*, *37*(1), 113–138. https://doi.org/10.18045/zbefri.2019.1.113

[32] Lyeonov, S., & Liuta, O. (2016). Actual problems of finance teaching in Ukraine in the post-crisis period. In *the Financial Crisis: Implications for Research and Teaching,* 145–152. Springer International Publishing. https://doi.org/10.1007/978-3-319-20588-5_07

5

Green Investment as an Economic Instrument to Achieve SDG

Olena Chygryn[1], Tetyana Pimonenko[2] and Oleksii Lyulyov[3]

[1]Ph.D., Associate Professor, Economics, Entrepreneurship and Business Administration Department, Sumy State University, Sumy, Ukraine
[2]Ph.D., Associate Professor, Economics, Entrepreneurship and Business Administration Department, Sumy State University, Sumy, Ukraine
[3]Ph.D., Associate Professor, Economics, Entrepreneurship and Business Administration Department, Sumy State University, Sumy, Ukraine

5.1 Introduction

The ongoing snowballing economic, social, and ecological development from one side provokes the improvement of the countries' welfare, on the other side leads to increase of the negative anthropogenic effect on the environment. The world community has already made a lot in that direction. Thus, a lot of regulations and stimulation laws, action plans and agendas, protocols, and instruments have been accepted and implemented. Thus, the latest document was "The 2030 Agenda for Sustainable Development" (The 2030, 2018) which indicated 17 sustainable development goals for achieving in 2030, where highlighted the importance of renewable energy, as it will support the achievement of climate policy goals. Noted, that the characteristics of renewable energy investment have received greater attention from energy policymakers and academics. In 2015, the share of renewables in total electricity generation was 23%, while a record amount of 285.9 billion US dollars was invested in renewables (IEA, 2016). Therefore, all abovementioned problems and goals require putting a powerful investment for their solutions and achievements. It should be underlined, that it is not so a big issue for the developed countries, but a lack of financial resources is a huge issue for the developing countries (Chygryn, 2018). On the other side, investment in green projects and activities are not attractive to investors. Firstly, it is the consequences of the existing stereotypes that investment in green industries are non-profitable. Secondly, such investments have a huge payback period which negatively influences making the decision if to invest or not. Thus, the increasing number of ecological problems contributes to accepting prompt actions for developing and implementing the corresponding instruments, which could resolve abovementioned issues and not retain the economic development in the country. Moreover, these actions should consider the current modern trends in the world economy. Lastly, it is necessary to indicate and summarize the benefits and positive effects of instruments from different point of views for all stakeholders with the purpose to promote and popularize such instruments. Despite ample global savings and record-low long-term interest rates, infrastructure investments in sustainable development projects are often unable to attract long-term private financing, and the costs of financing are relatively high—in some cases prohibitively so. While the volume of private finance including cross-border finance has grown rapidly over the past two decades, very little of this capital is being directed toward long-term investment, and even less is being made available for infrastructure financing. Improving access to and reducing the cost of private capital for sustainable infrastructure will require concurrent actions on several fronts: deepen domestic capital markets, enhance and scale up risk mitigation instruments, develop infrastructure as an asset class, expand the range of financial instruments, and greening of the financial system. In this direction, the traditional economic and ecological instruments should be adopted and modernized accordingly to the ongoing market economy. Thus, a very promising instrument is the green investment, which integrated the main aspects of the traditional investments and features of a green economy.

5.2 Method

Under this investigation, the authors used the traditional and modern methods of scientific knowledge: analysis and synthesis—in identifying trends of green investment in the developed countries; comparison and compilation—to analyze the experience of developed countries to support and develop green investment with the purpose to allocate resources for achieving the Sustainable Developments Goals 2030; the statistical and mathematical methods—in analysis of the economic, social, and ecological benefits of green investment for stakeholders; and the scientific support methods—to summarize and to formulate conclusions on perspectives to develop green investment in Ukraine. These approaches allow allocating the challenges and opportunities for Ukraine to develop green investment as additional financing for achieving the Sustainable Development Goals 2030. In addition, it allows taking into account the best EU practice on supporting green investment in Ukraine conditions. The main purpose of this paper is to analyze the potential of green investment in the EU and Ukraine. In this case, it is necessary to highlight the ecological, economic, and social benefits of green investment for stakeholders.

5.3 Results

The results of the analysis showed that green investment is a complex definition, which involved green and economic aspects. Noticed, that "green" is the general and broad term which defined by the scientists from the different points of views: philosophy, social, technical, and economic. The experts in the work "Defining and Measuring Green Investments" indicated that definitions of "green" can be explicit or implicit. Some are very broad and generic; others are more technical and specific. Some are investment-driven; others come out of ecological or ethical discussions (Inderst et al., 2012).

On the other side, in general, investment is time, energy, and assets; resources were spent with the purpose to receive future benefits. From the economic point of view, investment is the purchase of goods that are not consumed today but are used in the future to create wealth. Most spread definition in finance sense, investment is a monetary asset purchased with the idea that the asset will provide income in the future or will later be sold at a higher price for a profit (Investment, 2018).

It should be noted, that among scientists there are some narrow interpretations of the concept "green investment" and the nature and coverage of a wide meaning. In addition, a mostly green investment is associated with socially responsible investing, environmental, social and governance investing, sustainable, long-term investing or similar concepts (Inderst et al., 2012).

Traditionally, green investment is directed to manufacture products specializing on cleaning equipment and waste management, environmental control facilities and monitoring devices, etc. The authors Eyraud L., Martin P., and Moser D. in the papers (Eyraud et al., 2013; Martin & Moser, 2016) defined green investment as the investment which is directed to the reduction of CO2 emission. Noted, that in the paper (Adeel-Farooq et al., 2018) the authors defined green investment as greenfield investment and associated it with the capital which finances the green growth. The group of the scientists in the papers (Hagspiel et al., 2018; Cebula et al., 2015; Yevdokimov et al., 2018; Prokopenko et al., 2017) defined green investment as an investment in renewable energy. Mielke & Steudle, 2018 analyzed green investment as funding in technologies and projects for climate change mitigation (Pimonenko et al., 2017).

The experts of Triodos Bank identified green investment as the financial products which guarantee not only financial benefits but also environmental and social effects (Triodos Bank, 2018). During Stream "Our Green Future: Green Investment and Growing Our Natural Assets" the scientists underlined that green investing is the concept which involves social and environmentally responsible corporate governance (Djalilov et al., 2015; Chigrin, 2014). In addition, they highlighted that green investment is a way to decrease investment risk and at the same time, to promote green development (Summary, 2018).

The obtained results of analysis of approaches to define green investment among Ukrainian scientists showed that green investment defined as a capital which finances green projects (Andreeva, 2005; Vasylyeva et al., 2014), economic activities for developing ecosystems (Arestov, 2010), and for achieving of green and sustainable growth (Vyshnitskaya, 2013; Kvaktun, 2014). The consolidated approaches to define the term green investment are presented in Table 5.1.

Thus, the green investment could be defined as a monetary asset purchased on green goals with the purpose to achieve income and positive green effect in the future for achieving sustainable and green growth. In this case, green goals are: mitigate climate change, develop alternative energy resources, develop clean technology, etc.

Accordingly, the complexity of the term green investment determines the following features:

- orientation to use, protection, and reproduction conditions in support of natural resources;
- the investee has the general public character for many consumers and users and often the problem cannot be solved by a separate subject, region, or country;
- the need to consider different sources of investment, combined in time and space, their forms and types;

Table 5.1 Approaches to define "green investment"*.

Main goal	Contents
Prevention and liquidation pollution	Green investment is all types of property and intellectual values invested in order to mitigate climate changes and decrease the CO_2 (Andreeva, 2005; Eyraud et al., 2013; Mielke & Steudle, 2018).
The development of ecosystems	Green investment is not only environmental investments, and any investment aimed at developing ecosystems (Arestov, 2010).
The social, ecological, and economic effect	Eco investment is the all types of property and intellectual values which invested in economic activities with goals: to reduce negative anthropogenic impact on the environment, eco-destructive impact of the product and service during life circle; conservation, effective using of natural resources and improving natural resources areas; to ensure environmental security of the country (Vyshnitskaya, 2013).
The implementation of environmental protection measures	Green investment is funds investing only into environmental protection measures (Anischenko, 2007).
Creating special funds	Green investment is capital that is aimed to develop of profitable assets in the production and exploitation which, firstly, lead to reduce the utilization of natural resources and, secondly, to soften (or liquidated) negative impact on the environment and human health. (Kvaktun, 2014, Chygryn, 2016).
Green growth	Green investment is an investment in green and sustainable growth (The Green, 2013; Adeel-Farooq et al., 2018)
Investment in green projects and clean technology	Green investment is essential investment activities that focus on companies or projects that are committed to the conservation of natural resources, the production and discovery of alternative energy sources, the implementation of clean air and water projects, and/or other environmentally conscious business practices (Investopedia, 2018; Martin & Moser, 2016; Hagspiel et al., 2018)
Environmental and social performance	Green investment is the financial products that take into consideration issues wider than purely financial performance, such as environmental and social concerns (Triodos, 2018).
Promoting green development	Green investment is a concept involves socially and environmentally responsible corporate governance as a way to reduce investment risk and at the same time, to promote green development (Summary, 2018).

(*Continued*)

Table 5.1 (*Continued*)

Improving the environment	Green investment is traditional investment instruments (such as stocks, exchange-traded funds and mutual funds) in which the underlying business(es) are somehow involved in operations aimed at improving the environment. This can range from companies that are developing alternative energy technology to companies that have the best environmental practices (Green, 2018).

*Compiled by the authors

- natural system (assimilations potential), its elements cannot be discounted, although they may reduce, lose original properties under the influence of anthropogenic factors;77
- differ from the investment that provides the state, interstate, own, mixed forms of organization of social and economic activities of natural users;
- considering specific properties of self-regulation and heal itself ecosystems its individual components (Krasnyak et al., 2015).

It should be underlined, that additional costs of greening growth are insignificant compared with the costs of fuel savings compensating in large part for the investment requirements. Nevertheless, there are a huge number of barriers which should be overcome through the reorientation of the mind of the business vision from the overusing to green.

In addition, a lot of existing policies, instruments, and mechanisms should be modernized and adapted accordingly to the new market economy. The principal scheme of transformation from traditional business to green growth is shown in Figure 5.1. It should be underlined, that green investment has its own infrastructure, sources, and channels (Figure 5.1).

Green investment has distinctive characteristics compared to investments in other economic spheres. It is due to the fact that green investment is not directed to a profit. Although in the long-term, appropriate investment projects can be profitable or create conditions for a high level of profitability of other investment projects. Thus, worldwide experience often shows the absolute economic efficiency of investments. Modern innovative production and business activities are impossible without saving technologies and monitoring of environmental cleanliness products. Along with it, an important role in boosting green investment belongs to the government which at the international, national, regional, and local levels should develop appropriate conditions for promoting and encouraging the green investment. In this perspective, public administration, green investment should not be regarded

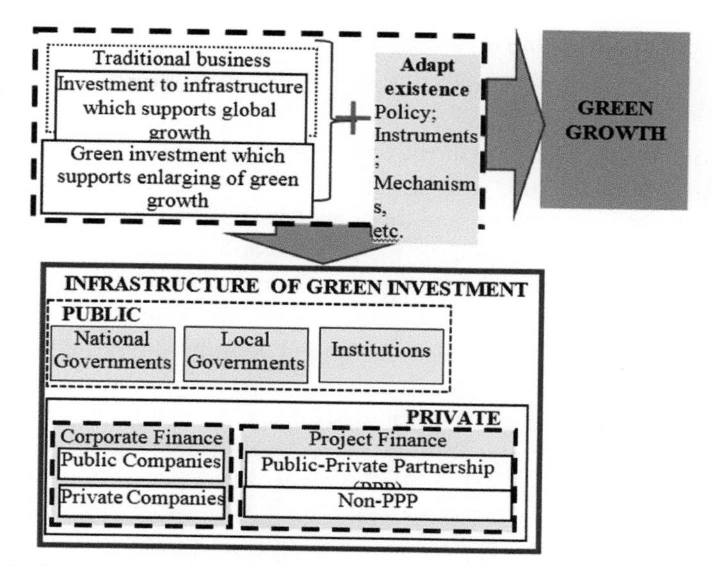

Figure 5.1 Principal scheme of transition from the existing model of business to green growth*.
*Developed by the authors.

only as investment in regeneration, protection and development of the environment, but also in creating conditions, motivation and encouragement to attract capital in the economic spheres, the reproduction and development of environmental consciousness of society as a whole (Chygryn, 2015).

The results of the analysis showed that the classification of green investment isn not so different from general investment. The scientists in the report (Chygryn, 2017) highlighted that green investment is closely connected with other types of investing and allocated the main approaches to classify green investment as follows:

- Green (eco-friendly, climate change, etc.) investing;
- the "E" in ESG (environmental, social, and governance) investing;
- thematic investing (in green sectors or themes such as water, agriculture);
- SRI (socially or sustainable responsible investing);
- RI (responsible investing);
- SI (sustainable investing), sustainable capitalism;
- Impact investing (including microfinance);
- Long-term investing;

- Universal ownership concept;
- Double- or triple-bottom-line investing (with financial, social, and ecological goals) (Inderst et al., 2012).

On the other side, the authors in their papers (Vyshnickaya, 2013; Kvaktun, 2014; Heinkel, 2003) highlighted that green investment involves all abovementioned investment, and these are types of green investment.

However, the experts from the EU commission assumed that green investment could be classified by the green assents. The authors in the paper (Ibragimov et al., 2019) proved that defining the meaning of green investment related to the assets type. In this case, the EU commission is going to develop the classification of the green assets with the purpose to avoid "greenwashing." This classification allows systematising the main approaches to classify green investment by groups follows as the level of investment; stakeholders; spheres and timeframe. Each group involves different types of investments. The summarising result is showed in Figure 5.2.

The experts in the paper (Bhattacharya et al., 2016) proved that developing green infrastructure requires attracting additional finance resources as the green investment. Thus, according to the Report "Better Finance, Better Development, Better Climate" the energy sector needs $40 trillion, transport $27 trillion, water supply and sanitation—$19 trillion. Furthermore, the authors highlighted the huge investment gap by sectors and level of countries' development. The detailed information on infrastructure needs is presented in Table 5.2.

The results of the analysis showed that in the world practice green investment is the most popular in renewable resources and financial sectors. In

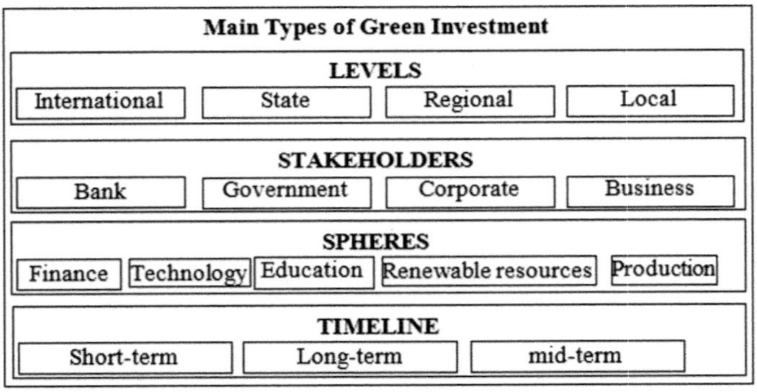

Main Types of Green Investment

LEVELS
International | State | Regional | Local

STAKEHOLDERS
Bank | Government | Corporate | Business

SPHERES
Finance | Technology | Education | Renewable resources | Production

TIMELINE
Short-term | Long-term | mid-term

Figure 5.2 The main approaches to classifying green investment*.
*Developed by the authors.

Table 5.2 Estimation of green investment in infrastructure for SDG in the world, (2015–2030), USD trillions.

Source	Energy	Transport	Water Supply & Sanitation	Telecom	Total
OECD, 2006	3,9	6	17	5,9	32,8
OECD, 2012	–	9,6	–	–	–
Boston Consulting Group (2010)	4	7	14,3	9,2	34,5
McKinsey Global Institute (2013)	13,2	25,8	12,7	10,3	62
World Bank (2013)	4	4,1	3,2	2,8	14,1
International Energy Agency (2014)	37,6	–	–	–	–
NCE, 2014 (Total Needs—BAU)	50,4	14,8	23,1	7,7	96,1
NCE, 2014 (Total Needs—Low Carbon)	–	–	–	–	101,6
NCE (Core Infrastructure BAU)	11	14,8	23,1	7,7	56,7
NCE (Core Infrastructure—Low Carbon)	–1,1	–	–	–	55,6
NCE (Energy - Primary Generation and Use)	39,4	–	–	–	39,4
UNCTAD (2014)	9,5–14,4	5,3–11,7	6,2	–	24,5–38,9
UN Sustainable Development Solutions Network (2015)	5,3–5,4	9,5	0,03	–	19,8–19,9
Brookings (2016)	20,5–23,9	27,2–31,4	12,7–14,7	–	74,7–86,6

Recourse: Bhattacharya et al., 2016.

addition, according to the survey of corporate pension funds reports (Eurosif, 2011) the equities and bonds are well ahead of other asset classes for green investment. The experts in the paper (Inderst et al., 2012) wrote that the same results were shown in another survey by (Novethic, 2011). It questioned 259 institutional investors of Europe with assets of EUR 4.5 tn. Thus, green investment is most popular among equities (40% fully implemented), followed by corporate bonds (31%), government bonds (24%), real estate (19%), private equity (15%), money market funds (14%), and commodities (8%) (Inderst et al., 2012). Therefore, green bonds market is increasing from year to year in the world (Figure 5.3). Similar conclusions are made by the group of scientists in the paper (Chygryn et al., 2018).

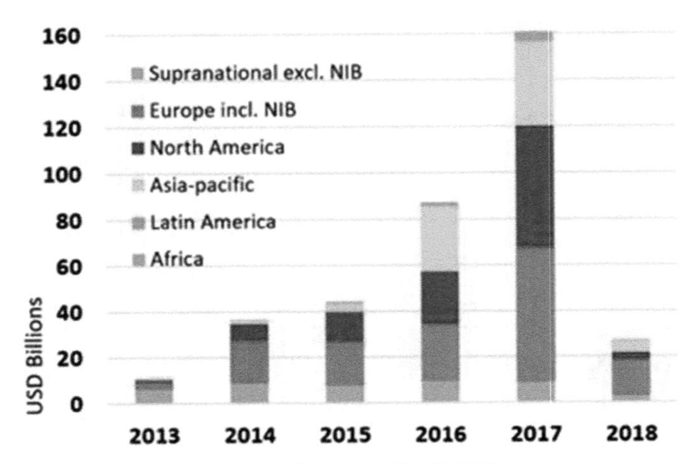

Figure 5.3 Dynamic of the green bond market in the world.
Recourses: The green bond, 2018.

The share of green bond in the debt market is 1% among G20 countries. However, the growth is significant: worldwide annual issuance rose from just US$ 3 billion in 2011 to US$ 95 billion in 2016 (Climate, 2017).

According to the official dataset (Figure 5.4), in 2016, the biggest progress at the green bond market had the following countries: France, Germany, Mexico, and South Africa. France issued the first green sovereign bond in early 2017—the largest green bond issued to date at EUR 7 billion—increasing the overall French green bond market by approximately 25% (Climate, 2017).

China is just behind the EU with regard to market penetration. The first Chinese green bonds were issued in the late 2015, but substantial growth had been made by China as in 2016 it became the largest single green bond issuing country (Climate, 2017).

According to the official report (Green bonds, 2018) developed by Climate Bond Initiatives, the goal is US$1 trillion of annual green bond issuance by 2020. According to Bloomberg New Energy Finance estimation, in 2017, investment in clean energy reached $333.5 billion, up 3% from a revised $324.6 billion in 2016, and only 7% short of the record figure of $360.3 billion, reached in 2015 (Bloomberg, 2018). In addition, the Chinese invested in 2017 to the clean energy more than $132.6 billion, up 24%, setting a new

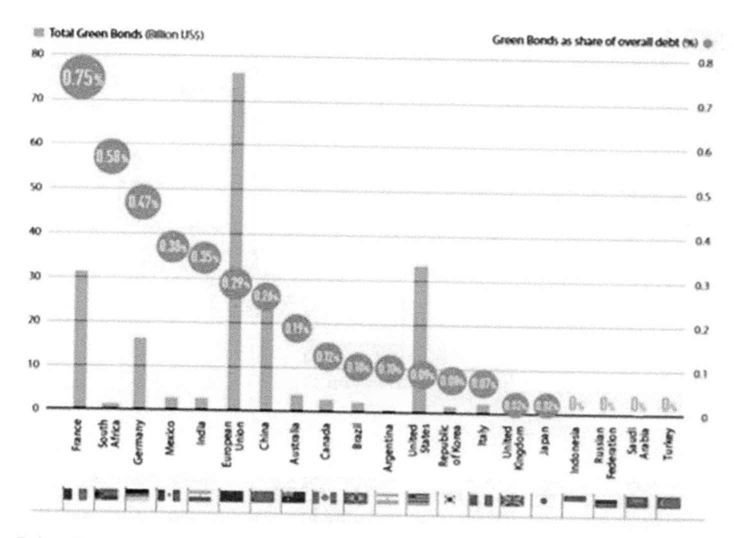

Figure 5.4 Green bonds market among G20 countries in 2016.Recourses: Climate, 2017.

all-time record. The next biggest investing country was the United States, at $56.9 billion, up 1% in 2016. In third place, Japan saw an investment decline by 16% in 2017, to $23.4 billion (Bloomberg, 2018).

The results of the analysis showed that among Asia Pacific (APAC), Europe, Middle-East and Africa (EMEA), and North, Central and South America (AMER) regions, the biggest sum of green investment in clean energy was in APAC region—$187 billion (Figure 5.5).

A lot of countries in the world have already taken a number of steps to align their financial systems with sustainable development and address risks related to climate change (Chygryn, 2017). Also given the diversity of financial systems across the countries, it is clear that measures need to be tailored to the specific needs and circumstances of each country. The compilation of experience on the green finance policy for countries with different financial systems is shown in Table 5.3.

According to the findings, Ukraine has already started to develop the relevant mechanism for supporting green investment market development. Thus, at the national level, the Government declared the concept of green bond market development. Green financial instruments with the correct target use and risk assessment could solve the range of issues, in particular: expanding funding for energy projects, strengthening the country's economic potential, and further integrating into the global economic environment. The results of

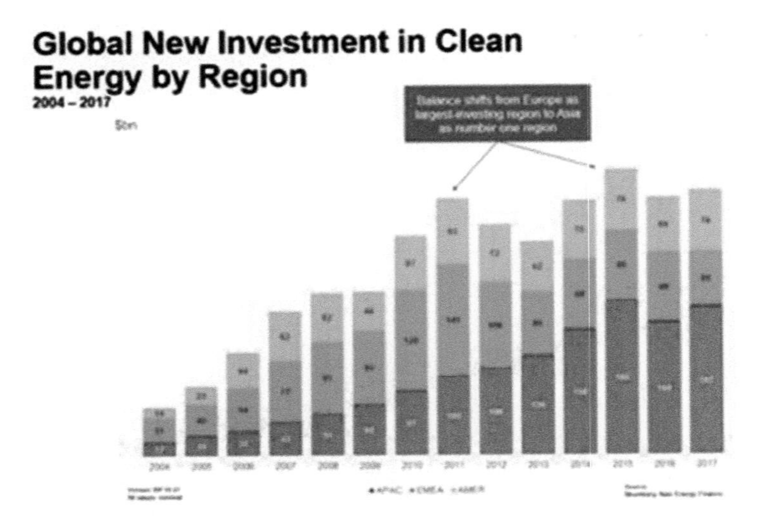

Figure 5.5 Dynamic of investment in clean energy by region.
Recourses: Bloomberg, 2018.

analysis allow allocating the main spheres for green investment in Ukraine (Figure 5.6).

In addition, the G20 experience showed that green investing has a range of opportunities and challenges for Ukrainian investors which justified the developing of supporting mechanism to overcome these barriers and challenges with the purpose to achieve green growth (Figure 5.7). Ukraine has focused on attracting investments to projects related to renewable energy and energy efficiency both on a national and regional level.

Thus, the Ukrainian Government has to develop priorities and choose strategies of investment in order to stimulate the involvement of green foreign investment and develop a favorable climate for domestic investors.

An important aspect of the effective functioning of green investment mechanism is the system of the motivation of the investors' involvement. The attitude of green investment is determined by many motives in their various combinations, which in their turn determine the system of their economic interest. Thus, it is necessary to emphasize and explain the main benefits of green investment for Ukrainian business sector. According to the EU experience, green investment allows receiving not only ecological effects but also social, political, and economic effects. So, the main economic benefits could be as follows:

Table 5.3 The green finance policy implementation.

Country	Sphere of Implementation	Content
Brazil	BOVESPA Stock Exchange	Has set up a Corporate Sustainability Index as early as 2005
	Brazil Central Bank	Has introduced requirements for banks to monitor environmental risks, building on a voluntary Green Protocol from the banking sector
	Brazil's Banking Association	Is developing a standardized assessment methodology and automated data collection system to monitor flows of finance for green economy sectors
China	People's Bank of China	Introduced green bond standards and green banking regulation
France	French Government	Introduced mandatory climate change–related reporting for institutional investors starting in January 2016.
Germany	German National Development Bank	Emitting green bond issuers worldwide
United Kingdom	Bank of England's Prudential Regulation Authority	Published a report on the impact of climate change on the UK insurance sector
India	The Reserve Bank of India	Lending to small renewable energy projects
South Africa	Johannesburg Stock Exchange	Environmental, social, and governance disclosure indicators have been introduced

*Compiled by authors on the base of Bhattacharya et al., 2016).

- increase productivity through the use of innovative and environmental technologies and equipment;
- reducing costs and product cost based on the reduction of energy intensity and resources;
- increase the competitiveness business entity and the possibility of entering new markets, etc.

From the political point of view: reducing the level of political dependence on foreign suppliers' resources; widening the opportunities to use of international agreements for activation quota trading, green production, etc.

The scientists (Ambec et. al., 2008) describe channels through which green investments can raise the benefits of firms or cut their costs: better access to markets; possibility for differentiation of products; commercialization of pollution-control technology; savings on regulatory, material, energy, and services; capital, and labor costs.

Figure 5.6 The main options for green investment in Ukraine.
*Compiled by the authors on the basis of (Krasnyak et al., 2015; Inderest et al., 2012; Bhattacharya et al., 2016).

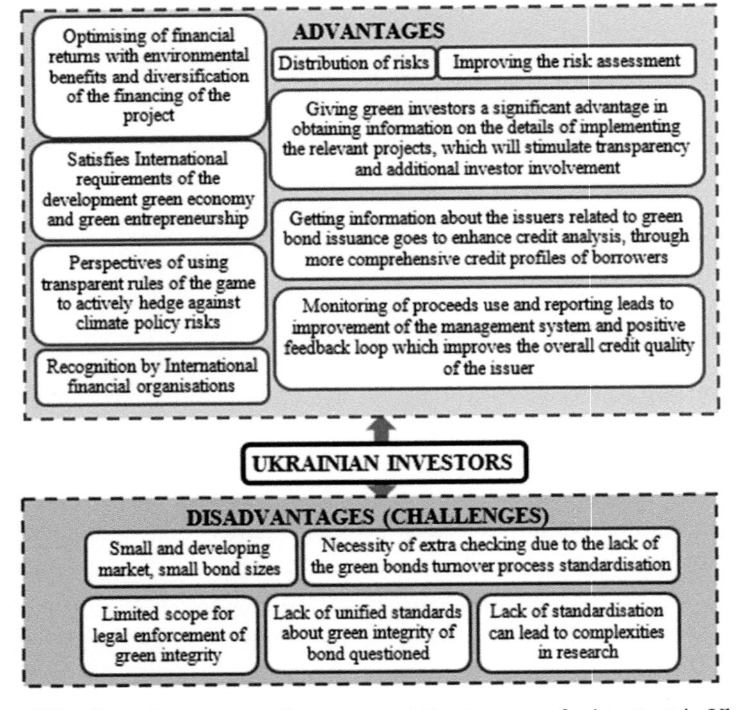

Figure 5.7 Green investment: advantages and disadvantages for investors in Ukraine.

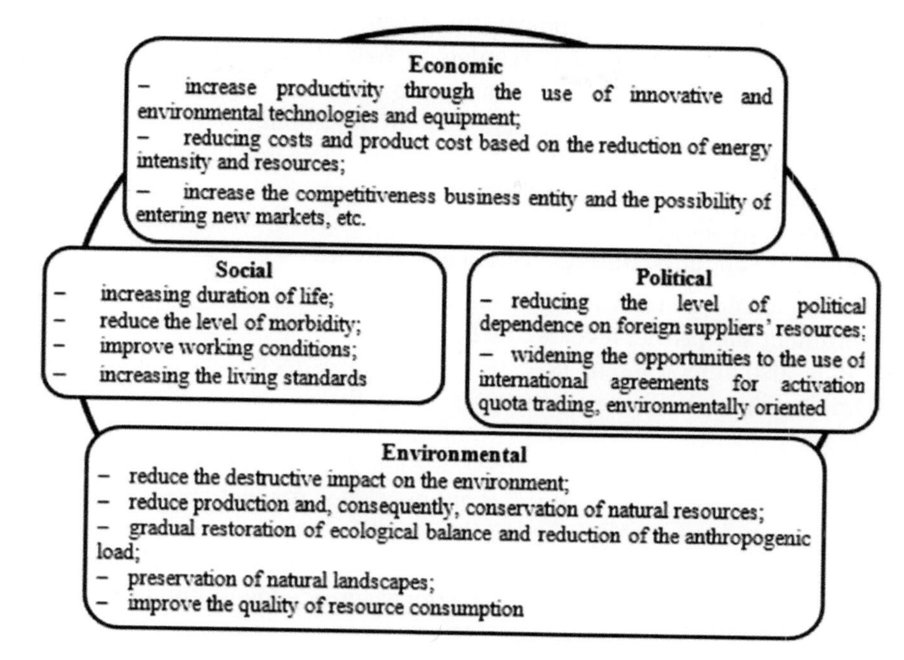

Figure 5.8 The benefits for issuers (who will disburse green investment)*.
*Developed by the authors on the basis of (Chygryn, 2015).

The consolidated benefits for green investors are presented in Figure 5.8.

It is necessary to emphasize policy spheres where the main changes should be done.

1 Promotion of the standardization of green finance practices, which includes recognizing the diversity of financial systems, establishing markets for green financial assets, and developing principles and guidelines for green finance for all asset classes, including bank credit, bonds, and secured assets.
2 Enhance the transparency of information by promoting disclosure standards for carbon and environmental risks: widening disclosure standards for carbon and environmental risks and related information flows for addressing the problem of information asymmetry. Because investors often do not know to what extent specific sectors and companies have been affected by climate change.

3 Support market development for green investment: advancement of financial institutions and regulatory frameworks which can play in developing new markets (the market for green bonds).
4 Maintenance developing countries in developing and implementing nationally sustainable finance roadmaps: devising their own national green finance practices and frameworks. Developing sustainable finance roadmaps should be geared to country-specific circumstances and needs and be supported through technical assistance.

The systemized general instruments of encouraging green investment for government and private level are in Table 5.4.

It is also important to note that green investment has a wide sphere of implementation in the different level of the economy. The worldwide experience (Green, 2017; Heinkel et al., 2001; Eyraud et al., 2013) proved the efficiency of developing the green financing market as a way to attract the additional finance recourse to the economy. Thus, for that purpose, the findings showed, that most effective activities are: to organize equity investment funds, to invest in clean cities, to support energy efficiency partnerships with financial institutions, to warehouse energy efficiency loans, bond issuance by

Table 5.4 Instruments of encouraging green investment*.

Level	
National Government	Private
1. Eliminating subsidies for fossil fuels and internalize their social and environmental costs.	1. Providing a one-year grace period for principal repayment, and to reduce transaction fees paid by private developers for project assessment
2. Establishing long-term binding targets for renewable energy with explicit paths to achieve them to make the opportunity for investment clear	2. Encouraging local banks to lend to renewable energy projects, provide credit guarantees, risk sharing, and long-term lines of credit
3. Using international best practices in economic assessments and offering tenders	3. Providing technical assistance to develop public underwriting criteria for renewable energy projects
4. Communicating with investors and other stakeholders to ensure policies and processes address barriers and encourage green investment	4. Providing accessible and inexpensive country risk insurance to increase the number of private international investors
	5. Developing structures that encourage new players to finance renewable energy in emerging markets

*Developed by the authors on the basis of (Chapman et al., 2016).

development authorities for energy efficiency, to promote innovation at early-stage companies and projects, to overcome financing barriers for residential solar project developers, public lending to facilitate commercial financing for biogas, to finance waste-to-energy, to extend the green bank model to international activities, etc. Moreover, the contribution to the growth of the green economy could be realized by cross-industry collaboration, evolving standards, and promoting green finance. All green investment initiatives should contribute to a recognized green purpose; reduction of global greenhouse gas emissions, enduring green impact, clear investment criteria implementation, robust green impact evaluation, effective monitoring and engagement, and transparent reporting.

5.4 Conclusion

The general increased interest in climate change and environment problem has increased attention to investments in green technologies and sustainable practices. For this reason, the important economic relationships, challenges, perspectives, and investment opportunities related to renewables and other green technologies were investigated. The findings allow making the conclusion that green investment is a very wide category and it is being used at all levels: the investment in primary technologies and projects and also to green companies and financial assets. Green investment can be independent, a sub-set of a broader investment theme or closely related to other investment approaches such as socially responsible investing, environmental, social, and governance investing, sustainable long-term investing, and others. The investors' and financial institutions' attention to climate change and environmental problems in general has been rising in recent years and investor financial initiatives in this respect are also growing. It is important to note that energy efficiency and environment protection represent a significant largely untapped opportunity for meeting the dual goals of risk-adjusted financial return and environmental protection. The world financial institutions are interested to invest in renewables as a cost-effective opportunity to reduce the carbon emissions and to prevent environmental damage. Thus, essential to develop specific recommendations for government and private investors for "green investing," which should include: encouraging consideration of green standards for all levels of the investment decision-process; transparency in "green issues" and strengthening disclosure to consumers and investors; encourage capacity building and development of internal and external "green audit" as well as raising "green" awareness and education. Ukraine will need to prioritize activities in investing in renewables and green technologies by selecting its own criteria, perhaps in conjunction with potential investors of a project-by-project basis.

**This research was funded by the grant from the Ministry of Education and Science of Ukraine (№ g/r 0117U003932)*

References

Adeel-Farooq, R. M., Abu Bakar, N. A., & Olajide Raji, J. (2018). Greenfield investment and environmental performance: A case of selected nine developing countries of Asia. Environmental Progress & Sustainable Energy, 37(3), 1085–1092.

Ambec, S., Lanoie, P. (2008). Does it pay to be green? A systematic overview. Academy of Management Perspectives 22(4): 45–62.)

Andreeva N.M. (2005). The theoretical basics of ecologization of investment activity. Research Bulletin of the National Forestry University, 15, 314–320

Anischenko V.O. (2007). To the issue of improving the theoretical and methodological principles of environmental investment. Actual problems of economics, 8, 175–183.

Arestov S. V. (2010). Foundations of ecosystem transfer in environmental investment. Retrieved from: http://www.nbuv.gov.ua/portal/soc_gum/en_re/2010_7_2/2.pdf.

Bhattacharya, A., Meltzer, JP., Oppenheim, J., Qureshi, Z. & Stern, N. (2016). Delivering on sustainable infrastructure for better development and better climate. Brookings Institution. Available: https://www.brookings.edu/wp-content/uploads/2016/12/global_122316_delivering-on-sustainableinfrastructure.pdf Bloomberg New Energy Finance. Retrieved from: https://about.bnef.com/clean-energy-investment/

Cebula, J., and Pimonenko, T. (2015). Comparison Financing Conditions of The Development Biogas Sector in Poland and Ukraine." *International Journal of Ecology & Development*™ 30.2, 20–30.

Chapman, Sarah., Inouye, Lauren., Smith, James., & Toriello, Carlos. (2016). Infrastructure for sustainable development: Central America renewable energy case study. IWANA.

Chigrin, O., and T. Pimonenko. (2014). "The ways of corporate sector firms financing for sustainability of performance." *International Journal of Ecology & Development*™ 29.3 (2014): 1–13.

Chygryn O.Y. (2015). Ways to financing environmental and recourse saving activity in Ukraine. Sustainable Human Development of local community. civil society P. 278–284.

Chygryn, O. (2016). The mechanism of the resource-saving activity at joint stock companies: The theory and implementation features. *International Journal of Ecology and Development*, 31(3), 42–59.

Chygryn, O. (2017). Green entrepreneurship: EU experience and Ukraine perspectives. *Waste management*, 243.5, 146.

Chygryn, O., Petrushenko, Y., Vysochyna, A., Vorontsova, A. (2018). Assesmentof fiscal decentralization influence on social and economic development. Montenegrian Jornal of economics, № 14 (4), 69–84.

Chygryn, O., Pimonenko, T., Lyulyov, O., and Goncharova, A. (2018). Green Bonds Like the Incentive Instrument for Cleaner Production at the Government and Corporate Levels: Experience from EU to Ukraine. Journal of Environmental Management and Tourism, (Volume 8, Winter), 9(17): 105-113. DOI:10.14505/jemt.v5.2(10).01

Climate Transparency (2017). Brown to Green: The G20 Transition to a Low-Carbon Economy, Climate Transparency, c/o Humboldt-Viadrina Governance Platform, Berlin, Germany, www.climate-transparency.org.

Djalilov, K., Vasylieva, T., Lyeonov, S., & Lasukova, A. (2015). Corporate social responsibility and bank performance in transition countries. Corporate Ownership and Control, 13(1CONT8), 879–888.

Eurosif (2011). Corporate Pension Funds & Sustainable Investment Study.

Eyraud, L., Clements, B., & Wane, A. (2013). Green investment: Trends and determinants. Energy Policy, 60, 852–865. doi:10.1016/j.enpol.2013.04.039

Hagspiel V, Dalby PAO, Gillerhaugen GR, Leth-Olsen T, Thijssen JJJ, (2018). Green investment under policy uncertainty and Bayesian learning, Energy (2018), doi: 10.1016/ j. energy.2018.07.137.

Heinkel, R., Kraus, A., & Zechner, J. (2001). The Effect of Green Investment on Corporate Behavior. Journal of Financial and Quantitative Analysis, 36(4), 431-449. doi:10.2307/2676219 https://www.oecd-ilibrary.org/docserver/e3c2526c-en.pdf?expires=1552237619&id=id&accname=guest&checksum=39502C79D3DA4B93EF51D97E0771A868

Ibragimov Z., Lyeonov S., Pimonenko T., (2019). Green investing for SDGS: EU experience for developing countries. Economic and Social Development (Book of Proceedings), 37th International Scientific Conference on Economic and Social Development - "Socio Economic Problems of Sustainable Development", Baku, P. 867–876.

Inderst, G., C. Kaminker and F. Stewart (2012). "Defining and Measuring Green Investments: Implications for Institutional Investors' Asset Allocations", OECD Working Papers on Finance, Insurance and Private Pensions, No. 24, OECD Publishing. http://dx.doi.org/10.1787/5k9312twnn44-en

Inderst, Georg. (2016). Infrastructure Investment, Private Finance, and Institutional Investors: Asia from a Global Perspective. ADBI Working Paper Series, No. 555, January. Manila: Asian Development Bank Institute

International Energy Agency, (2016). International Energy Outlook 2016. https://www.eia.gov/outlooks/ieo/pdf/0484(2016).pdf. Investment. Retrieved from: https://www.investopedia.com/terms/i/investment.asp#ixzz5LVfUJt3b

Investopedia. (2018) Green Investing. Retrieved from: https://www.investopedia.com/terms/g/green-investing.asp#ixzz5La3ZIXli

Krasnyak V., Chygryn, O. (2015). Theoretical and applied aspects of the development of environmental investment in Ukraine. *Marketing and Management of Innovations*. 3, P. 226–234.

Kvaktun O.O. (2014). The real green investments as an effective tool for sustainable design and construction of Ukraine's regions. Retrieved from: http://ecoukraine.org/_ld/0/7_ecpros_2014_83_.pdf.

Lyeonov, S. V., Vasylieva, T. A., & Lyulyov, O. V. (2018). Macroeconomic stability evaluation in countries of lower-middle income economies. *Naukovyi Visnyk Natsionalnoho Hirnychoho Universytetu*, (1), 138–146. doi:10.29202/nvngu/2018-1/4

Martin, P. R., & Moser, D. V. (2016). Managers' green investment disclosures and investors' reaction. Journal of Accounting and Economics, 61(1), 239–254. doi:10.1016/j.jacceco.2015.08.004

Mielke, J., & Steudle, G. A. (2018). Green Investment and Coordination Failure: An Investors' Perspective. Ecological Economics, 150, 88–95. doi:10.1016/j.ecolecon.2018.03.018

Novethic (2011), 2011 Survey, European Asset Owners" ESG Perceptions and Integration Practices.

Novethic (2012), Green Funds. A sluggish Market.

OECD (2017). Green Investment Banks Innovative Public Financial Institutions Scaling up Private, Low-carbon Investment. (2017). OECD Environment Policy Paper No. 6 January. Retrieved from: Green Investment. Retrieved from: http://wgeco.org/green-investment/

OECD. (2006). Infrastructure to 2030: Telecom, Land transport, Water and Electricity. Paris: OECD

OECD. (2008). Is it ODA?. Paris: OECD. Available at: http://www.oecd.org/dac/stats/34086975.pdf

OECD. (2012). Strategic Transport Infrastructure Needs to 2030. International Futures Programme. Paris: OECD.

OECD. (2013). Long-Term Investors and Green Infrastructure: Policy Highlights. Paris: OECD. http://www.oecd.org/env/cc/Investors%20in%20Green%20Infrastructure%20brochure%20(f)%20[lr].pdf

OECD. (2013). What do we know about Multilateral Aid: The 54 billion dollar question. OECD Policy Brief. Paris: OECD. http://www.oecd.org/development/financing-sustainable-development/13_03_18%20Policy%20Briefing%20on%20Multilateral%20Aid.pdf

OECD. (2014). OECD Science, Technology and Industry Outlook 2014. Paris: OECD.

OECD. (2014). Pooling of Institutional Investors Capital – Selected Case Studies in Unlisted Equity Infrastructure. Paris: OECD. OECD. (2014). Private Financing and Government Support to Promote Long-Term Investments in Infrastructure. P. 36. Paris: OECD.

OECD. (2014). Report on Effective Approaches to Support Implementation of the G20/OECD High-Level Principles on Long-Term Investment Financing by Institutional Investors. P.30. Paris: OECD.

OECD. (2014a). Infrastructure to 2030: Mapping Policy for Electricity, Water and Transport. Paris: OECD.

OECD. (2015-2). Climate Finance in 2013-15 and the USD 100 billion goal. Report by OECD in collaboration with CPI, p. 10. Paris: OECD.

OECD. (2015a). Towards a Framework for the Governance of Public Infrastructure. OECD Report to G20 Finance Ministers and Central Bank Governors. Paris: OECD.

OECD. (2015b). G20/OECD Report on Investment Strategies. Rreport prepared for G20 Finance Ministers and Central Bank Governors. Paris: OECD.

OECD. (2015c). Size of Public Procurement. In Government at a Glance 2015. Paris: OECD.

OECD. (2015d). Smart Procurement: Going Green – Better Practices for Green Procurement. GOV/PGC/ETH(2014)1/REV1. Paris: OECD.

OECD. (2015e). Taxing Energy Use: OECD and Selected Partner Countries. Paris: OECD.

OECD. (2015f). Mapping of Instruments and Incentives for Infrastructure Financing: A Taxonomy. OECD Report to G20 Finance Ministers and Central Bank Governors, September 2015. Paris: OECD.

OECD. (2015g). Effective Approaches to Support Implementation of the G20/OECD High-Level Principles on Long-Term Investment Financing by Institutional Investors. Report to G20, November 2015. Paris: OECD.

OECD. (2015h). Mapping Channels to Mobilize Institutional Investment in Sustainable Energy. Paris: OECD.

OECD. (2016). DAC Members List. Available at: http://www.oecd.org/dac/dacmembers.htm.

Pimonenko, T., Prokopenko, O., & Dado, J. (2017). Net zero house: EU experience in ukrainian conditions. *International Journal of Ecological Economics and Statistics*, 38(4), 46–57.

Prokopenko, O., Cebula, J., Chayen, S., & Pimonenko, T. (2017). Wind energy in israel, poland and ukraine: Features and opportunities. *International Journal of Ecology and Development*, 32(1), 98–107.

FAO (2018). Summary of Stream 5: Our green future: green investment and growing our natural assets. Retrieved from: http://www.fao.org/fileadmin/user_upload/rap/Asia-Pacific_Forestry_Week/doc/Stream_5/Stream_5_Summary.pdf The 2030 Agenda for Sustainable Development. Retrieved from: https://sustainabledevelopment.un.org/content/documents/21252030%20Agenda%20for%20Sustainable%20Development%20web.pdf

The Green Investment Report. (2013) World Economic Forum. www3.weforum.org/docs/WEF_GreenInvestmentReport_ExecutiveSummary_2013.pdf

Triodos Bank (2018). Green investment – what does it actually mean? Retrieved from: https://www.triodos.co.uk/en/personal/ethical-investments/green-investments/ Unlocking the green bond potential in India. Retrieved from: https://archive.nyu.edu/bitstream/2451/42243/2/Unlocking%20the%20Green%20Bond%20Potential%20in%20India.pdf Green bonds as a bridge to the SDGs. (2018). Retrieved from: https://www.climatebonds.net/files/files/CBI%20Briefing%20Green%20Bonds%20Bridge%20to%20SDGs%281%29.pdf The green bond market in Europe 2018. (2018). Prepared by the Climate Bonds Initiative. Retrieved from: https://www.climatebonds.net/files/files/The%20Green%20Bond%20Market%20in%20Europe.pdf

Vasylyeva, T. A., & Pryymenko, S. A. (2014). Environmental economic assessment of energy resources in the context of ukraine's energy security. Actual Problems of Economics, 160(1), 252–260.

Vyshnitskaya O.I. (2013). Environmental investments: essence, classification, principles and directions of realization. - Bulletin of Sumy State University. Economy Ser., 2, 51–58 320.

Yevdokimov, Y., Chygryn, O., Pimonenko, T., & Lyulyov, O. (2018). Biogas as an alternative energy resource for ukrainian companies: EU experience. *Innovative Marketing*, 14(2), 7–15. doi:10.21511/im.14(2).2018.01

6

The Effect of Shadow Economy on Social Inequality: Evidence from Transition and Emerging Countries

Sergij Lyeonov[1], Tatiana Vasylieva[2], Inna Tiutiunyk[3] and Iana Kobushko[4]

[1]Doctor of Economics, Professor of the Economic Cybernetics Department, Sumy State University, Sumy, Ukraine
[2]Doctor of Economics, Professor of the Finance and Entrepreneurship Department, Director of Oleg Balatskyi Academic and Research Institute of Finance, Economics and Management, Sumy State University, Sumy, Ukraine
[3]PhD in economics, Associate Professor, Senior Lecturer of the Finance and Entrepreneurship Department, Sumy State University, Sumy, Ukraine
[4]PhD in economics, Senior Lecturer of the Management Department, Sumy State University, Sumy, Ukraine

Abstract

In the context of the economic crisis in the country, the decline in the level of material well-being and social protection of the population, the problem of income shadowing, as the main component of reducing the level of social protection in the country, increasing the level of its economic inequality, becomes especially relevant. Nowadays, the shadow economy is one of the threats of economic development of the country and one of the main obstacles of its social development. One of the biggest negative effects of the shadow economy on the level of social development of the country is the evasion of social security contributions, which leads to a significant decrease in funding for programs and measures of social security of the population, problems with social benefits, pensions, unemployment insurance, health insurance, etc. Shadowing the economy significantly reduces the pace of reform of health care, education, social protection, etc. This chapter examines the relationship between shadow economy and the level of the social inequality compare magnitudes of the effects of its interaction among transition and emerging countries. In order to test the hypotheses about the negative impact of the shadow economy on the level of social inequality in the country, empirical models with interaction terms were presented: with Gini index, the ratio of the average income of the richest 10% to the poorest 10%, the ratio of the average income of the richest 20% to the poorest 20% as dependent variables, and inequality-adjusted Human Development Index. Based on the comparison of indicators of social inequality and the shadow economy, the authors distinguish clusters of countries by the nature and direction of the relationship between the analyzed indices. Empirical calculations have proved the high impact of the shadow economy on the indicators of social development of the country. All indices are statistically significant at the level of 1%, 5%, and 10% respectively. Based on the panel data regression models, groups of indicators, which have a linear and non-linear relationship with shadow economy, have been defined. The scientific contribution of the research consists in the fact that the current studies regarding the impact of the level of shadow economy on the level of the social inequality indicators are fragmentary. This study allowed the authors to conclude on the need to take into account the level of shadowing of the economy in the process of developing programs and measures to improve the level of social development of the country, reduce the volume of economic and social inequality of the population, etc.

Data set

This study examines the impact of shadow economy on social inequality by using a panel data set of 32 transitions and emerging countries in the period 2005–2018. The information base of the study is the data of the World Bank and the International Monetary Fund.

Methods & Methodology

In order to test the hypothesis of the effect of shadow economy on the level of social inequality in countries we employ a time-series analysis with the Johansen cointegration test, a vector error correction model, and the Granger causality tests.

6.1 Introduction

Current trends in world economic development indicate the presence of a significant number of factors that negatively affect the economic, political, and social development of the country. One of the sources of constant negative multidimensional influence on public policy is the shadow economy. Fragmentary understanding of the real scale and qualitative characteristics of the systemic phenomenon, as the shadow economy, have a very negative impact on the functioning of the national economy. Failure to take into account the actual scale of shadow economic activity leads to a distortion of macroeconomic indicators, which ultimately affects the reliability of the processes of forecasting and modeling the socio-economic situation in the country. Without a thorough understanding of the volume of shadow financial transactions, it is virtually impossible to assess existing trends and real processes taking place in the formal sector of the economy.

The constant growth of the world level of the shadow economy harms the country's development indicators. According to experts, its average level is 32% and is gradually growing every year (Medina and Schneider, 2018). This problem is especially acute for Emerging countries in which the institutional component of the mechanism for combating illegal financial transactions is in its infancy and therefore, unable to completely do its functions.

The shadow economy is a complex phenomenon whose consequences are observed in many areas of society. The majority of scientists study the shadow economy through the prism of its impact on the country's economic development (Makarenko and Sirkovska, 2017). At the same time, its social component, which is affected by shadow operations on a scale no less than the economic one, is often ignored.

One of the negative effects of the shadow economy on social development is the social contributions evasion. It leads to a decrease in funding for social programs and benefits, problems with the state pension system, unemployment, and health insurance, etc. The pace of reforming health care, education, and social sectors is declining.

In addition to its economic consequences, it is important to link the level of shadow economy with indicators of social development. The deficit of budget funds leads to lower social living standards, deteriorating the level of their material well-being (Kouassi, 2018). Thus, for some countries (Bulgaria, Sweden, Italy, etc.) there is a gradual increase in the Gini coefficient during the analyzed period, which is evidence of rising poverty in the country, low efficiency of state institutions, and reduced quality of life. One of the

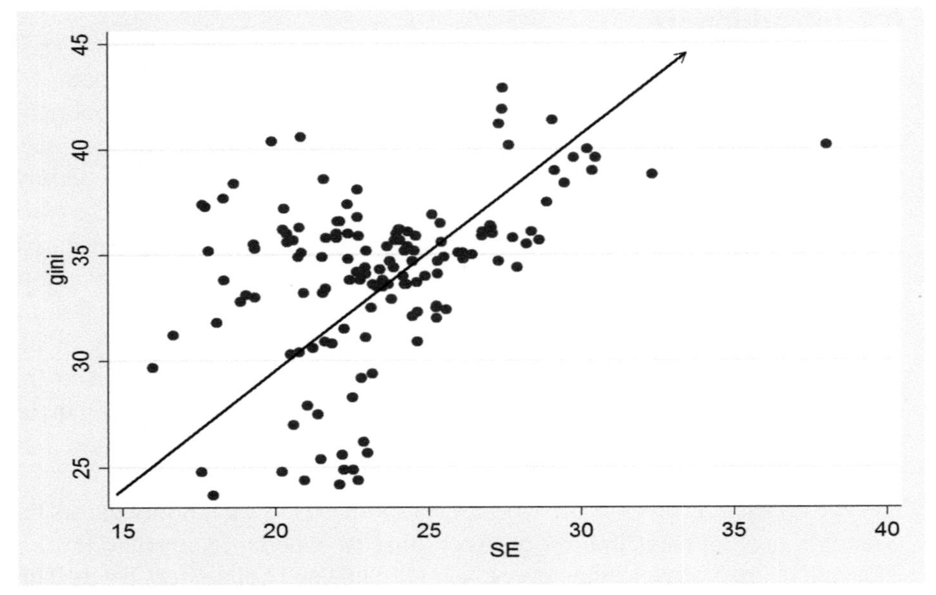

Figure 6.1 The level of shadow economy and Gini coefficient in European countries in the period 2005–2018.
Source: World Bank, 2020.

reasons is the existence of the shadow economy, which is accompanied by the enrichment of participants in shadow operations and the lack of funds from the official sector of the economy. The following scatter plot depicts a positive correlation between the share of the shadow economy in GDP and income inequality in European countries in the period 2005–2018. The growth of the shadow economy is accompanied by an increase in the Gini coefficient.

6.2 Literature Review

In assessing the relationship between the informal economy and social development, some authors use indicators of the quality of life of society, such as the number of secondary and higher education institutions, the density of hospital beds, kindergartens, education, and others. (Balas and Kaya, 2019; Kuzmenko and Roienko, 2017; Singh, 2018). However, these indicators characterize the social development of the country as a whole and do not fully reflect the change in the level of quality of life under the influence of the shadow economy. For example, the density of hospital beds cannot fully characterize the level of social security of the population. In countries with low levels of material development, the imperfection of the medical system, sufficient provision of hospital beds may occur in conditions of low access to medical facilities due to the inability to pay for treatment, etc.

Katrechka and Dahlberg (2014) studied the relationship between these indicators in the long run by four econometric models, in which the level of the shadow economy used as a factor, and as a result—life expectancy, HIV prevalence, school enrollment, and mortality up to 5 years.

Empirical studies for 31 countries with a high level of economic development and 27 countries with a medium level of development proved that the negative effect of the shadow economy outweighs the positive one in the social sphere for all analyzed countries. A change in the level of the shadow economy by 1% is more threatening for low- and middle-income countries than for developed ones.

Rosser et al. (2003) hypothesized a positive effect of shadowing on income inequality. On the example of 16 countries with economies in transition in two periods (1987–1989 and 1993–1994), the authors conclude that an increase in inequality causes more informal activities due to the decline in social solidarity and trust, and expanding informal activities lead to more inequality because of falling tax revenues and weakened redistributive policies.

The opposite view is held by Schneider and Dominik, 2000, who argue that two-thirds of the income earned in the shadow economy is immediately spent in the official economy. Smith, 2002 claims, that the shadow economy creates new jobs and increases the level of an individual's income. Moreover, Valentini, 2009 argues that the income evasion attached to an unobserved economy tends to reduce inequality measured by regular wages in the case of Italian private sector employees.

Kim (2005) argues that the main motive for participating in shadow financial transactions is the desire to escape poverty. Thus, according to the author, informal economic activities are used as salvage for the poor to lessen poverty.

Huynh and Nguyen (2019) examined the impact of the shadow economy on income inequality by using a panel data set of 19 Asian countries in the period 1990–2015 using the fixed effect, random effect, and SGMM method. The results of the analysis showed that the shadow economy reduces income inequality. Thus, the shadow economy significantly increases the income share held by the lowest quintile and decreases the income share held by the highest quintile.

Saibal and Shrabani (2013) by Panel Least Square and Fixed Effects Models proved that if the level of the informal economy grows, corruption is less harmful to inequality. Authors found that in the absence of the shadow economy, corruption increases inequality. However, with larger shadow economies in South Asia, income inequality tends to fall.

Musaeva (2017) justified the positive impact of the shadow economy on the material well-being of an individual and negative in the case of participation in the shadow operations of two or more people. The author argued that the high level of shadow economy intensifies the social inequality and discrepancies associated with the income and property of its population.

In general, all theories describing the relationship between the shadow economy and indicators of social development can be divided into three types: the theory of dualism, voluntarism, and legalism. The theory of dualism considers the shadow economy as a set of economic activities carried out to generate income for poor members of society. The voluntarism and legalism theories argue that the existence of the shadow economy creates unfair competition between formal and informal business. Thus, the shadow economy leads to a decrease in the share of income of rich members of society and as a result to reduces inequality in society.

However, despite the presence of a significant number of approaches to assessing the impact of economic shadowing on the level of social inequality (Boiko and Samusevych, 2017; Cseh-Papp et al., 2017; Dean et al., 2017; Poliakh and Nuriddin, 2017; Bhandari 2020; Bhandari and Shvindina 2019) the complex study of the relationship between these indicators for European countries is absent, and existing developments are fragmentary.

6.3 Methods

In the research, we based social inequality on the World Bank's approach, according to which it can consider as a deviation of the distribution of income or consumption expenditure among individuals or households within the country from their uniform value.

This study aims to confirm the hypothesis of the negative impact of the shadow economy on social inequality in European countries in both the short and medium-term. As for indicators of social development, we have chosen the Gini index, the ratio of the average income of the richest 10% to the poorest 10%, the ratio of the average income of the richest 20% to the poorest 20% as dependent variables, and inequality-adjusted Human Development Index. The annual values of the above indicators were used for calculations.

We will test the hypothesis using the VAR/VEC model, which describes the relationship between the indicators. The choice of model depends on the characteristics of the indicators (stationary, cointegrated):

1 In the case of non-stationarity and existing cointegration between indicators will be used VAR model in difference;
2 In the case of non-stationarity and cointegration of indicators will be used vector error correction model (VEC model);
3 If all indicators are stationary we will apply the VAR model which has the following form:

$$y_t = a_0 + \sum_{m=1}^{p} A_m y_{t-m} + \sum_{n=0}^{q} B_n x_{t-n} + \varepsilon_t \tag{6.1}$$

The information base for the indicators of the level of shadowing of the economy is the data of the International Monetary Fund, obtained based on the calculations of Medina and Weiss. Information on social inequality is based on OECD and World Bank data.

6.4 Results

At the first stage, we will investigate the static indicators that characterize the stability of the analyzed indicators: standard deviation, coefficient of variation, maximum, and minimum values. The calculations (Table 6.1) indicate a significant variability and deviation of the indicators over the years. The standard deviation of the analyzed countries characterizes by a significant scale. The greatest variability has indicators for countries with a shadowing rate of more than 30%. If for countries with a level of shadow economy up to 10% the scope of variation of the human development index is 0.0046, then for those with the highest level of shadow economy its value is 0.018 and is almost three times higher than the previous value. A similar situation is followed by other indicators of social development. The highest levels of variation are in the shadow economy and the Gini index—3.13 and 0.5, respectively.

The analysis of the correlation between the analyzed indicators carried out using the multiple regressions method showed the influence of the shadow economy on the indicators of social development in terms of groups of countries. Most results are statistically significant at 0.1%. For almost all groups of countries, the relationship between the level of the shadow

Table 6.1 Descriptive statistics of variables for the period from 2005–2018.

Groups	Variable	Mean	Std. Dev.	Max	Min
Countries with the level of shadow economy 0–10%	H0	0.8275	0.0046291	0.82	0.83
	IN200	5.697692	0.0870971	5.58	5.83
	IN100	4.340769	0.2343239	3.82	4.52
	DG0	32.5675	1.054965	31.2	34.58
	SE0	8.663077	0.7133183	7.75	9.83
Countries with the level of shadow economy 10–20%	H10	0.79875	0.0099103	0.78	0.81
	IN2010	4.59	0.1410082	4.41	4.91
	IN1010	3.666154	0.0888386	3.4	3.77
	DG10	30.54201	0.611261	29.41	31.26
	SE10	13.79	0.7961157	13.02	15.39
Countries with the level of shadow economy 20–30%	H20	0.7475	0.010351	0.76	0.73
	IN2020	5.77	0.3021865	5	6.15
	IN1020	4.668462	0.1352064	4.48	4.88
	DG20	34.17459	0.4961968	33.42	34.8
	SE20	23.36846	1.216647	21.62	25.41
Countries with the level of shadow economy 30–40%	H30	0.66775	0.0187197	0.64	0.691
	IN2030	12.46423	0.4944561	11.72	13.2
	IN1030	10.46423	0.4944561	9.72	11.2
	DG30	31.92143	0.8087773	30.8	33.35
	SE30	50.45964	3.133153	46.30927	55.995

economy and the Gini index has not been confirmed. At the same time, a statistically significant relationship between the level of shadowing and the Human Development Index was established for almost all groups. The results of the calculations are shown in Table 6.2.

To build a model of the dependence of social development indicators on the level of the shadow economy, we will check the all-time series for stationarity using the Dickey–Fuller test (Table 6.3). According to the results, most indicators are non-stationary. The absolute value of the calculated value is less than the critical value at 1%, 5%, and 10% levels of significance. For example, the value of the ADF test statistic for IN10 for group of countries with the level of shadow economy less than 10% is 0.105 is less than the critical value for sample 25, which is -2.66 and indicates the non-stationarity of the data analyzed. Similar results were obtained for all group of countries and indicators. For HDI0, IN200, DG0, HDI10, IN2010, DG10, HDI20, IN2020, IN1020, and DG20, the obtained values are more than critical. The results of the Philips Perron Test Statistics allow us to reject the unit root null hypothesis for stationarity of all indicators within all groups of countries at the 10% level of significance.

Table 6.2 Multiple regressions (OLS) for shadow economy and indicators of social inequality in European countries.

	HDI	IN20	IN10	DG
0–10	–4.058174	2.943048*	0.913874*	0.0156895
	(4.344109)	(1.348477)	(0.3361108)	(0.7376207)
cons	–3.25767**	–7.56996**	–3.789497***	–2.431589*
	(0.8229164)	(2.346371)	(0.4932852)	(0.8279568)
10–20	–2.754574**	1.145244*	0.9681602	–0.021955
	(0.6312634)	(0.4400251)	(0.6186238)	(0.8141087)
cons	–2.621488***	–3.727442***	–3.240226**	–2.008774
	(0.1420812)	(0.6704797)	(0.8036432)	(0.9662717)
20–30	–2.800827**	–.6746249**	–0.3525456	–0.6759321
	(0.6668371)	(0.2057948)	(0.5304645)	(1.029482)
cons	–2.288353***	–.2735224	–0.911952	–2.181309
	(0.1943133)	(0.3605776)	(0.8172833)	(1.106264)
30–40	–1.47924**	1.501344***	1.25941***	1.603194*
	(0.2826594)	(0.1196806)	(0.1006663)	(0.5452747)
cons	–1.319279***	–4.472356***	–3.641509***	1.142803*
	(0.1144879)	(0.301885)	(0.2363011)	(0.6220718)
***p*<.05 ***p*<.01 ****p*<.001. Standard errors within parentheses**				

For most indicators, p-value (probability) does not allow to reject the null hypothesis about the presence of a single root in the time series. The first difference of the series (Table 6.4) is fixed, which indicates that they are all integrated degree 1 (I (1)). The absolute value of the calculated value of t-statistics in the first differences exceeds the critical values for the significance level of 1%, 5%, and 10%. The p-value for all analyzed indicators is less than 10%, which allows us to reject the null hypothesis about the non-stationarity of the first differences of the data series with a minimum probability of error (almost 0% of cases with 100%). Thus, the series in the first difference is stationary and has the order of integration 1.

In the next stage, to select a model of the relationship between the shadow economy and the social development we will test the hypothesis of the cointegration of data series by Johansen tests. This method is to test the hypothesis of the cointegration of indicators from rank 0 to rank k-1. If the hypothesis is not rejected for rank 0, then the rank is considered to be null (no cointegration). And so on to k-1. In the latter case, the alternative hypothesis is the not cointegration of the original series. If the trace statistic is more than 5% critical value it allows us to accept the alternative hypothesis of data cointegration. The model assumed the presence of a constant and the absence of a trend in the cointegration ratio.

Table 6.3 The results of testing the data group for stationarity by the Dickey–Fuller test.

Groups	Variables	ADF Test Statistics*			Philips Perron Test Statistics		
		Prob.	lag	Test Statistic	Prob.	lag	Test Statistic
0–10	HDI0	0.0104	0	–3.416**	0.0104	0	–3.416**
	IN200	0.0741	2	–2.700**	0.7133	0	–1.104
	IN100	0.9664	0	0.105	0.9664	0	0.105
	DG0	0.0001	0	–4.634***	0.0001	0	–4.634***
	SE0	0.4215	4	–1.719	0.4945	0	–1.578
10–20	HDI10	0.0078	1	–3.508**	0.0423	1	–2.927*
	IN2010	0.0859	0	–2.635*	0.0468	1	–2.887*
	IN1010	0.9839	0	0.470	0.9839	0	0.470
	DG10	0.0285	0	–3.075**	0.0200	3	–3.200**
	SE10	0.1508	0	–2.369	0.1403	1	–2.405
20–30	HDI20	0.0002	2	–4.478***	0.0325	5	–3.027**
	IN2020	0.0008	0	–4.142***	0.0008	0	–4.142***
	IN1020	0.0165	1	–3.266**	0.4250	3	–1.712
	DG20	0.0110	3	–3.400**	0.3806	5	–1.800
	SE20	0.2305	0	–2.136	0.2305	0	–2.136
30–40	HDI30	0.7210	0	–1.085	0.0104	0	–1.085
	IN2030	0.8700	4	–0.604	0.9107	0	–0.397
	IN1030	0.9138	0	–0.377	0.9169	2	–0.358
	DG30	0.3970	4	–1.767	0.4227	0	–1.716
	SE30	0.6191	4	–1.322	0.7598	1	–0.982

As shown the results in Table 6.5, for all group of analyzed countries, the obtained values for rank 0 are more the critical values and allow to accept the hypothesis of cointegration of the analyzed data series. For example, for countries with the level of shadow economy less than 10% of GDP, the value of trace statistic for IN20 is 22.91 and exceeds 5% (15.41) and 1% critical value (20.04), for HDI it is 33.76, for DG—17.81. At the same time, rank 1 is lower than the 5% critical value for DG0 for countries with the level of shadow economy less than 10%, 20% to 30%, HDI for countries with the level of shadow economy 20% to 40%.

The stationary of the first differences and the cointegration of the data indicate the expediency of using the VAR model to formalize the relationship between the level of the shadow economy and social development indicators.

The cointegration of the data indicates that there is a relationship between them. The next step in formalizing the relationship between indicators is to determine the time lag through which this effect is maximum. For this purpose, we will determine the optimal structure of lags for the VAR model using tests for maximum lag and exclusion.

Table 6.4 The results of checking the first differences of the data series for stationarity by the Dickey–Fuller test.

Groups	Variables	ADF Test Statistics		
		Prob.	lag	Test Statistic
0–10	HDI	0.0104	0	–3.416**
	IN20	0.0081	0	–3.496**
	IN10	0.0266	0	–3.099**
	DG	0.0001	0	–4.634***
	SE	0.005	0	–4.253***
10–20	HDI	0.0078	1	–3.508**
	IN20	0.0859	0	–2.635*
	IN10	0.9839	0	0.470
	DG	0.0285	0	–3.075**
	SE	0.0014	1	–4.006***
20–30	HDI	0.0002	2	–4.478***
	IN20	0.0008	0	–4.142***
	IN10	0.0492	1	–2.868*
	DG	0.0110	3	–3.400**
	SE	0.0025	0	–3.844***
30–40	HDI	0.0302	0	–3.053**
	IN20	0.0439	1	–2.913*
	IN10	0.0459	1	–2.895*
	DG	0.0000	1	–5.042***
	SE	0.0433	1	–2.918*

Table 6.5 Johansen tests for cointegration.

	Rank	5% Critical Value	1% Critical Value	Trace Statistic			
				HDI	IN20	IN10	DG
0–10	0	15.41	20.04	33.7582	22.9130	20.3813	17.8120
	1	3.76	6.65	9.7638	9.7638	7.9099	0.8482
10-20	0	15.41	20.04	30.0722	37.0565	15.7717	12.2981
	1	3.76	6.65	9.4941	8.5321	4.8302	5.6012
20–30	0	15.41	20.04	16.5710	36.4055	15.5623	16.3661
	1	3.76	6.65	1.9361	12.4537	1.8800	2.6640
30–40	0	15.41	20.04	15.9277	27.7348	27.5639	18.1221
	1	3.76	6.65	1.6979	8.4290	8.3708	7.5471

The results in Table 6.6 show that for countries with a shadow economy of 10% to 20%, the maximum lag is 5 years, 20% to 40% is a maximum lag of 3 years, and 0% to 10% is a maximum of 6 years. The VAR model with these lags has the best values for the Akaike, Hannan-Quinn, Schwarz Bayesian criteria among other considered model specifications.

Table 6.6 The maximum lag of the impact of the shadow economy on social development for countries with a level of shadowing from 0% to 20% of GDP.

lag	LL	LR	df	p	FPE	AIC	HQOC	SBIC
dH0, dSE0, dIN100, dIN200, dDG0								
0	99.6299		25		1.3e-18	–27.0371	–27.5146	–27.0758
1	–	–	25		–1.6e-91*	–	–	–
2	1327.26	–	25	–	–	–369.216	–372.559	–369.486
3	1320.94	–12.638	25	–	–	–367.41	–370.753	–367.681
4	1324.97	8.062	25	0.999	–	–368.562	–371.905	–368.833
5	1338.39	26.843	25	0.364	–	–372.397	–375.74	–372.667
6	1347.36	17.945	25	0.845	–	–374.96*	– 378.303*	– 375.231*
dH10, dSE10, dIN1010, dIN2010, dDG10								
0	103.936		25		3.7e–19	–28.2675	–28.745	–28.3061
1	670.761	1133.6	25	0.000	1.5e–84*	–183.075	–185.94	–183.306
2	1334.52	1327.5*	25	0.000	–	–371.292	–374.635	–371.562
3	1340.67	12.292	25	0.984	–	–373.048	–376.391	–373.318
4	1332.46	–16.423	25	–	–	–370.702	–374.045	–370.972
5	1341.75	18.586	25	0.817	–	– 373.357*	– 376.699*	– 373.627*
6	1339.74	–4.0129	25	–	–	–372.783	–376.126	–373.054

Based on the determination of the number of lags and cointegration relations in the paper, a VAR model was built. It describes and confirms the existence of a connection between the level of the shadow economy and indicators of the social development of countries.

This model reflects the dependence of the values of the differences in the values of the shadow economy on the lag differences in the values of both the same indicator and other parameters. The model can be written as follows:

$$D(SE) = f\,(D(SE\ (L)), D(H\ (L)), D(IN10\ (L)),\\ D(IN20\ (L)), D(DG\ (L))) \tag{6.2}$$

where D (SE)—the value of the differences in the series shadow economy;

D (SE (L))—values of lag differences of the shadow economy series;
D (*H* (L))—values of lag differences of the inequality-adjusted Human Development Index series;
D (*IN10* (L))—values of lag differences of the ratio of the average income of the richest 10% to the poorest 10% series;
D (*IN20* (L))—values of lag differences of the ratio of the average income of the richest 20% to the poorest 20% as dependent variables series;
D (*DG* (L))—values of lag differences of the Index Gini series.

All model variables are endogenous. Exogenous variables were not included. To visualize the simulation results, graphs of responses of the model parameters to single and accumulated shocks of the shadow economy were constructed. The model assumes that other parameters do not change at this time. Graphs of single shocks are bestowed in Figure 6.2.

The most sensitive to changes in the shadow economy are the indicators of the social development of countries with a shadow economy of 0% to 10% and 30% to 40%. At the same time, for countries with a shadow economy of 10% to 30% of GDP, a change in its level has an insignificant effect on the Gini index.

Testing the model for normality, autocorrelation, stability, using the Lagrange-multiplier test, Jarque–Bera test, and Eigenvalue stability condition confirmed the reliability of the results. Based on the result of the stability test of the underlying VAR model, it is evident that as required, all the eigenvalues lie inside the unit circle. This implies that the estimated model is dynamically stable.

Since the estimated VAR passed all the diagnostic tests, we re-estimated the VAR by the Granger causality test. The results show that bi-directional

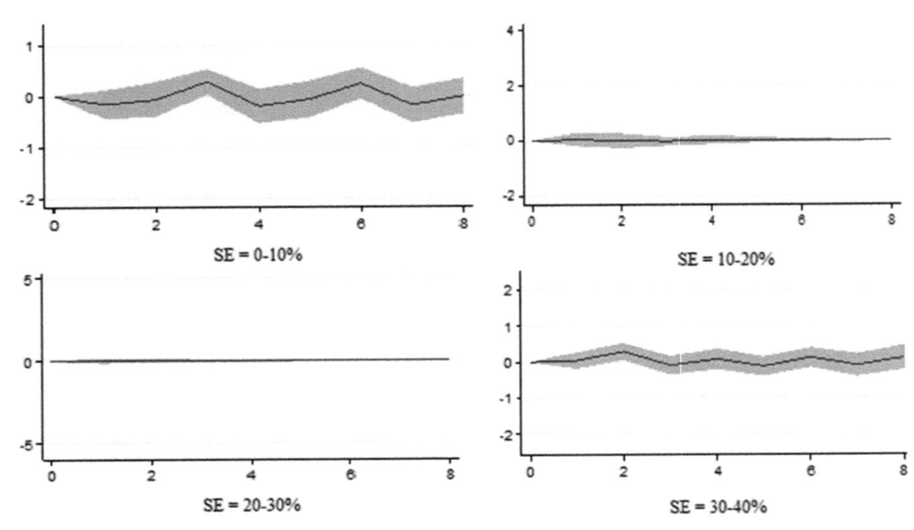

Figure 6.2 Impulse function of the Gini index response to the shocks of the shadow economy.

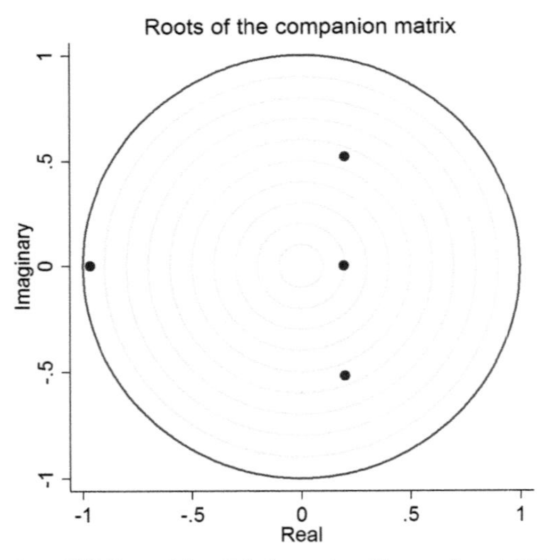

Figure 6.3 Results of VAR model validation using Eigenvalue stability condition test.

causality exists between shadow economy and indicators of social development of the countries. This means that shadow economy causes decrease in the level of social development of the country, and low level of social development causes shadow economy.

6.5 Conclusion

Inefficiency and non-transparency of the state tax, investment, and budget policies deepen imbalances in the economy and intensifies the macroeconomic instability. Since the significant amount of shadow capital is formed through tax evasion, it is important to manage tax gaps, improve and develop the current system of economy de-shadowing (Logan and Esmanov, 2017; Louis, 2017; Morscher et al., 2017; Ch and Semenog, 2017; Dave, 2017a, 2017b).

The shadow economy has become a complex systemic phenomenon that requires the introduction of coordinated, balanced, and effective instruments for de-shadowing economic activity. To solve the problem of attracting additional resources to the economy, it is important to implement a policy to withdraw funds permanently from the shadow sector of the economy to the official one.

This research investigates the existence and essence of a causal link between the level of the shadow economy and the country's social development. Based on the economic and mathematical modeling, the influence of shadow economy on the Gini index, the ratio of the average income of the richest 10% to the poorest 10%, the ratio of the average income of the richest 20% to the poorest 20% as dependent variables, and inequality-adjusted Human Development Index are proved. Based on the Granger test, the obtained results indicate a bi-directional causality between the shadow economy and social development.

Thus, the analysis allows us to conclude that the social component is quite sensitive to the growing share of the shadow economy in the country. At the same time, the level of social development is one of the main determinants of shadow activities by economic entities (Tiutiunyk et al., 2019; Kostyuchenko et al., 2018). This causes significant harm to the state, the main consequences of which are: the decreased financial capacity of public authorities to finance social programs due to reduced tax revenues to the budget; violation of the right of employees to decent working conditions, lack of control over non-compliance with labor discipline in informal employment; impossibility for shadow employees to use the social protection package in case of illness, accident, unemployment, etc.; non-compliance with the principles of equality and fairness of taxation, which leads to higher tax rates and becomes a burden only for officially employed persons; reducing the reliability of forecasting the economic and social development of the country, and the effectiveness of the development programs of the state.

References

Balas, A. N., Kaya, H. D. (2019). The Global Economic Crisis and Retailers' Security Concerns: The Trends. SocioEconomic Challenges, Vol. 3, Issue 2, pp. 5–14. doi: http://doi.org/10.21272/sec.3(1).5-14.2019.

Boiko, A., Samusevych, I. (2017). The role of tax competition between the countries of the world and the features of determining the main tax competitors of Ukraine among the European countries. Financial markets, institutions and risks, Vol. 1, Issue 1. pp. 72–79. doi: 10.21272/fmir.1(1).72-79.2017.

Bhandari, Medani P. (2020) Second Edition- Green Web-II: Standards and Perspectives from the IUCN, Policy Development in Environment Conservation Domain with reference to India, Pakistan, Nepal, and Bangladesh, River Publishers, Denmark / the Netherlands. ISBN: 9788770221924 e-ISBN: 9788770221917

Bhandari, Medani P. (2020), Getting the Climate Science Facts Right: The Role of the IPCC (Forthcoming), River Publishers, Denmark / the Netherlands- ISBN: 9788770221863 e-ISBN: 9788770221856

Bhandari, Medani P. and Shvindina Hanna (2019) Reducing Inequalities Towards Sustainable Development Goals: Multilevel Approach, River Publishers, Denmark / the Netherlands- ISBN: Print: 978-87-7022-126-9 E-book: 978-87-7022-125-2

Ch, A. R., Semenog, A. (2017). Non-bank financial institutions activity in the context of economic growth: cross-country comparisons. Financial Markets, Institutions and Risks, Vol. 1, Issue 2, pp. 39–49. doi: 10.21272/fmir.1(2).39-49.2017.

Cseh-Papp, I., Szira, Z., Varga, E. (2017). The situation of graduate employees on the Hungarian labor market. Business Ethics and Leadership, Vol. 1, Issue 2, pp. 5–11. doi: 10.21272/bel.1(2).5-11.2017.

Dave, H. (2017). An Inquiry on Social Issues - Part 1. Business Ethics and Leadership, Vol. 1, Issue 2, pp. 78–87. doi: 10.21272/bel.1(2).78-88.2017.

Dave, H. (2017). An Inquiry on Social Issues - Part 2. Business Ethics and Leadership, Vol. 1, Issue 3, pp. 45–63. doi: 10.21272/bel.1(3).45-63.2017.

Dean, J., Syniavska, O., Minenko, S. (2017). Using economic-mathematical modeling in the study of the economic component of terrorism. *SocioEconomic Challenges*, Vol. 1, Issue 2, pp. 103–109. doi: 10.21272/sec.1(2).103-109.2017.

Huynh C. M., Nguyen T. L. (2019). Shadow economy and income inequality: new empirical evidence from Asian developing countries.

Journal of the Asia Pacific Economy, Vol. 25, pp. 175–192. doi:10.1080/13547860.2019.1643196.

Katrechka A., Dahlberg S. (2014). The effect of the shadow economy on social development. A comparative study on advanced and least developed countries. Thesis (Master). University of Gothenburg.

Kim, B. Y. (2005). Poverty and Informal Economy Participation: Evidence from Romania. *The Economics of Transition*, Vol. 13(1), pp. 163–185. doi:10.1111/j.1468-0351.2005.00211.x.

Kostyuchenko N., Starinskyi M., Tiutiunyk I., Kobushko I. (2018). Methodical Approach to the Assessment of Risks Connected with the Legalization of the Proceeds of Crime. *Montenegrin Journal of Economics,* Vol. 14, No. 4, pp. 23–43.

Kouassi, K. (2018). Public Spending and Economic Growth in Developing Countries: a Synthesis. *Financial Markets, Institutions and Risks*. Vol. 2, Issue 2, pp. 22–30. doi: 10.21272/fmir.2(2).22-30.2018.

Kuzmenko, O., Roienko, V. (2017). Nowcasting income inequality in the context of the Fourth Industrial Revolution. *SocioEconomic Challenges*, Vol. 1, Issue 1, pp. 5–12. doi: 10.21272/sec.2017.1-01.

Logan, W., Esmanov, O. (2017). Public financial services transparency. *Business Ethics and Leadership*, Vol. 1, Issue 2, pp. 62–67. doi: 10.21272/bel.1(2).62-67.2017.

Louis, R. (2017). A new economic order for global prosperity. *SocioEconomic Challenges*, Vol. 1, Issue 2, pp. 52–58. doi: 10.21272/sec.1(2).52-59.2017.

Makarenko, I., Sirkovska, N. (2017). Transition to sustainability reporting: evidence from EU and Ukraine. *Business Ethics and Leadership*, Vol. 1, Issue 1, pp. 16–24. doi:10.21272/bel.2017.1-02.

Medina, L., Schneider, F. (2018). Shadow Economies Around the World: What Did We Learn Over the Last 20 Years? IMF Working Paper WP/18/17. International Monetary Fund. doi:10.5089/9781484338636.001.

Morscher, C., Horsch, J., Stephan A. (2017). Credit Information Sharing and Its Link to Financial Inclusion and Financial Intermediation. *Financial Markets, Institutions and Risks*, Vol. 1, Issue 3. pp. 22–33. doi: 10.21272/fmir.1(3).22-33.2017.

Musaeva, K. M. (2017). Impact of the shadow activity on social and economic relationships, income level and welfare of the population. Social *Behavior Research* and *Practice*, Vol. 2(1), pp. 27–32. doi:10.17140/SBRPOJ-2-109

Poliakh, S., Nuriddin, A. (2017). Evaluation Quality of Consumer Protection by Financial Markets Services. Financial Markets, Institutions and Risks, Vol. 1, Issue 3, pp. 75–81. doi: 10.21272/fmir.1(3).75-81.2017.

Rosser, J. B., Rosser, M. V., Ahmed, E. (2003). Multiple Unofficial Economy Equilibria and Income Distribution Dynamics in Systemic Transition. Journal of Post Keynesian Economics, Vol. 25, pp. 425–447.

Saibal, K., Shrabani, S. (2013). Corruption, Shadow Economy and Income Inequality: Evidence from Asia. 9th Australasian Development Economics Workshop 2013, Australia. pp. 1–29.

Schneider, F., Dominik, H. E. (2000). Shadow Economies: Size, Causes, and Consequences. Journal of Economic Literature, Vol. 38(1), pp. 77–114. doi:10.1257/jel.38.1.77.

Singh, S. N. (2018). Regional Disparity and Sustainable Development in North-Eastern States of India: A Policy Perspective. SocioEconomic Challenges, Vol. 2, Issue 2, pp. 41–48. doi: 10.21272/sec.2(2).41-48.2018.

Smith, R. S. (2002). The Underground Economy: Guidance for Policy Makers? Canadian Tax Journal/RevueFiscale Canadienne, Vol. 50, pp. 1655–1661.

Tiutiunyk, I., Kobushko, I., Ivaniy, O., Flaumer, A. (2019). Innovations in the Management of Tax Gaps in the Economy: Foreign Economic Component. Marketing and Management of Innovations, Issue 3, pp. 112–125. http://doi.org/10.21272/mmi.2019.3-09.

Valentini, E. (2009). Underground Economy, Evasion and Inequality. International Economic Journal, Vol. 23(2), pp. 281–290. doi:10.1080/10168730902903433.

7

Cultural Dynamics and Inequality in the Turkish Sphere Concerning Refugees

Kemal Yildirim

European School of Law and Governance, Kosovo
E-mail: conflictresearch@yahoo.com

Abstract

Cultural Dynamics of Refugees in Turkey:

Every society has the same needs and that every society finds different solutions to these needs according to one of the founders of cultural antrhopology, Bronislaw Malinowski I also believe that Culture is dynamic because it represents the experiences, beliefs, norms, etc., of living people. A good definition for culture is, "Anything that can't be explained by DNA." So the culture is defined always evolving, being pushed, stretched and changed. What was a forbidden belief in the past becomes a norm... what was a norm becomes arcane and sometimes offensive.

Culture is the solution to that. Societies of all the world has developed cultural solutions that have different forms depending of historical, political, economical, and social differences in the people that developed them. When a new need arise or when the society has changed, for internal or external reason, culture also changes to meet the new request. In this sense, we may assume that there are cultures that change their political or economic system because someone gains enough power to take control of a previously unruled society. Meanwhile, there are also cultures where the conversion to a new religion changes the morality of the people, with the abolition of some prescriptions and the evolutions of some rituals into different forms of organization.

Cultures are not rare animals to be preserved like pandas in a reserve but they are changing because their very basic nature is to change and to be adaptive. Preserving a culture considering it like a fixed, solid institution is sometimes like a nostalgic attempt to preserve the idea we have of some cultures. Safeguarding the cultural diversity of humans is important but this is not the same as considering cultures as static, old-fashioned, and "traditional". This research paper will find out the Dynamics of Syrian refugees in Turkey and within transition to Asylum within EU.

Finally, Cultural Dynamics of Syrian refugees describes how to deal with the complexities of human interactions within the framework of social systems where they belong to. Their Cultural Dynamics should be viewed from the Post Modernist multiplicities. Not being able to create a ground. Yet couldn't accept.

Keywords: Culture, Dynamics, refugees, Human rights, Turkey.

7.1 Argumentation and Claim

The main purpose of this chapter is to argue the current position and role of the Turkish civil religion in historial perspective on political agenda with affects on social and cultural agenda on the society. *In the late 1970s, the Turkish political scene was characterized by a thorough ideological polarization between right-wing ultranationalists (ülkücüs—idealists) and radical left-wing groups, along with a lack of decisive authority on the part of the government.* In fact, the important phase of the Turkish political islam in late 19th century exploited by islamist political forces in the 1990s had its origins in a crucial policy choice made by military government that came to power on September 12, 1980.

The coup aimed to end an extraordinary outbreak of extremist politics and to the attendant political violence between radical leftists and ultranationalists. The coup was initiated primarily against the leftists.The military's strategy for legitimizing the Turkish state and securing popular support for it involved a radical departure from the Kemalist secularism that had defined Turkey until then.

The Turkish-Islamic Synthesis (i.e., a mixture of Sunni Islam and Turkish nationalism), adopted and implemented by the military and maintained by the center-right Motherland Party (MP) rule (1983–91), opened the door to organizational and framing activities by Islamist forces—activities reinforced by such external factors as the Islamic Revolution in Iran and the financial support of Saudi capital, and supported by an emerging Islamist business class in Turkey.

These activities laid the foundation for the entry of political Islam into electoral competition and its eventually successful bid for power in the 1990s. In any case, it is likely that the Committee of Union and Progress was stuck to the ideology and discourse of Ottomanism, so they started to highlight Turkishness and rising Turkish nationalism following the footsteps of the "Turkish-Islamic Synthesis" with a special stress on military that appeared during the reign of Abdülhamid II.

However, on the other hand Turkish nationalism did not become a coherent full-fledged ideology. The Committee of Union and Progress utilized the unifying and mobilizing force of Islam particularly during the Balkan Wars and World War I. This also contributed to the rise of "Turkish-Islamic Synthesis". So, the paper will thus explain the religious identity with features on national identity with Laicism and official religion as Islam.

7.2 Juxtaposing the Turkish Civil Religion with the Muslim Perception of the Political Community

Unionists refined and used official Islam policy of Abdulhamid II. Although they did not lean toward religion, they did not hesitate to use the power of religion in the mobilization of masses. One of the most distinctive characteristics of the education in terms of ideology is that it used Islamic and Turkish identity together, with the priority of Turkish one. After the Balkan Wars in particular, it could be observed that there was an impatient and inexperienced effort to create a Turkish-Islamic synthesis. Its primary reason was to take advantage of the power of Islam for mobilizing society. It has two aspects: first, gaining support of Muslims living in Ottoman Empire and other countries; second, sanctifying nationalist values like homeland, nation, and the state with Islamic values like holy war (cihat), martyrdom, and being war veterans (gazilik) with militaristic purposes. They aimed to have men accept to do their military service, make them obey their commanders and reduce desertions. Likewise, the First World War led Unionists to create a Turkish-Islamic synthesis despite their nationalist tendencies. The declaration of holy war is the most typical example. Turkish-Islamic synthesis created by the political power aimed to raise "pious, patriot, nationalist people" in primary schools.

Students who had been forced to fulfil their religious duties in Hamidian era were forced to do their militaristic nationalist duties in the Second Constitutional Period. Students who had shouted as "Long Live Sultan" in Hamidian era began singing national anthems as well as religious chants in the new period. While forming associations was forbidden in Hamidian era, people had to become members of power, youth, and scoutcraft associations under the strict control of political power. Even military students had to do their trainings without bullets and weapons in Hamidian era; however, in the Constitutional era both civilian and military students took military lessons and did target practice with real weapons. For the first time, a "national festival/holiday" was accepted—"10th of July National Festival". In this era, the obligation to worship at schools was abrogated. They changed the contents of religious lessons and reduced their hours. Ethics lessons had religious contents in Hamidian era, but in this period they acquired "national" contents. For instance, Knowledge of Ethics was taught as a secular/social and political ethics, one hour a week. New topics like "idea of entrepreneurship" and "savings" were added to ethics. Religion became the sub-title of ethics rather than the opposite; in addition they argued that ethics could be taken as a different science. It was indicated that ethics was discussed as distinguished from religion since Socrates up to Kant; therefore, it was taken as a social fact. In this new lesson differently entitled as "knowledge of ethics, civilization, religion or economy" different sub-titles like civil ethics, political ethics, and religious ethics were discussed. Opponents were swift to protest against this new lesson added with the name of "religious ethics". For them ethics and religion were the same, ethical principles were not contrary to religion and ethics was the subsection of religion. In religious books, it was still emphasized that official sect was Hanafiyyah school of Sunnism. Yet, the most critical change was that there was no emphasis on loyalty and obedience to Sultan any more. Finally, holy war had a sacred and refined meaning.

Turkish nationalism that became the official ideology of this period was swift to take place in textbooks. Therefore, dominant elements of the political culture in the era appeared in all textbooks including reading ones. The nationalist characteristic of the government became apparent in the First World War. One of its typical examples is Köprülü Mehmed Fuad's reading book that came first in the contest of the Ministry of Education. Reading texts were chosen from among the works of eminent, patriot and nationalist writers. Even in grammar books, "enemies" were identified in terms of "us" and "others" within the scope of nationalist ideology. "Others" were particularly Greeks and Bulgarians; there are various illustrations depicting their tortures and persecutions in Balkan Wars. This expression is particularly striking: "We have too many enemies but the most treacherous ones are Bulgarians and Greeks". Both in lessons and physical trainings, there are subjects implicitly related to military service in order to glorify it and get people to love the service and prepare for it.

Major phenomena inherited by the Republican Turkey are official ideology and official institution of history itself. The "six arrows" of the Republican People's Party became the principles of the state/official ideology. Revolutionism, populism, nationalism, statism, republicanism, and laicism gained official status in the constitutional level. Republican rulers/elites evaluated these principles with different comments in different periods. The official ideology turning into "Turkish-Islamic Synthesis" in the Second Constitutional Period was crystallized as Turkish nationalism in the Republic. Now the nationalism of the Republic is processed within the nation–state system. "Turkism" was reformed as the mixture of legal, cultural, ethnic, and religious qualities. The definition of "Turk" was interpreted with different contents according to different periods; some elements dominated the scene while others were forgotten about. These preferences were reflected on historical approaches and textbooks. Islam continues to be an element used by official ideology when necessary. Republican laicism became the main characteristic of the political system. To sum up, in the Second Constitutional Period, Ottoman Commission of History would be renamed as Turkish History Commission; then in the place of it Turkish Historical Society would be established. "History shows how absolutist regime caused great evil, while constitutionalist regime led to fortunate outcomes; it tells us zealous struggles and challenges by which we acquired our liberty and constitutionalism. Thus history serves to give a more evident idea about the legitimacy of the constitutionalism and creates a more intimate allurement and attachment towards it. In brief, history constitutes a powerful foundation for civil information and political culture". The emphasis on "Turkishness" outweighed as it can be seen in textbooks being read in military schools of the Hamidian era. Now Ottomans were also called Turks.

The most serious break point between Hamidian era and the new one is about the history and (therefore) the origin of the state. Origins of Ottomans no longer dated back to the emergence of Islam; rather it started with the ancient Turkish history going back a long way to the first age. Religious explanations about the creation were narrated with few sentences; after religious creation passage titled as "Nature of Universe", the other one whose title is "Creation" states that Earth is a distinct part split from Sun.

National identity began to be narrated within a militarist characteristic. During lectures, subjects like military achievements of Turks, their heroic actions, and janissaries' achievements were taught. Great political figures from Islamic, Turkish, and Ottoman history were taught, the story of their tombs (if possible) were narrated and visited. After Abdulhamid's deposition, new subjects like "Era of Despotism", "His Reactionism, Domestic and Foreign Affairs in Hamidian Era", "Young Turks", "Great Transformation", "Despotic Hamidian Era", "Struggles for Liberty", and "July the 10th" were added to syllabuses. French Revolution was another important topic.-Indeed, Unionists (İttihatçılar) considered themselves as Turkish agents of French Revolution and the Declaration of Liberty as French Revolution of Turkey.

The concepts "national" and "nation" have more than one meaning and Constitutional Turkism (nationalism) made use of this fact. First, it was associated with the multinational Ottoman nation and Ottomanism denoting the unity of different ethnic groups and religions. Secondly, it evokes a religious community (ummah) with its religious and traditional meaning, thus equals to Islamic nation. Thirdly, nationalism that was the dominant movement of the era was used as the equivalent of Turkism. Therefore, "national manners" could bear different meanings according to target groups. Indeed, in the constitutional era, the concepts such as "National Economy", "National Literature", – and "National Manners/Culture" – had national emphasis. The motto of education in the era was "national manners/culture". Lots of books and articles were written on this concept. Texts in Köprülü Mehmed Fuad's reading book that won the first place in the contest of the Ministry of Education and that was used in lessons were written by nationalist, eminent, and influential literary figures.

> *Köprülüzade Mehmed Fuad; Milli Kıraat Mekâtib-i İbtidaiyenin Son ve Sultanilerin İlk Sınına Tedris Edilmek Üzere Maârif Nezareti Tarafından Bilmüsabaka Birinciliğe Kabul Edilmiştir. Beşinci Kısım* (Istanbul: Kanaat Kütüphanesi ve Matbaası, 1331).

Köprülü Mehmed Fuad's book clearly reveals the nationalist tendency in the Ministry of Education. The first text of the book was the poem "Turan" written by Ziya Gökalp in his Kızıl Elma (Red Apple). Other examples written by Turkish nationalists like Mehmed Emin, Mehmed Ali Tevfik, Hamdullah Subhi, Ahmed Hikmet, and Ziya Gökalp, İsmail Safa, Halide Edip, and Ali Canib followed the first one.

The discourse of "Turkish-Islamic Synthesis" contains a militaristic nationalism in itself. When possible military service was glorified and legitimized with religious values. Texts and illustrations that would exhilarate students' religious and national affections in order to reclaim lost territories started to take place in textbooks. At school, students began taking physical training and doing target practice with real weapons. Unlike Hamidian era, students did not shout as "long live Sultan" at the beginning and the end of their lectures; rather they began singing national songs, marches, chants, and "nationalist" poems. In textbooks, recent political history was narrated in such a way that Young Turks and CUP rulership were legitimized; Abdulhamid II was portrayed as a "despotic" sultan. The restoration of the constitution, also called as "the declaration of liberty", for the first time started to be celebrated as a national holiday and schools were suspended on that day. Conceptions like motherland, nation, state, constitution, assembly, election, law and court that were banned in Hamidian era were sanctified. Education turned into a process in which a participative political culture came into being with more or less laic/national values.

7.3 Juxtaposing the Moral and the Social Claims and Counter-Claims, (supporting for being against) for the Presence of Refugees in Turkey

Most of the research examine people's (prejudicial) beliefs and feelings toward refugees and immigrants and do not consider behavioral intentions. Yet, these intentions are closest to people's actual behavior, and research has demonstrated, for example, that protest intentions and actual behavior tend to be associated Refugees receive various sorts of assistance, but often they also face discrimination and social exclusion. Thus, both positive and negative behavioral intentions are important to study, and people might demonstrate a mixture of both. The positive–negative asymmetry in intergroup relations indicates that positive evaluations and intentions differ from negative evaluations and intentions. A less positive orientation toward an outgroup compared with the ingroup is consistently found on evaluation dimensions and behavior with positive connotations, but not on negatively valued dimensions or negative behavior. One reason for this is that, in general, the differential evaluation of negative traits and behavior is socially less acceptable than the differential evaluation of positive traits and behavior showed that negative valence increases the social concern with the legitimacy and appropriateness of unequal group distinctions. Furthermore, the domains of positive and negative actions and behavioral intentions have been found to involve different moralities with distinct motivational and regulatory systems. Positive behavior that focuses on advancing other's well-being raises questions of prescriptive morality that indicates what one should do, whereas negative behavior involves proscriptive morality that indicates what one should *not* do. Prescriptive morality is abstract, commendatory, and discretionary, whereas proscriptive morality is concrete, condemnatory, and duty-based resulting in greater moral blame. This means that negative actions and behavioral intentions against disadvantaged people are likely to be morally more difficult than not assisting or helping them. In the present study and considering the normative and moral implications, we expected that Turkish respondents make a distinction between positive and negative behavioral intentions toward Syrian refugees.

People can not only feel threatened and demonstrate prejudicial reactions toward refugees, but they also can act favorably toward this group. There are many examples of assistance and help being provided to refugees and, as indicated by the Turkish opinion polls discussed, these acts can be based on humanitarian concerns. Humanitarian concerns involve a sense of compassive care and moral responsibility for the welfare of fellow human beings, especially when they are in need. These concerns have been found to be associated with stronger support for refugees and are based on a shared humanity. The human level of identity defines Turks and Syrian refugees as forming part of the same humanity. The common in-group identity model suggests that a superordinate identity makes subgroup boundaries less salient and that former outgroup members will be part of the in-group resulting in more favorable attitudes and behaviors. There is extensive empirical evidence supporting this model in a range of settings and among various groups, including the positive effect of shared humanity for attitudes toward asylum seekers. Humanitarian concern reflects an identification with other human beings with the related moral responsibility to help them in times of need. Shared humanity has been found to be associated with the endorsement of human rights, intergroup empathy, and providing humanitarian aid and relief. Thus, it can be expected that stronger humanitarian concern is associated with stronger positive behavioral intentions toward these refugees and weaker negative behavioral intentions.

In addition to humanitarian concern being expected to be associated with behavioral intentions toward refugees, we also examined whether these concerns moderate the association between feelings of outgroup threat and positive and negative behavioral intentions. In relation to the so-called refugee crisis, host societies often struggle with finding a balance between humanitarian considerations and societal interests. People might not only be concerned about the threats that refugees can pose to the unity and safety of society but can also feel a sense of compassion and moral responsibility toward refugees. This could mean that the expected link between perceived threats and behavioral intentions depends on the level of humanitarian concerns. The perception of threat might be less strongly associated with behavioral intentions among individuals with stronger humanitarian concerns. Research has demonstrated that moral norms can influence the expression or suppression of prejudices. When people feel threatened by an outgroup but also consider members of this outgroup as fellow human beings, this might increase the intention to act positively toward them and suppress the intention to act negatively. Humanitarian concerns make it possible that feelings of threat are less likely to translate in lower positive behavioral intentions and in higher negative behavioral intentions. Thus, the negative association between threat and positive behavioral intentions can be expected to be weaker for Turkish participants with stronger humanitarian concerns. Correspondingly, for these respondents, the positive association between threat and negative behavioral intentions can be expected to be weaker.

I have also observed whether the perception of threat mediates the association between national identification and negative and positive behavioral intentions of Turkish citizens toward Syrian refugees and whether the role of threat depends on the level of humanitarian concern. First, Turkish participants responding the queries were expected to differentiate between positive and negative behavioral intentions. Second, higher national identification was expected to be associated with more negative and less positive behavioral intentions toward Syrian refugees, and perception of threat was expected to mediate this relationship, because higher identifiers will perceive Syrian refugees more strongly as a threat to Turkish identity and security. Third, Turkish respondents with stronger humanitarian concerns were expected to have more positive behavioral intentions and less negative behavioral intentions toward Syrian refugees. Fourth, these concerns were expected to moderate the association between perception of threat and behavioral intentions toward Syrian refugees.

In contrast to the existing research on (negative) beliefs and feelings toward refugees, we focused on both positive and negative behavioral intentions. Furthermore, it would be necessary to examine these intentions in relation to national identification and perceptions of threat as well as humanitarian concerns. We have to conduct our study in an underresearched national context that is highly relevant for understanding how people react toward the arrival of refugees.

I am satisfied that it will be found that stronger national identification will most probably be associated with more negative behavioral intentions and less positive behavioral intentions toward Syrian refugees via perceived threat. Additionally, stronger humanitarian concern will be associated with a stronger intention to help and support refugees and a weaker intention to protest against them. Furthermore, the findings may as a fact suggest that the combination of perceived threat and humanitarian concern can perhaps work out differently for positive and negative behavioral intentions toward Syrian refugees. This could mean that when people feel threatened by refugees, an emphasis on humanitarian concerns might not always have beneficial consequences for refugees: It might reduce negative behavioral intentions but also the inclination to offer help and support. This possibility may also have practical implications. Public campaigns and social policies that appeal to humanitarian concerns for improving intergroup relations between Syrian refugees and Turkish people should be managed cautiously. People might understand these campaigns and appeals as ignoring their genuine feelings of threat and as implying a moral accusation of failing to meet humanitarian standards. To them, these campaigns and appeals might be threatening to their sense of moral self, which leads to justifications of their behavior. This means that people's feelings of threat should be taken seriously and not dismissed as being misguided and prejudicial. An appeal to humanitarian concerns might be most effective when feelings of threat are considered and reduced. This means that future studies should investigate the correlates and causes of feelings of threat and the ways in which these hamper the inclination to offer help and support. In doing so, it is important to develop a more detailed understanding of the similar as well as different processes involved in positive and negative behavioral intentions.

7.4 Comments on the Claims Symbolically Connected the Principles of the Turkish Kemalist Civil Religion, Muslim Religion, Erdogan's Civil Religion, etc.

European Union members' pushing back against burden-sharing measures has led to what can be described as burden-shifting, that is, placing the task of hosting and providing for the care of refugees onto Middle East countries such as Lebanon, Jordan, Egypt, and of course, Turkey. While the AKP has taken steps such as pushing for safe zones to prevent greater numbers of refugees flowing into the country and the consolidation of a Kurdish autonomous zone within Syria—its view of Syrian refugees in Turkey seems to be that they are more of a boon than a burden. Indeed, AKP leaders welcomed Syrian asylum seekers as their "brothers", spending billions of dollars to set up "perfect . . . five-star" camps to house them and even insisting that they should be offered citizenship. Work permits were issued to those who fulfilled the requirement qualifications; ID cards facilitating access to free health care services were distributed.

Behind these seemingly benevolent outreaches, however, lie at least some less than altruistic motives. Indeed, President Recep Tayyip Erdoğan's threats this past November to open Turkey's borders and allow masses of refugees to flow into Europe if the European Union (EU) freezes talks on Turkey's membership accession—threats he has made repeatedly—suggest refugees represent more of a tactical boon than a brotherly one. Further, an in-depth report released by the International Crisis Group on November 30, 2016 indicates that despite its welcoming rhetoric the Turkish government has a very long way to go in developing plans to accommodate and integrate Syrians into Turkey's society. Only 0.1% of Syrian refugees have received work permits, and while health care services are free, medicines are not, leaving refugees paying an exorbitant share of their often meager earnings for medication. Some positive outreaches notwithstanding, the AKP's handling of Syrian refugees when examined more closely reveal that the party is utilizing the massive numbers Turkey hosts to its own political advantage.

To avoid losing educated new business owners to better economic prospects in Europe, the Turkish government is creating incentives for Syrians to stay and contribute to Turkey's struggling economy. Turkey has also gone a step further by refusing exit permits to Syrians with university qualifications. As a *Guardian* report notes, families such as that of Loreen and Shero waited years for their US resettlement application to be approved, only to find the Turkish government had blocked their exit because Loreen had a degree in banking.

A *Politico* report tellingly titled "Turkey Hoards Well-Educated Syrians" recounts that Sameer and his family suffering the fate because his wife has a university degree. A spokesperson for Germany's Interior Ministry told a reporter in June that more than 50 cases of Turkey refusing exit permits to Syrians already given visas for Germany had been reported in just a few weeks, while 292 refugees had been allowed to enter Germany. Although publicly denying the application of any selective criteria to Syrians' exit permit requests, one Turkish official speaking on the condition of anonymity stated that the government believed "the most vulnerable need to be helped before others" in justifying why educated refugees were not being allowed to leave. Whatever the actual intention behind this policy of retaining skilled Syrians, the detrimental effects on Syrian families who have been wading through the bureaucratic process of resettlement for years—many of whom do not even have work permits in Turkey and thus are underemployed if employed at all—are starkly clear.

All of the steps the Turkish government has taken to host Syrians fleeing violence in their homeland notwithstanding, these effects highlight the everyday challenges and daily setbacks facing Turkey's population. Families that ended their leases, sold their furniture, and purchased plane tickets are being told they are not allowed to leave. Others reading the news may see that the Turkish president who welcomed them with open arms now threatens to send them on to Europe by the busload. Still others may be experiencing the societal backlash against Erdoğan's political gambit in announcing a citizenship process for Syrians; the hashtag *#ÜlkemdeSuriyelilerİstemiyorum*—"I don't want Syrians in my country"—was trending on Twitter, and Syrian-operated shops were attacked. Such mixed and hurtful messages only compound the physical, emotional, financial, and other stresses that come along with displacement in conflict.

7.5 Inequality Issues

Literature on poverty offers many definitions, each highlighting the standpoint of its user (UNDP, 2006) inequality itself is a broad term, for many intellectuals inequality is when people are not treated as equals, with the same privileges, status, and rights due to their common humanity.

To address inequality, it is important that the disadvantaged are supported with appropriate resources to level the playing field alongside provision of equal opportunities (Oxfam, 2012). We are certainly aware that while a number of people are fully incorporated into the advanced industrial economy of the emerging global system, while others are marginalized.

Poverty is regional at some parts of the world. What we notice is that highly developed nations may accept immigrants who bring human capital and their investment while most others are excluded even when there are political as well as economic reasons for their migration and humanitarian reasons for their admission. However, it is clear that humanitarian concerns and obligations under UN conventions are of more an importance while refugees and human rights oblige wealthy countries to accept refugees and asylum seekers who are deemed to be genuine victims of persecution. However, "economic migrants" may time to time be excluded or perhaps accepted on a non-permanent basis to pursue the poorly paid heavy manual and service occupations that the indigenous population do not wish to undertake.

There may be several exceptions for a number of immigrants with capital to invest, or human capital in the form of vocational qualifications, which may or may not be recognized in the receiving country. Most of the refugees and economic migrants are likely to experience discrimination in wealthy lands. However, the hardest of cares for the internally displaced and refugees probably rests on less developed countries in regions where mostly armed conflicts transpire.

Most of the Syrian and other refugee women in Turkey encounter by a greater distance disadvantages while entering the labor market and display lower labor market outcomes, mostly because of lower levels of educational attainment acquired in their countries of origin, which arguably attributed to culturally embedded gender inequalities, which may continue to their disadvantage in the countries of destination. Meanwhile, because of further roles attained such as child-care, refugee women may not always continue relevant trainings, which would contribute to their labor market participation.

I believe that it would be necessary to take into account that further qualitative research is a necessity with the refugee women residents in Turkey and/or Europe, who are either legally or illegally employed in their countries of destination. First to see if their current professions are in line with their attainments, talents, and furthermore with their expectations and aspirations; and identify further, what additional challenges refugee women face, when are employed in the labor markets of their host societies. Likewise, as another option, measures already taken or being discussed as a response to the aforementioned challenges refugee women encounter from the view of adaptation into labor markets in their particular lands of destination, can either be equate to or in opposition to either local, international, supra-national, or European levels. However, according to international sources, there are 1.5 million school-aged Syrian refugee children living in Turkey, Jordan, and Lebanon, but about just half of them do not have access to formal education. Host countries may probably have taken generous steps to increase student enrollments of primary, secondary, and high school students, such as offering free public education and opening afternoon "second shifts" at schools to accommodate more children. But barriers such as child labor, enrollment requirements, language difficulties, and a lack of affordable transportation are keeping children out of the classroom. Children with disabilities are of an another greatest problem faced and most of secondary school age children are also likely to be in health hazard condition. Although refugee-related organizations are working to ensure that all of these children can realize their right to education. Especially most of Syrian children with disabilities may face particularly daunting challenges. Public schools, which have a poor record including Turkish and Lebanese children with disabilities, often reject them, because they lack the resources or skills to educate them. While Syrian refugee children with disabilities are able to enroll, public schools do not ensure they receive a quality education on an equal basis with others. All of the Turkish students are legally required to complete 12 years of schooling; school attendance for Syrians is voluntary, further complicating the mission to integrate Syrians into Turkish society via the educational system. Reportedly, the Turkish government is considering a plan to withhold monthly ESSN support allowances—120 Turkish lira, or about $22, per family member—from Syrian families whose children do not attend school. This would be a separate measure, beyond the withholding of the CCTE monetary inducement specifically tied to school attendance (Demirtaş 2018).

> *"Most of the Syrian kids' education has gone backward. Because I dont believe most of these children have a brilliant future since they left their country and their homes and presently they don't even have an education or a future"* (Demirtaş 2018).

"To Grow Up Without getting any formal Education" is the second of a three-part series addressing the urgent issues of access to education for Syrian refugee children in Turkey, Lebanon, and Jordan. Most of Syrian families can't afford to put their children in school at countries such as Turkey, Lebanon, or Jordan. They believe that all their children were studying in Syria regularly, but if they would put them in school at countries they fled to, how would they live then? Because they would have to buy them clothes and pay for transportation. Even if everything was free, the children couldn't go to school according to major number of Syrian refugees. However, Turkish Government thinks that grant of Turkish citizenship to Syrian refugees does not begin to resolve the fundamental questions about the future disposition of Syrian refugees in Turkey. They believe that they should determine whether to acknowledge that the vast majority of Syrians will likely remain in Turkey and, if so, it must consider how to integrate these refugees into wider Turkish society. Given the size of the Syrian refugee community, its lack of obvious alternatives to Turkey, and the potential consequences for Turkey of ignoring

the problem, efforts at integration appear to be Turkey's only logical solution. Failure to integrate the Syrians could create perhaps new divisions in Turkish society as well as deepen pre-existing economic, religious, and ethnic divisions. But I don't think granting citizenship in its current, limited scope and pace would likely have little impact on the problem, but there can be better alternatives toward integration that do not involve mass conferral of citizenship.

Questions

Q.1 Why inequality is the cause which creates the conflict and people have to leave their loved place?

Inequality between countries essentially means looking at poverty, strong link between the wealth of a country and the probability of it suffering from civil war (Collier and Samban (2004). From other hand, Frances Stewart looks not only at economic inequalities between groups, but also inequalities in social, political and cultural dimensions (Stewart 2010).

- *"Economic inequalities include access to and ownership of financial, human, natural resource-based and social assets. They also include inequalities in income levels and employment opportunities.*
- *Social inequalities include access to services like education, healthcare, housing, etc.*
- *Political inequalities include the distribution of political opportunities and power among groups, such as control over local, regional and national institutions of governance, the army and the police. They also include inequalities in people's capabilities to participate politically and express their needs.*
- *The dominant group capture the political oppurtunities and power of goverence, so that marginalized group always remain at the lower stratas of political, social, and economic system.*
- *Cultural inequalities include disparities in the recognition and standing of the language, religion, customs, norms and practices of different groups"* (Stewart 2010).

For mobilization of people to go to war, there must be an issue around which they can be organized. Economic or political inequalities have the potential to be such an issue, but it is much easier to organize people around it when they are already part of a group and inequalities can be interpreted as a consequence of conscious discrimination against this group. A good example is provided by Joshua Gubler and Joel Sawat Selway (2012), who describe two rebel leaders trying to organize a rebellion.

One of the leaders is able to appeal to an ethnic group. The group already has a shared history and the leader does not face the problem of having to convince a set of individuals that they are a group. Moreover, this ethnic group may have its own language and norms, which facilitates in-group communication. Once the group has been mobilized, it is difficult for members to leave it, as they cannot simply change their skin color, family ties or cultural heritage. This increases the rebel leader's capacity to exercise social control. The second rebel leader wants to rally a lower economic class. Creating a shared history and organizing and convincing individuals that they are part of a distinct group is a lot harder in his case. They share their history and language with members of higher economic classes and there is much greater mobility between the classes.

A focus on inequalities between different groups has another advantage. As the figure below shows, the levels of inequalities between groups may be overlooked by data examining the inequalities across a population as a whole. A country may have a highly equal distribution of income overall, but it may be divided very unevenly between particular cultural groups. In this light, focusing on inequalities between different groups draws attention to discriminatory relationships between groups in a society (Cramer, 2005).

7.6 A Short Approach Toward Syrian Citizens in Turkey Migration and Refugees

Within the scope of the study, regardless of the bureaucratic sense of what does a refugee mean in a scientific concept, we need to identify those people who have been displaced by any sort of reason and how they have taken refugee status in another country so they are classified as refugees.

Number of those fleeing from Syria to Turkey as refugees since the beginning of crisis of the war in Syria has rapidly increased in Turkey. In this context, number of those refugees from Syria to Turkey has gradually increased. This has been noted by United Nations High Commission of Refugees, so the agency's figures as of February 7, the number of total Syrian refugees in Turkey indicates that it is 3 million and 644 thousand. Of the refugees in Turkey, most of them populated first in cities such as Hatay, Gaziantep, and Sanliurfa as well as Izmir and Istanbul. It is observed that major cities such as Ankara, Bursa, and Konya have also accomodated a large number of Syrian refugees (UNHCR Turkey, 2019). It is possible to state that factors such as existing social capital play an important role. On the other hand, Turkey is the country with the highest Syrian refugee population when compared with rest of the nations, although the neighboring countries have not accomodated that much refugees in their own territories. (The United Nations Refugee Agency 2019) data of the population of Syrian refugees in Turkey at the end of 2014, quoted to be 1 million 622 thousand; at the end of 2016 it was quoted to be 2 million.)

It is important to look at the number of Syrian refugees taken in by neighboring countries of Turkey such as Iraq. They received a total number of 232 thousand 317 at the end of 2014, 229 thousand 774 at the end of 2016 and 252 as of January 2019. It is observed that there are 451 thousand. 1 million 145 at the end of 2014 in Lebanon.The Syrian refugee population, which was one thousand thousand, 1 million 69 thousand at the end of 2016, January 2019 It is observed that it has decreased to 947 thousand as of. At the end of 2014 in Egypt, 138. Two-thirds of the participating Syrian refugees in Turkey wanted to stay longer.. This situation, pointed out that the Syrian refugees in Turkey want to stay long in Turkey.

However, it reveals the necessity of developing an effective integration process at this point, in order to develop a two-way communication process. It is also important to make it open to this communication. The analysis of the profile of the migration typologies of Syrian refugees in Turkey would be an important case to discuss in the context of war of Syrian refugees in the first place, individuals forced to undertake international migration due to various reasons. It is possible to state that this migration depends on the situation of the war and the region of residence. It can be carried out reactively or proactively. In the framework of the research, the concept of refugee displaced in a sociological context, not as a legal definition but it is used to describe a person who has taken refugee status in another country. In the legal context it would be perhaps necessary to explain the status of Syrian citizens in Turkey, then it would be possible to consider such a proposition. Border entry without discrimination of religion, language, and race within the framework of "open door" policy that they did not send any Syrians back and told them "temporary protection status".

Turkish authorities expressed that refugees were given shelter and hosted on temporary status defining the temporal trends of migration also touches on the importance of their approach to perform the open-door policy that has since 2011 that accepts refugees. Thus Turkey's open door policy that has been implemented and explained as the temporary protection law is as follows:

"Temporary Protection" in Article 7 of the "Temporary Protection Regulation"

Forced to leave, unable to return to the country of departure, emergency and temporary protection massively or individually during this period of mass influx to find International protection request coming to or crossing their borders protection provided protection to the refugees. International legal system helps Syrians to get temporary protection status in Turkey. , By applying an open door policy, accepting the refugees unconditionally, social and economic support of the refugees with modern methods. Whether or not he/she has a passport from the first day of his/her commencement, sect, gender, and acceptance of anyone in need of help, regardless of race, and meeting their needs is called an open door policy

It might thus be possible to say that it is within a successful integration of refugees fleeing from Syria into Turkey as it also points to the existence of an institutional basis for the process. While the success of this integration process with local people is the outcome of open door policy, Turks see the case for the success of the integration process and as the outcome of the intercultural relations to accept them as parameters that will probably not hurt the establishment of communication channels.

Finally, the Turkish community hosting Syrian refugees should be encouraged to integrate in order to get a success of this public relations approach, in which it aims to develop the needs to be measured for a normative public relations at the point of measurement.

I believe a reference point for the outputs of the theory of Perfect Public Relations should be taken into consideration. In this respect, the Turkish society providing qualified information about refugees and about Turkish society to refugees thus is not transferred, this flow of information includes only the truth for both parties and whether it is carried out by a method with codes of ethics or not, it should therefore is an ideal procidure that includes empathy, consensus, and social peace. But it is not implemented, and the problems of the public are approached in an impartial manner. It should be therefore checked that it is approached carefully.

On the other hand, any sort of problems raised against refugees during the research phase whether it can be detected correctly or not should be carefully reviewed, and meanwhile it is worth to know which one of those occuring problems spread at which locations, and which social actors are considered as interlocutors or which expressions, etc. Finally we need to know whether the interaction is considered to be intense thus all should be questioned.

7.7 Conclusion

The largest refugee population hosted by Turkey amounts to the highest in the world. The Government of Turkey (GoT) appraise the total number of overall registered Syrians under Temporary Protection (SuTPs) at 2,225,147. Meanwhile, World Bank policy outline estimates, Turkey's Response to the Syrian Refugee Crisis is huge and as a result of this I am satisfied personally that road ahead is likely closed at the moment. Turkey's unique response to its displacement crisis is understandable; but also the challenges in managing the socio-economic dimensions of displacement should be considered since the crisis in the Middle East (on Syrian conflict) created the refugee situation. The most important factors are likely to be political tensions with armed conflict as well as ethnic intolerance with religious fundamentalism are among those facts influencing increase in refugees from Syria. Although the UN Note presently highlights remaining critical policy issues for Turkey, we may then question what lessons could be drawn from the Turkish hosting experience for other countries' refugee response efforts. Certainly there are so many other social issues distinguished from economic issues (such as immigration) have both social and economic aspects.

However, it would be clear that a wider lens may probably needed than the purely economic perspective taken by Collier and colleagues. So, most of the evidence indicate that when inequalities intersect with ethnic, religious, or regional divides, thus it may create a high risk of civil conflict. Inequalities in societies therefore should be taken seriously into account in which it includes looking at the ways where inequalities are addressed, devoting attention to citizenship and property rights and inclusive peace processes, and making sure that reconstruction efforts in fragile post-conflict situations do not ignite renewed inequalities. That will not only lead to more equal societies, but can also prevent war. In this sense, I believe refugees and economic migrants alike experience discrimination in wealthy countries but not in Turkey. As a result, the heaviest burden of care for the internally displaced and refugees rests on less developed countries in regions where armed conflicts have occurred.

It should be explained on a suitable ground that integration of Syrian refugees into society will bring fruitful and random able workforce into the society. On the other hand, Syrian refugees can protect their cultural values, but the need for Turkish society to adapt to cultural and social values. A message strategy that encourages people should be formulated. Thus, they do not lose their cultural values, I think it is possible to reduce the negative effect of fear on social integration.

Finally, the Turkish community hosting Syrian refugees should be encouraged to integrate. The success of this public relations approach, which aims to develop the needs might be measured time to time for a better outcome. A normative public relations as a reference point for the outputs of the theory of Perfect Public Relations at the point of measurement should be taken into account. Providing detailed information with refugees about Turkish society as well as in return to provide detailed information about the refugees to the Turkish society would be important for both parties to be familiar with each others.

In this context, informing about the refugees to the Turkish society in a most accurate way will contribute to the integration process. On the other hand, refugees are also needed to have a strong identity within the Turkish society that hosted them for a long time. As a result of this, it is therefore important that they are willing and positive toward the adaptation process.

References

Cederman, L. Weidmann, N. and Gledditsch, K.S. (2010). Horizontal Inequalities and Ethno-Nationalist Civil War: a Global Comparison. Paper prepared for presentation at Yale University, p. 29.

Collier, P. and Hoeffler, A. (2004). "Greed and Grievance in Civil War". in: Oxford Economic Papers. 56(4): 563–595, p. 577.

Collier, P. Hoeffler, A. Rohner, D. (2008). Beyond Greed and Grievance: Feasibility of Civil War. Department of Economics, University of Oxford.

Cramer, C. (2005). Inequality and Conflict: A Review of an Age-Old Concern. United Nations Research Institute for Social Development, Programme Paper 11, p. 11.

Demirtaş, Serkan (2018). "Turkey wants EU aid as long as Syrian crisis lasts", Hürriyet Daily News, December 1, 2018, available at http://www.hurriyet dailynews.com/opinion/serkan-demirtas/turkey-wants-eu-aid-as-long-as -syrian-crisis-lasts-139344. Turkish agencies administer both programs, ESSN and CCTE, out of funds provided by the European Union. See text box "Some key sources of funding for Syrian refugees", elsewhere in this paper.

Gubler, J.R. and Selway, J.S. (2012). Horizontal Inequality, Crosscutting Cleavages, and Civil War. In: Journal of Conflict Resolution. 56(2): 206-232, p. 209.

Holmqcist, G. (2012). Inequality and Identity: Causes of War? Discussion Paper 72, Nordiska Afrikainstitutet, Uppsala, p. 11.

Jok, J.M. (2007) Sudan. Race, Religion, and Violence. Oxford: One World Publications, p. 115.

Kärki T, Napoli C, Riccardo F, et al (2014). Screening for infectious diseases among newly arrived migrants in EU/EEA countries - varying practices but consensus on the utility of screening. Int J Environ Res Public Health;11:11004-11014.

Morin, R. (2012). Rising Share of Americans See Conflict Between Rich and Poor. Washington D.C.: Pew Research Center.

OECD (2012). Perspectives on Global Development 2012: Social Cohesion in a Shifting World. OECD Publishing, p. 58.

Østby, Gudrun (2008). Polarization, Horizontal Inequalities and Violent Civil Conflict. In:Journal of Peace Research. 45(2), p. 155.

Oxfam (2012). No Accident: Resilience and Inequality of Risk, Oxfam International

Pareek M, Watson JP, Ormerod LP, et al (2011). Screening of immigrants in the UK for imported latent tuberculosis: a multicentre cohort study and cost-effectiveness analysis. Lancet Infect Dis;11:435–444.

Sambanis, N. (2004) Poverty and the Organization of Political Violence. In: Globalization, Poverty, and Inequality. Brookings Trade Forum. Washington D.C.: Brookings Institution Press, pp. 165–211;

Sambanis, N. (2004). Poverty and the Organization of Political Violence. In: Globalization, Poverty, and Inequality. Brookings Trade Forum. Washington D.C.: Brookings Institution Press, pp. 165–211;

Stearns, Jason (2012). From CNDP to M23: The Evolution of an Armed Movement in Eastern Congo. London/Nairobi: Rift Valley Institute, p. 20.

Stewart and Brown, (2010). Horizontal Inequalities as a Cause of Conflict: A Review of CRISE Findings. World Development Report, Background Paper, p. 3.

Stewart, F. (2000). Crisis Prevention: Tackling Horizontal Inequalities. Oxford Development Studies. 28(3): 245–263;

Stewart, F. (2008). Horizontal Inequalities and Conflict: Understanding Group Violence in Multiethnic Societies. New York: Palgrave Macmillan.

UNDP (2006). International Poverty Centre, What is Poverty? Concepts and Measures, Poverty in Focus, United Nations Development Program

UNHCR (2006). UNHCR Master Glossary of Terms, United Nations High Commissioner for Refugees (UNHCR); Department of International Protection (DIP); Protection Information Section (PIS): Refugees. Rev. 1. June 2006 *www.refworld.org/docid/42ce7d444.html* (last access: June 16, 2017).

United Nations (1951). UN General Assembly: Convention Relating to the Status of Refugees. July 21, 1951. *www.refworld.org/docid/3be01b964.html* (last access: June 16, 2017)

Veldhuijzen IK, Toy M, Hahné SJ, et al (2010). Screening and early treatment of migrants for chronic hepatitis B virus infection is cost-effective. Gastroenterology 138:522–530.

WHO (2016). World Health Organization. Regional Office for the Eastern Mediterranean: Eastern Mediterranean Region. Framework for Health Information Systems and Core Indicators for Monitoring Health Situation and Health System Performance. *http://applications.emro.who.int/dsaf/EMROPUB_2015_EN_1904.pdf?ua=1* (last access: May 8, 2017).

8

Genesis and Reasons for Social Inequality in the Globalization Era

Teletov Aleksandr Sergeevych[1] and Teletova Svetlana Grigorievna[2]

[1]Professor of the Department of Public Management and Administration, Sumy National Agrarian University, Sumy, Ukraine
[2]Associate Professor of the Department of Russian Language, Foreign Literature and Methods of their Teaching, Sumy State A. S. Makarenko Pedagogical University, Sumy, Ukraine

Abstract

Since older days inequality existed between people, although its best representatives always tried to reduce it. The study demonstrates that globalization has not reduced social differentiation between people. Moreover, globalization deepened it in some way since the scope of global inequality, reflecting the difference in welfare between the richest and the poorest countries in the world, exceeds the inequality within a particular country. At the same time, inequality indices in individual countries continue to increase. There are many examples, including those that confirm the efforts of the civilized world to reduce inequality between people. In general, the problem of inequality between people, the current rate of which is a major threat to stability in the whole world, has not been solved yet. This section deals with some challenges of existing social differentiation.

8.1 The Reality of Social Inequality

It is a known fact that, from time immemorial, man has become accustomed to living at the expense of another. Recently, inequality between people, countries, and peoples of different identity (sex, sexual orintations, race, class, gender, etc.) have intensified with the development of globalization, uneven distribution of resources, and growing intercultural tension, which creates social tension. Inequality initially implies different opportunities and unequal access to material and intangible benefits: resources, education, power, prestige. Most people believe that they do not participate adequately in society. In recent decades, significant social shifts have occurred, and interethnic and interregional conflicts are intensifying. The peculiarity of the global social imbalance is that its causes lie not only in individual states, but also at the supranational level, which leads to the fact that there is no dominant strategy in the world to overcome urgent social problems.

Inequality in human society acts as one of the urgent objects of sociological research.

For a long time, many scholars have wondered whether a society can exist in principle if it does not trace inequality or hierarchy. In order to answer this question, it is necessary to understand the causes of social inequality. The problems of inequality arise immediately in the two most developed areas of society: economic and social. Consider inequality in education. Economic growth brings new opportunities. But new opportunities are not enough. Access to education is expanding, and this is happening all over the world, but even if what a person acquires as knowledge should open up opportunities for him, then this is also not enough. Equality is justice in the distribution of such opportunities.

A certain degree of inequality, of course, is inevitable, but it is necessary to constantly strive to reduce it, and not vice versa. For example, inequalities arising from the development of technology or based on economic growth from Spain with Portugal in the 15th century to China in the 21st century are an engine and historically inevitable.

In the countries of the former USSR, the specifics are not associated with economic growth, but with the processes that took place in their economy in the 90s, so inequality here is different.

The information society has exposed the problem of inequality. Over the past 20 to 30 years, there has been more transparency—social media has allowed us to see more inequality, for example, after learning about the salaries of many VIP officials, the society in Ukraine was horrified. Many countries are trying to reduce inequality in society. For example, legislative regulation of the equitable distribution of resources and capital takes place: in France—a bill against the waste of food and products, Japan—a bill for the conservation of resources, etc. Since various consumer goods are produced one and a half times more than people need, there are many products that are simply thrown away, one and a half times more resources are spent and one and a half times more waste is generated, and in the European Union, 100,000 people go hungry, in the rest of Europe–more than a million, in the North America and South America—tens of millions, in Asia and Africa—hundreds of millions, and in total—more than 700 million people.

That is, to reduce different types of inequalities, specific measures are needed in each of the branches of human activity, the most important of which is the fight against poverty on the one hand and the fight against squandering on the other. An attempt to answer some of the challenges of deepening social differentiation, which largely determines the stability of the entire global social space, is devoted to this article.

8.2 Introduction

The world is constantly being changed. Today, humanity faces the choice once again. Globalization regime is in a severe crisis for the second time after the COVID-19 pandemic, and most countries have to solve their problems alone. Thanks to the globalization processes in the second half of the 20th century, the world has succeeded in directing efforts to unite people. The globalization regime, according to S. Pereslegin [1], has been in a major crisis since the fall of the Twin Towers in New York on September 11, 2001. The economic crisis of 2008–2009 and the political crisis of 2013–2014 marked the beginning of the transition from global to post-global regimes, about which the leading international experts mentioned at that time. It is commonly known that globalization is based on the free movement of people, finances, and goods and services. The beginning of the global pandemic caused the situation that this movement has been limited. The borders between the United States and other countries, within the EU, between the EU and other countries, are closed. A system will not be quickly renewed. It means that all countries in the world will not be united. Then, this situation can lead to a transition from global to local world subsystems, markets, etc.

Globalization has forced each country to develop that sector, which it can do best. For example, it is the raw material for Russia, consumer goods for China, and financial management for the United States. When the COVID-19 pandemic finishes, countries will have to develop all capacities, for example, as in such cases, when sanctions are imposed on a particular state [2]. It affects more likely the middle class, as it is usually after crises. Huge business wins here, again increasing inequality between people. Financial flows and asset ownership will be changed. Financial assets will be redistributed in favor of large funds, transnational business, etc. The problem will appear regarding changes in the world financial system, the world economy with the extension of the sixth technological way and the service economy liquidation. The global financial and economic crisis will exacerbate the social inequality problem. The social inequality is one of the objects of philosophical, sociological, and marketing research. For a long time, scientists have raised the question of whether society can exist without inequality or hierarchy. That is why it is necessary to understand the reasons for social inequality.

8.3 Methods

The study uses historical and logical approaches, retrospective analysis, statistical methods, the comparative and contrastive method for identifying distinctive features in different countries and different historical periods.

8.4 Results and Discussion

Modern globalization is the process of transforming regional social and economic systems into a single world system. Analyzing globalization from a historical point of view, one should notice that a gradual change in social styles defines its every stage—from the Mediterranean trade to the present days. It is common knowledge that all industrialized countries have undergone three stages of scientific and technological progress: industrialization, computerization, and informatization. At the industrialization stage, the fuel and energy resources and production technologies prevailed. They formed the ground for the creation of an elemental base and the emergence of electronic computing machines. During the computerization stage, the development of materials science and technology to improve the production of electronic computers prevailed. In the third stage of informatization, intelligence and information technologies, characterizing the transition to a new qualitative state—the information society, played the leading role, see Table 8.1 [3].

By the way, the transition from socio-economic structure to socio-economic structure went through a series of wars or including the World war. The transition to the fifth socio-economic structure was after the Cold War. Today, the COVID-19 epidemic is a world war. Until now, the war often caused inequality between people; today, the epidemic is the reason. On the one hand, the epidemic is real, on the other—it is being promoted as a convenient ground for certain social classes to solve their problems [4].

Table 8.1 Historical and economic development of humanity.

		Historical Periods of Development			Population	
No.	Historical Classification	Timeline	Duration	Transition to the Next Period	Number of People, Million	The Principle of Settlement
1	Harvesting and hunting	From 40,000 BC	About 30 thousand years	The beginning of the ancient world	Up to 10	Decentralized resettlement of people
2	Horticultural	From 10,000 BC to the beginning of AD	About 8 thousand years	The heyday of ancient civilization	50	
3	Agrarian	20th century BC–16th century AD	About 2.5 thousand years*	The discovery of America	400	Concentration of employed population in cities
4	Industrial	17th century–early 20th century	300–350 years	World War II	3000	
5	Information technology	Mid-20th century–2000	About 60 years	The terrorist act on September 11, 2001.	About 7000	Decentralization production, export of capital, etc.
6	Globalization	Late 20th century–2013	15–20 years	World crisis, events in Ukraine, the Middle East, Eastern Europe		
7	Global instability	Since 2014	5 years	COVID-19 The transition of humanity to a new state?		

*the period from the 5th to the 15th century AD (middle ages) is ruled out as ineffective in terms of innovation: decentralization of population displacement, practical cessation of its growth, etc.

The factors that influence the deepening of social differentiation in countries may vary. In the United States, inequality is a result of high immigration; in China, inequality is based on a significant increase in the number of millionaires and billionaires; in the former Soviet Union, inequalities arose from the processes that took place in the 1990s, when countries transitioned from socialism to wild capitalism, that caused the impoverishment of the greater part of the society.

During the globalization, inequality between people, countries, and nations has increased due to the unequal distribution of resources, growing intercultural tension, which together form social tension. Due to the uneven distribution of income, many people find themselves deceived. The influence of the elites on the state policy in favor of their interests increases the dissatisfaction of people who practically do not participate in the life of society, country, or region. Thus, significant social shifts occur and international and interregional conflicts continue. The state and society have joint efforts to solve inequalities. There is no single solution here since it is desirable to obtain a result that will satisfy the maximum population. The following measures can be implemented to reduce the gap between the poor and the rich: to introduce progressive taxation, to apply rigid tax evasion measures to large corporations, to protect the poor, providing them with a guaranteed minimum income, to invest in social services, including free education and health care.

The peculiarity of the global social imbalance is that its reasons are not only within individual states but also at the supranational level. It complicates the study of the reasons and the structure of social inequality. Multipolarity has caused the fact that there is no dominant strategy in the world to solve lately-stored problems. Inequality implies different opportunities and unequal access to available tangible and intangible goods. The reasons for inequality in society lie in the following basic aspects: (1) *Income* is the ownership of a "working" property or wages for physical or mental work done by a person (the lower the income, the lower the person in the hierarchy of society); (2) *Education* is a complex of knowledge and skills acquired by a person during his or her education in educational institutions; (3) *Power* is the individual's ability to impose his or her worldview, point of view, opinions, regardless of the public desire. The level of power is measured by the number of people to whom it applies (the President or Prime Minister has one level of power and the Mayor of a small town has another one); and (4) *Prestige* is the position in society and its assessment, formed by applying public opinion.

Social inequality has been accumulating for centuries. Besides, it makes a constant strain in society. Certainly, there are some attempts to reduce inequality, such as increasing access to education in the whole world. Access to a good education increases a person's chances to get a decent job. Even if the knowledge acquired during training is supposed to open up new opportunities for a person, it is not enough. Equality is fairness in the distribution of such opportunities.

Absolute equality is certainly impossible. The inequality degree depends not only on the person's belonging to one or another social class but also on the abilities, perseverance, working capacity of the person, family relationships etc. Therefore, the state has to help people at different stages in the three abovementioned components: education, health care, and obtaining elementary accommodation. A child from a poor family has the same right to an education with the wealthy citizen's child, starting from primary school to higher education. Everyone at any age is entitled to free treatment for illnesses, epidemics because most people do not suffer from their whims, and even those who take their health more seriously will need more and more help at the end of their lives. Thus, it can be assumed in theory that on average each person will waste the same amount of money.

Not all people can have their accommodation, so the government should assist in it by providing municipal housing or paying excess for renting it. Since the possibility to possess own home and the size of a pension depends also on the individual's abilities, we can see fair or stimulating inequality based on the assumption that the social inequality source should be sought in the political and economic social system [5, p. 31]. Although it is difficult to make people agree with even more or less fair conditions. For example, everyone has got accustomed to the fact that employees do not know the real wages of their colleagues. If they knew exactly each other's income, they would certainly deal with the following questions: "We do the same work, why does he (she) get higher wage?" or vice versa "I work better than my colleague, why do we get the same salary?"

The analysis of numerous concepts of social inequality, developed by the classics of sociological science and modern sociologists, enabled the Russian researcher N.G. Osipova to give a comprehensive definition of this phenomenon: social inequality is a reflection of the complex social structure of society and the specific form of social stratification, in which certain individuals, social groups, classes, and layers are on different steps of the vertical social hierarchy and have unequal opportunities to meet their material, social, or spiritual needs [6].

It is commonly known that from olden times, one person used to live at the expense of another one. Crone J. notes that "people have created the inequality they have, but they are also able to mitigate it. It is impossible to reconcile with inequality, it is important to determine to what extent it is necessary to have it" [7, p. 47]. So, when is inequality allowed and when is not? The social state can answer this question.

We compare the "sociality" degree of the states in different periods on the example of Ukraine, which until 1991 was a part of the USSR and is independent now, with the United States and other countries. In this comparison, we use the theory created in the 1940s by the American economist Abraham Maslow (1908–1970), according to which all human needs are divided into five levels. The so-called Maslow's hierarchy of needs (the triangle of needs by A. Maslow) is widely used in the practice of marketing and management. Needs are not only classified there but also arranged in a clear hierarchical sequence. As a rule, a person starts with an attempt to meet the priority needs, and higher-order needs are the motives of his or her behavior only when the needs of the lower level are satisfied, the needs of the next level become a powerful factor that identifies people's behavior. From a marketing standpoint, the Maslow's hierarchy of needs helps define with what products, for which classes (layers) of the population and in which countries one can enter the target markets.

Contrary to the conventional approach, we start from the lower or the average level of needs, if we do not classify the pyramid into primary and secondary needs. This is convenient because in the late '70s to the early '80s of the last century the majority of the population in Ukraine was able to spend considerable money on goods from this level. The primary needs were practically satisfied, although there was no modern variety of food products, the general output of clothing, shoes, and furniture was sufficient, but the population no longer wanted to "just dress" or have "just furniture". They wanted to have products that fit the fashion for that period, fit a particular person, differ from others. The needs for physical and psychological security were met to some extent. Housing and utilities were scanty and had little or no impact on the family budget. Each person was provided with free medical care, had the opportunity to get an education (including higher education) for free, and most importantly—was confident in getting a job. Upon reaching retirement age, a person was able to retire quietly, which was sufficient not only to meet physiological needs but also social needs, contacts, mutual assistance, etc., see Table 8.2.

Table 8.2 Comparison of some goods in the third level of Maslow's hierarchy of needs.

Name of Need	Opportunity to Satisfy Needs			
	USSR, 1986		Ukraine, 2019	
	Price, SUR	Percentage of Average Salary	Price, UAH	Percentage of Average Salary
Textbooks, medium quality art books, etc.	1–2	1	200–300*	5
Visiting a world-class museum, zoo, etc.	to 1	up to 0.5	up to 200	to 4
Mid-range accommodation on the seaside (per day)	to 5	2	600	10–15
Vacation on a permit	200–500	100	20,000–30,000	300–400
Playing technically challenging sports	Free		up to 10,000	150

*In 2019, 100 UAH (Ukrenian Currency) at a cost of approximately equal to 1 rubles in 1986.

At that time, goods and services that met such needs included attending theatres, concert programs, sports competitions, and the opportunity to play sports. Tickets to the theatre, concert, football or hockey (other competitions were often free of charge) cost 1% to 3% of the family's monthly budget, that is, everyone could attend them. The youth could go in for sports, including biathlon, archery, academic rowing, and canoeing. Currently, these and some other sports have reduced their popularity to a great extent, and the minimum price for tickets to a pop star concert is 20% to 30% of the family's monthly budget.

Another set of goods at this level comprised the possibility of active rest, traveling or treatment in the sanatorium. For example, a trip to a mountaineering camp costs 50% to 70% of the average monthly wage (the rent of appropriate equipment, work of an instructor included), but it was sold to employees for 30% and was generally given to students free of charge. The entire USSR was entangled with a network of various tourist routes, houses, and recreation centres. The tickets for different modes of transport cost for the bus—three times, for train—four times, for the plane—seven or ten times cheaper by consumer value than now. At that time, there was a large social group, which did not aspire to a career, and considers its earnings as a living means and dealing with favourite business: tourism, music, etc.

Social needs include also the ability to buy a book. When you look at the copies of the books of that time, you do not believe that they had hundreds of thousands of copies. Almost everyone was able to purchase books. Now, most people cannot buy books at all: art and non-fiction books are 5 to 7 times more expensive, educational and scientific books—10 times than 30 years before. Prices for children's books have increased 30 times. Nowadays, Ukraine has been driven out of the countries community with priorities for social needs, mutual assistance, and love. In Europe, they include Norway, Sweden, Finland, Austria, etc., 83% of total family's income is spent on nutrition (physiological needs), though Ukrainians consume only more potatoes, and bread as much as in the '80s. Outbreaks of chicken flu in poor countries have reduced the price of chicken by only 5% to 15%, in developed countries more than doubled, and in some countries—consumers have virtually abandoned this type of food by the time of its guaranteed safety.

According to the A. Maslow's hierarchy of needs, in the lower countries, most people consume largely the "second hand", which help them dress, receive state subsidies for housing, energy, heat, and others. Countries with similar priorities include a large part of Africa, some Asian countries—Afghanistan, Bangladesh and some countries in the Southeast, as well as some countries in Central and Latin America. In Europe, typical representatives of such countries are Albania, Moldova, and Georgia. Certainly, in developed countries, there are social groups for which the priority is the consumption of products that meet physiological needs. As a rule, it is a lower-level group, the number of which in the total population does not exceed 20%. These are unemployed, migrants, and displaced persons with no benefits, hourly-paid workers, and persons oppressed by the state on a national feature. In Ukraine, not only the majority of workers and peasants belong to such people, but also a considerable part of the intelligentsia, employees, etc. The security issues as well as social needs were previously solved and are now being solved only partially for them that greatly increases inequality in society, see Table 8.3 [8, p. 26].

In the '70s to '80s of the last century, there were no needs from 4 and 5 levels, that is, self-esteem, personal dignity, and recognition needs in Ukraine as part of the Soviet Union. The needs of 1 to 3 levels could satisfy almost everyone. In developed countries, such people relate to the middle class.

Table 8.3 The general level of wealth in some countries.

Characteristic	Development level		
	Minimum	Middle	Maximum
Average salary	Up to € 300 most African countries, some Asian countries. **Ukraine** about **300 €**	In the world, the average salary is 2000 €	More than 4000 € average salary in Switzerland
Number of dollar billionaires per 10 million population	Not at all-Baltic countries, Balkan countries (except Greece), Belarus, Portugal	On average, 5 people in the world; in **Ukraine**-also **5**	Maximum number (20–60)-Cyprus, Switzerland

In the United States, vice versa, most people belong to the top two levels of Maslow's hierarchy of needs. Higher authorities, the third-rate businessmen, top executives, top scholars of various fields, belong to the 4th level of the pyramid. Products from this group worldwide include prestigious travel cars and tours, multi-meter downtown apartments, and large suburban homes, fine dining restaurants with sophisticated taste, education at famous educational institutions, health care by the latest scientific and practical achievements, etc. For example, historically, despite the incomparable living standards of these countries, visiting an opera in Ukraine is a need for contacts, and in the United States, it is for recognition. Countries, where the most population need recognition, include the so-called "advanced civilized countries": France, the United Kingdom, Germany, Canada, Japan, some of the Arab oil-producing countries, etc. Although much of the highly skilled migrants to these countries from the USSR live financially better than their remaining counterparts, they suffer from the lack of social status they are accustomed to. They do not occupy a decent place in the hierarchical structure of society, do not have the opportunity to spend money on appropriate goods, communicate with the colleagues of their level, and therefore lose their social status and, to some extent, their qualifications.

According to A. Maslow, the top of the dominant consumer priorities hierarchy is meeting the needs for self-actualization, self-realization, and development. It is an opportunity to purchase a multimeter limousine, a luxury yacht or island in the Pacific or the Atlantic Ocean, a sports club, etc., spend money on holidays and sponsor help. Only the elite can realize themselves in this way: billionaire businessmen and multimillionaires, bankers, owners and co-owners of well-known companies and brands, sports superstars and show business, world-famous models. It should be noted that certain types of goods or services, such as a bunch of flowers for $100 or a visit to a prestigious restaurant, can also be purchased by consumers from other levels of hierarchical needs for self-actualization, but this only happens in individual cases, whereas the upper classes constantly behave in such way. It is inequality.

The desire to receive as many social benefits as possible generates the main form of inequality—economic inequality. Based on unequal ownership of material goods and material resources, economic inequality divides society into rich and poor. The rich have high incomes and property that give them access to intangible assets: prestigious education, improved health care, exclusive recreation, owning football clubs, etc. The poor are constantly lacking in money and goods to support their livelihoods and to support their children, to have convenient accommodation, and they are constantly unemployed. The closure of many businesses and low wages make people leave their homes and seek employment in richer countries, de facto developing and supporting their economies. Inequality in the wealth distribution in the country increases the number of people thinking about emigration (e.g., in Ukraine in 2014 it was 15.3%, and in 2018 it is already 23.5%). The most capable, skilled and able-bodied of them can stay there forever, while others are not necessary, and return to their countries. It has happened recently during the worldwide epidemic COVID-19 ("the modern pandemic reveals the main contradiction of the post-industrial world: the country's inability with its political institutions, formed in the age of modernity to be opposed to the universal threat") [4]. It is the inequality and the lowest standard of living that drives residents of Libya, Syria, Iraq, and other African and Asian countries to risk their lives while illegally moving to European countries, Mexicans to the United States, etc.

Today, globalization is the one that maximizes economic inequality. Thus, the authorities of the global transnational business territories, as noted by S. Pereslegin, has already been tested by the experience of Ukraine [1]. In less than 30 years, Ukraine, which is trying to integrate itself into the globalization processes, has sharply worsened most economic indices. As an example, it is possible to compare current economic indices of Ukraine with those of 30 years before, see Table 8.4 [9, p. 9–10].

Ukraine becomes a raw material addition to developed countries owing to globalization processes. Table 8.4 demonstrates that among the economic indicators that show the volume of the produced goods, almost all declined by an order (10 times decreased) and only two times increased: (a) exports

Table 8.4 Comparison of some economic indicators.

Economic Indicators	Period	
	1990	2018
Territory (thousand km^2)	603,7	576,7
Population (million people)	51,8	37,9*
Produced		
electricity, billion kWh	298,5	159,3
coal, million tons	164,8	33,2
oil, million tons	5,3	1.6
natural gas, billion million m^3	28.1	21.0
gasoline, million tons	8.4	0.8
steel, million tons	48.5	21.2
pipe, million tons	6.5	1.1
tractors, thousand	106.0	3.4
cars, thousand	156.0	6.6
buses, thousand	12.6	0.8
textile, million m^3	1210,0	82,4
shoes, millions of pairs	196,0	24.5
cereals and legumes, million tons	51,0	66,0
meat, million tons	4.4	3.2
milk, million tons	24.5	10,0
sugar, million tons	6,8	1.8
flour, million tons	7.7	1.9
potato, million tons	16.7	22.1
Put into operation		
housing, million m^2	17.4	10.3
number of students for whom educational institutions were built, thousand	163,9	17.2
existing medical institutions, thousand	6.4	1.6

*Not including the population of Crimea, Donbass, emigrants and displaced persons

of grain and legumes, through the exploitation of land; (b) the collection of larger yields of potatoes, which are usually grown by citizens on their lands for personal consumption and the substitution of calorie and more necessary products for a person.

The asymmetric nature of globalization in favor of so-called "Golden Billion" countries (the countries of the "golden billion"-the USA, Canada, the countries of Western Europe, Japan, Australia, and some others) is mentioned in [10]. In [11, p. 97] it is emphasized that the leaders of globalization ideas include top management of multinational corporations, government officials, public relations experts, and right-wing politicians, who are called "agents of globalism". In particular, they argue that globalization is inevitable and promotes the spread of liberal values around the world and benefits the entire global community. These agents of influence try to introduce neoliberal values into the political, economic, and social life of countries that have not only permanent but also temporary economic difficulties and are de facto institutions of global social inequality in the world. The International Monetary Fund (IMF) is the main institution among them. When any country applies to the IMF to request a loan, the fund sets out the mandatory conditions for possible lending. For example, for some countries in Eastern Europe, the country's development concept called shock therapy has been developed. It aims at a complete structural restructuring of the country's economy, which is to privatize the public sector, drastically to reduce costs for the social sphere, to sell state-owned lands to the private sector, etc.

Sometimes such a structural adjustment reminds a witty marketing strategy. It is commonly known, according to the classic approach of F. Kotler, the last fifth marketing concept is the socio-ethical marketing concept in which society appears in the link manufacturer-consumer and in favor of which a particular marketing activity should be performed [12]. According to the author, this concept is the most modern. The difference between the IMF's proposed strategy is that it is opposed to the social and ethical marketing concept and directed against society, for example, free trade has nothing to do with anti-crisis measures. Adoption of such programs aggravates the condition of a country, increasing inequality both domestically and abroad. Rodrik points out that this approach forces the poor countries to adopt the programs they have chosen for them [13, p. 81–82]. As a result, one can observe rising unemployment, the rapid increase in inequality, and the ever-increasing division into the rich and the poor.

Thus, the globalization ideology increased inequalities based on the unfair distribution of material and human resources, both inside and outside the countries on a global scale. The list of countries affected by the ideology of globalization is increasing. It is well known that in such countries of Latin America as Argentina, Brazil, Ecuador, Peru, large lands are in the private property of a small number of owners. It does not always increase but constantly maintains inequality between people. For example, Kerbo [14, p. 33–34] notes that "in some Latin American countries, 10% of people possess 90% of the land. Landowners use their land to grow so-called "money plants" (coffee, sugar cane, etc.), which are to be sold to developed countries. People who do not have their land are not able to provide their families with food, resulting in famine and related illnesses in their environment". Despite this danger, in March 2020 the Verkhovna Rada of Ukraine adopted a law to open the land market in the country, against which, three-quarters of the citizens in the country oppose according to many inquiries.

This example confirms that not all laws that are appropriate for developed countries with a large number of resources are suitable for developing economies where there are cheap labor, unemployment, etc. An example of such a country is India, the world's second-largest population of more than 1 billion people. In his work [15], Vandana Shiva refutes the view that India's economic and some social achievements came at the expense of economic globalization and trade liberalization, rather than the liberation from colonial dependence on Britain after World War II. On the contrary, as a result of globalization, small farmers in India lost their ability to sell produced goods and land was expropriated by large corporations, as a result of which more than 160,000 farmers committed suicide [16]. It means that everything that was in the colonization of Asian countries, comes back in the worse form during globalization processes.

Ukraine is the country among the former Soviet Union republics which mostly suffered from the imposition of neoliberal values and Western patterns of economic and social behavior. Ukraine was provoked into new borrowings and became dependent on the international capital dictates against the background of the country's unprecedented de-industrialization. All this has led to the social inequality of Ukrainian citizens. If in 1989, there were no more than 2% of the population living below the poverty line, now it is about 60%. Today, the ratio of rich and poor in Ukraine (1:70) exceeds the level of this coefficient in Western European countries (1:6–7) by order and continues to grow to a threatening scale.

Today, virtually all mass media considers the question regarding the need to orient Ukraine to a small businessman (mainly the agricultural sector). This fact does not stand up under criticism, because, firstly, no country has advanced to the forefront of agriculture, but rather on the contrary: for example, agricultural Argentina was the seventh economy of the world at the beginning of the 20th century, and now it is in the sixth 10 countries by nominal GDP, and Canada, which has been the first in the world to grow cereals for almost the entire second half of the 20th century; secondly, people in the country, which took 6th place in Europe and 8th place in the world mainly due to industrial production, and, in the 1980s, had 4.7% of world GDP, would never agree to become a country laborer—they will better leave their homeland forever in search of a better fortune and "boost" the economy in other countries; thirdly, what assistance can be given to farmers in the possible purchase of their land when today there are 26,000 farmers in

Ukraine who cultivate 8% of agricultural land. It takes about 300,000 farmers to be sure that the land will reach its cultivator. Does Ukraine have innovative prospects in the agrarian sector and what are they? The United States and the EU insist that there is no other way out in our country. Their desire can be understood, but the country could flourish at the expense of agriculture during the agrarian regime, i.e., during the Ancient World, the principles of which prevailed prior to the beginning of industrialization [17, 18].

A weak capitalist state will always depend on the "raw" player and lose to stronger, as a rule, the so-called "Golden Billion" countries [6]. As can be seen from Table 8.5, in Ukraine GDP per capita is only 9% of the EU average index.

The social inequality problems, besides the economic sphere, also arise in the public sphere. Two polar poles can be proposed regarding the equality and inequality of people's lives in society. One pole explains the inequality phenomenon based on the diversity of social functions, which are typical for different classes, layers, and groups of society. Each group is engaged in its own business: the production of material goods, the creation of

Table 8.5 Country comparison by GDP for one year (period from 1.07.2016 to 30.06.2017).

Name of Country	GDP in USD, Billion	GDP in USD per Capita
First level development countries		
USA	18 986	58 323*
Japan	4 947	39 065
Germany	3 478	42 121
United Kingdom	2 535	38 407
France	2 452	36 516
Italy	1 845	30 466
Canada	1 572	42 940
South Korea	1 472	28 607
Australia	1 324	54 153
Spain	1 242	26 733
Second level development countries		
China	11 508	8 315
Brazil	2 002	9 565
Russia	1 461	10 130
Mexico	1 088	8 426
Third level development countries		
India	2 335	1 743
Ukraine	104	2 599

*It should be noted that in the US, GDP per capita is not the best. Thus, in Luxembourg it is $ 1 00573 per capita, in Switzerland–$ 79891, in Norway–$ 70912.

spiritual values, and management of the first and second group. The effective functioning of society requires a combination of these three types of human activity. Each person, occupying a certain place in the society, automatically receives a certain status, and it is quite difficult to move from it to another one. Opportunities that let a person reach a position in society, either in ownership of property or capital or in originating from a well-known or privileged family, belonging to a higher class of society or certain political forces. It means that social inequality is primarily an inequality of statuses. In the 17th and 20th centuries, the inequality of statuses led to revolutions aimed at changing the human status in society.

Another pole arose during the times of the USSR, in which the origin of man, for exceptions, had virtually no meaning, and on the contrary, youth from the countryside, after military service were given certain privileges when entering higher education institutions. It was possible to enter law HEIs (legal institution of higher education) only having 2 years of work experience. Certainly, these are polar statements, and theoretically justified optimal state for each country is different, according to the development of the economy, historical and demographic features, mentality, etc. The ability to identify and follow it is the task of further research on inequality. Some inequality degree is probably inevitable, but it is always necessary to try to reduce it. For example, inequalities between countries caused by the development of advanced technologies or based on economic growth—from Spain with Portugal in the 15th century to China in the 21st—are the engine and historically inevitable, see Figure 8.1.

The information-oriented society has exposed the problem of inequality because over the last 20 to 30 years the media has allowed us to see more inequality [19]. For example, Ukrainian society was shocked when they found out about the revenues of VIP-officials, top-managers of some state-owned companies, sports stars, and show business, while doctors, teachers, and scientists receive less. Social studies indicate an increase in public discontent

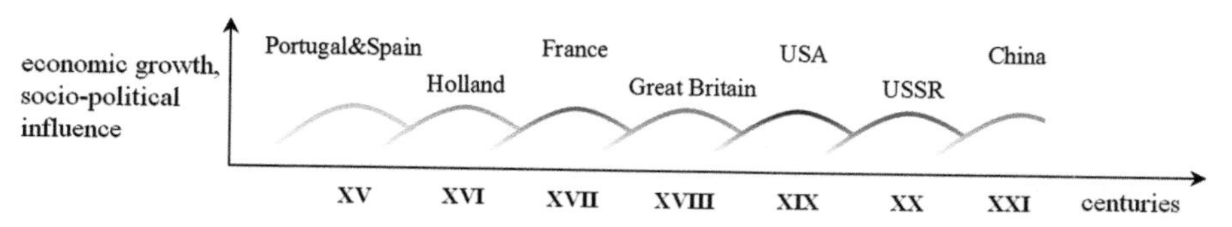

Figure 8.1 Countries that have made the most significant leap in development in each of the centuries.

with this situation, especially during the COVID-19 pandemic. Half a century ago, P. Bourdieu [20, p. 69] noted that "no science deals with social relations with the same obviousness as sociology, and at the same time, it touches the vital interests of people". Today, it is obvious that specific managerial decisions take into account economic, political, demographic, scientific and technical, cultural and some other factors. For example, combating inequality in large companies is also a business idea that should come from the heads of commercial companies to executives whose activities depend on the public.

There are countries (Scandinavian, Belarus, Japan, and some others) where inequality is minimized. Many countries are trying to reduce inequality in society. In some countries, there is a legislative regulation of the equitable distribution of resources and capital. For example, in France, there is a legislative draft against the waste of food and products (every Frenchman can form an association and demand for free food which was not sold at the supermarket). In Japan, there is a legislative draft to conserve resources—businessmen often independently offer to provide funds for the society (when one of the unsuccessful years of the state budget was not enough for one month, businessmen completely replenished it at their own expense). In the US, there are funds which manage most inheritance under the principle of trust but do not share between family members of the former magnate. The powerful marketing development in the last century confirms that since various consumer goods are produced one and a half times more than people need, and many products are simply thrown away—one and a half times more wasted resources are spent, and one and a half times more waste is generated. At the same time, one hundred thousand people are starving in the European Union, more than one million in the rest of Europe, tens of millions in North America and South America, hundreds of millions in Asia and Africa, and more than 700 million in total.

It means that reducing various types of inequalities in each sphere of human activity requires specific measures. The main measures include the fight against poverty, on the one hand (around 3.7 billion poor people in the world) and the fight against squandering, on the other.

In addition to income inequality, access to education, and health care, there are certainly other inequalities, such as race-ethnical, related to race and ethnicity; gender, related to women's oppression; age, which consists in discrimination based on age, etc. They are also rather derived from economic and social inequality. Thus, economic inequality in the globalized age has been significantly increased, and without a context of globalization, it is impossible to consider any phenomenon in society since these problems concern all humanity and cannot be solved at the local, regional, individual, or continent level.

8.5 Conclusion

Social inequality has a different nature and is the result of many contexts and systems that prevail in society. Inequality is based on the historical features of the country's development, the mentality of the population, the development of technology, economic growth, industrialization, political, demographic, scientific, technical, cultural, and some other factors. The reasons for inequality lie in such aspects as income, education, power, and prestige. It is necessary to give priority to more general interests of society than to the interests of the richest part of society to reduce inequality between citizens. Thus, one should, first, identify the type of social system that would be acceptable if not for everyone, at least for most people. In practice, states should strive to balance their citizens' living standard by eliminating economic, social, age, gender, and ethnic discrimination, regulating labor and capital markets, and introducing progressive taxation and social security policies. There is no doubt that the world will be changed after the COVID-19 pandemic: the economic crisis will be deepened, deglobalization will be observed, interstate relations will be partially limited, etc. It is also an opportunity to reduce inequality between people to some extent. The main task for further research on inequality and ways to overcome it in society is to find concrete steps to achieve this objective.

References

[1] Цыганов А. Маски сорваны (2020), Коронавирус – часть мирового заговора во главе с IBM : интервью с С. Переслегиным // Царьград. URL: https://tsargrad.tv/articles/maski-sorvany-koronavirus-chast-mirovogo-zagovora-vo-glave-s-ivm_247478

[2] Davis L., Engerman, S. Sanctions (2003), Neither War nor Peace. Journal of Economic Perspectives. Vol. 17. No. 2, P. 192.

[3] Телетов О. С. Економіка (2016), України в умовах можливого переходу суспільних укладів розвитку людства в нову якість // Економічний вісник НТУУ "КПІ".. № 13. С. 21–27.

[4] Alain Badiou. (2020), Sur la situation épidémique. *Quartier général.* URL: https://qg.media/2020/03/26/sur-la-situation-epidemique-par-alain-badiou/

[5] Ritzer G. (2000), Sociological theory. McGrow Hill,. P. 31.
[6] Осипова Н. Г. (2014), Проблемы современного глобального неравенства // Вестник Моск. ун-та. Сер. 18. Социология и политология. № 2. С. 119–137.
[7] Crone J. (2007), How can we solve social problems& Thousand Oaks. 2007. P. 47.
[8] Тєлєтов О. С., Летуновська Н.Є, Провозін М. В. (2019), Соціальна інфраструктура сучасних підприємств і територій [монографія] / за заг. редакцією доктора економічних наук, професора О. С. Тєлєтова. Суми : Триторія, 240 с.
[9] Bowel T. (2002), Hegemony and bifurcation points in world history // The Future of Global Conflict. N. Y., P. 282.
[10] Steger M. (2003), Globalization. N. Y., P. 97.
[11] Philip Kotler, (2016), Principles of Marketing. Prentice-Hall,. 672 P.
[12] Rodrik D. (1994), The rush to free trade in developing world: why so late? Why now? Will it last? Voting for Reform: Democracy, Political Liberalization and Economic Adjustment / Ed. By St. Haggard, St. B. Webb. N.Y., P. 81–82.
[13] Kerbo H. R. (2006), World poverty: global inequality and the modern world system. Boston, P. 33–34.
[14] Shiva V. (2006), India divided: diversity and democracy under attack. N. Y.
[15] Shiva V. (2009), Harvest of suicide.. URL: http://www.project-syndicate.org/commentary/harvest-of-suicide/russian
[16] Тєлєтов О. С. (2019), Іміджелогія як засіб випереджаючого інноваційного розвитку // Механізм управління формуванням стратегій випереджаючого інноваційного розвитку промислових підприємств»: монографія / за заг. ред. к. е. н., доц. Н. С. Ілляшенко. Суми : Триторія, С. 35–47.
[17] Lyulyov, O., Pimonenko, T., Stoyanets, N., and Letunovska, N. (2019), Sustainable development of agricultural sector: Democratic profile impact among developing countries//World Economy. 10(4), p. 97–105. doi:10.5430/rwe.v10n4p97.
[18] Тєлєтов О. С. (2018), Інноваційний розвиток маркетингової інформаційної системи та системи маркетингових знань // Інформація та знання в системі управління інноваційним розвитком / за ред. д. е. н. доц. Ю. С. Шипуліної. Суми : Триторія, С. 106–116.
[19] Bourdieu P. Metier (1968), de sociologue. pp. 69.
[20] Tsyganov A. (2020). Masks are ripped off: Coronavirus is part of a global conspiracy led by IBM : interview with S. Pereslegin. *Tsargrad.* Retrieved from: https://tsargrad.tv/articles/maski-sorvany-koronavirus-chast-mirovogo-zagovora-vo-glave-s-ivm_247478 .

[21] Davis L., and Engerman, S. (2003). Sanctions: Neither War nor Peace. *Journal of Economic Perspectives*. Vol. 17. No. 2, 192.

[22] Tielietov O.S. (2016). Ukraine's economy in the face of the possible transition of social setups human development in new quality. *Economic bulletin of National technical university of Ukraine "Kyiv polytechnical institute"*, 13, 21–27.

[23] Alain B. (2020). On the epidemic situation. *Quartier general*. Retrieved from: https://qg.media/2020/03/26/sur-la-situation-epidemique-par-alain-badiou/

[24] Ritzer G. (2020). Sociological theory. McGrow_Hill.

[25] Osipova N.G. (2014). Problems of modern global inequality. *Bulletin of Moscow University. Series 18. Sociology and Political Science*, 2, 119–137.

[26] Crone J. (2007). How can we solve social problems. Thousand Oaks.

[27] Tielietov O.S., Letunovska N.Ye., and Provozin M.V. (2019). Social infrastructure of modern enterprises and territories. Trytoriia.

[28] Mygovych I.I. (2020). The shackles are heavy.

[29] Bowel T. (2002). Hegemony and bifurcation points in world history. *The Future of Global Conflict*, 282.

[30] Steger M. (2003). Globalization. N. Y.

[31] Kotler Ph. (2016). Principles of Marketing. Prentice-Hall.

[32] Rodrik D. (1994). The rush to free trade in developing world: why so late? Why now? Will it last? Voting for Reform: Democracy, Political Liberalization and Economic Adjustment (Ed. by St. Haggard, St. B. Webb.). N.Y.

[33] Kerbo H. R. (2006). World poverty: global inequality and the modern world system. Boston.

[34] Shiva V. (2005). India divided: diversity and democracy under attack.

[35] Shiva V. (2009). Harvest of suicide. Retrieved from: http://www.project-syndicate.org/commentary/harvest-of-suicide/russian.

[36] Tielietov O.S. (2019). Imideology as a means of leading innovative development. *Mechanism of management of formation of strategies of leading innovative development of industrial enterprises*. (Ed. by N.S. Illiashenko). Trytoriia, 35–47.

[37] Lyulyov, O., Pimonenko, T., Stoyanets, N., and Letunovska, N. (2019). Sustainable development of agricultural sector: Democratic profile impact among developing countries. Research in World Economy, 10(4), 97–105. doi:10.5430/rwe.v10n4p97.

[38] Tielietov O.S. (2018). Innovative development of marketing information system and system of marketing knowledge. *Information and knowledge in the innovative development management system*. (Ed. by Yu.S. Shypulina). Trytoriia, 106–116.
[39] Bourdieu P. (1968). Metier de sociologue. Paris.

9

The Luxury Lockdown: Tackle to Covid-19, Fatal to Hunger—Widening Inequality*

Man Bahadur Bk[1] and Medani P. Bhandari[2]
[1]Former Chief Secretory, Bagmati Province, Government of Nepal, Nepal
[2]Professor and Advisor of Gandaki University, Pokhara, Nepal, Prof. Akamai University, USA and Sumy State University, Ukraine

Abstract

This chapter aims to present the humanitarian aspects of Covid-19, leading toward the more divisive and unequal society. Covid-19 is a global crisis and the crisis always hits the people who have no resources to alter with the situation. The Covid-19 regime is about a year old; however, it completely changed the global picture of human civilization. Firstly, it remained with the blame culture, the so-called conspiracies within nations and even among scientists and the world media big and small also played important roles to horrify the general public. The world's economic situation and political or power greediness divided world due to few tempered leaderships, whose role was basically to blame other nations, or in the worse cases own scientists and society. Until, very recently, world was in unprecedented crisis, however, there are lights of hope (the invention and execution of vaccinations). This chapter basically presents the opinion with the grounded factual evidence.

*The part of this opinion piece was published at, (Bhandari, Medani, 2019). In the Covid-19 Regime—What Role Intellectual Society Can Play. International Journal of Science Annals, 3(2), 5–7. doi:10.26697/ijsa.2020.2.1 https://ijsa.culturehealth.org/en/arhiv\https://ekrpoch.culturehealth.org/handle/lib/71 and (Bhandari, Medani, 2019), The Phobia Corona (COVID 19)—What Can We Do, Scientific Journal of Bielsko-Biala School of Finance and Law, ASEJ 2020, 24 (1): 1–3, GICID: 01.3001.0014.0769, https://asej.eu/resources/html/article/details?id=202946

[1]Dr. Bk, the Principal Secretary for the Provincial Government of Bagamati/Nepal, is former Fulbright scholar and the author of the book *Eradicating Hunger: Rebuilding Food Regime*.

9.1 Introduction

The world is facing one of the most difficult challenges of current time due to Covid 19, and the floating major message of this alarm "is no guaranty of anything" "beware, beware, and beware"; still we humans have not won the ultimate reality of life—the death (Bhandari 2020).

- Covid-19—- *What is it:*
- Coronaviruses are a type of virus. There are many different kinds, and some cause disease. A newly identified coronavirus, SARS-CoV-2, has caused a worldwide pandemic of respiratory illness, called Covid-19.
- Covid-19 is the disease caused by the new coronavirus that was first identified in December 2019.
- Covid-19 symptoms include cough, fever or chills, shortness of breath or difficulty breathing, muscle or body aches, sore throat, new loss of taste or smell, diarrhea, headache, new fatigue, nausea or vomiting, and congestion or runny nose. Covid-19 can be severe, and some cases have caused death.
- The new coronavirus can be spread from person to person. It is diagnosed with a laboratory test.
- There is no coronavirus vaccine yet. Prevention involves frequent hand-washing, coughing into the bend of your elbow, staying home when you are sick and wearing a cloth face covering if you can't practice physical distancing. (Lauren M. Sauer, M.S.) https://www.hopkinsmedicine.org/health/conditions-and-diseases/coronavirus

In reality, the human civilization history has seen many severe pandemics and to some extend there has been always some remedy or a way out to get rid of the pandemic (Watts, 1999; Green, 2002; Crutzen, 2006; WHO 2020; Bhandari 2020;). The table below shows the brief Pandemics History.

There are already thousands of papers and books available which explains the problems and consequences of the pandemic (Google, 2020; LePan, 2020; Sah et al., 2020; Silwal, Chalise and Limbu, 2020; Pooladi et al., 2020; Bhandari et al., 2020; Ardakani et al 2020; Egede and Walker, 2020; Patel et al., 2020; Bhandari 2020,A, B, C, D; Cucinotta and Vanelli, 2020; Pardo, 2020; OECD, 2020; Brandily, 2020; Chernick, Copeland and Reschovsky, 2020; Chung and Soh, 2020; COE, 2020; Council of State Governments, 2020; McKinsey and Cie, 2020; Muro and Maxim, 2020; Skjesol and Tritter, 2020; UN, 2020). More importantly, "*WHO is bringing*

Pandemics: History

Name	Time Period	Type/Pre-human Host	Death Toll
Antonine Plague	165–180	Believed to be either smallpox or measles	5 M
Japanese smallpox epidemic	735–737	Variola major virus	1 M
Plague of Justinian	541–542	Yersinia pestis bacteria / Rats, fleas	30–50 M
Black Death	1347–1351	Yersinia pestis bacteria / Rats, fleas	200 M
New World Smallpox Outbreak	1520 onwards	Variola major virus	56 M
Great Plague of London	1665	Yersinia pestis bacteria / Rats, fleas	100,000
Italian plague	1629—1631	Yersinia pestis bacteria / Rats, fleas	1 M
Cholera Pandemics 1–6	1817–1923	V. cholerae bacteria	1 M+
Third Plague	1885	Yersinia pestis bacteria / Rats, fleas	12 M (China and India)
Yellow Fever	Late 1800s	Virus / Mosquitoes	100,000–150,000 (U.S.)
Russian Flu	1889–1890	Believed to be H2N2 (avian origin)	1 M
Spanish Flu	1918–1919	H1N1 virus / Pigs	40–50 M
Asian Flu	1957–1958	H2N2 virus	1.1 M
Hong Kong Flu	1968–1970	H3N2 virus	1 M
HIV/AIDS	1981–present	Virus / Chimpanzees	25–35 M
Swine Flu	2009–2010	H1N1 virus / Pigs	200,000
SARS	2002–2003	Coronavirus / Bats, Civets	770
Ebola	2014–2016	Ebolavirus / Wild animals	11,000
MERS	2015–Present	Coronavirus / Bats, camels	850
Covid-19	2019–Present	Coronavirus—Unknown (possibly pangolins)	2541 K (Johns Hopkins University estimate

Coronavirus Cases: Coronavirus Cases: 114,606,873; Deaths: 2,541,302; Recovered: 90,159,180 as of February 28, 2021, 20:18 GMT, available at https://www.worldometers.info/coronavirus/ **Pathogen**: SARS coronavirus 2 (SARS-CoV-2)

Sources: Nicholas LePan, 2020. Visualizing the history of Pandemics, available at https://www.visualcapitalist.com/history-of-pandemics-deadliest/ https://www.worldometers.info/coronavirus/

the world's scientists and global health professionals together to accelerate the research and development process and develop new norms and standards to contain the spread of the coronavirus pandemic and help care for those affected". WHO is gathering the latest international multilingual scientific findings and knowledge on COVID-19? The global literature cited in the WHO COVID-19 database is updated daily (Monday through Friday) from searches of bibliographic databases, hand searching, and the addition of other expert-referred scientific articles. This database represents a comprehensive multilingual source of current literature on the topic. While it may not be exhaustive, new research is added regularly (WHO 2021). In addition to WHO, almost each and every International Agencies, have established a section to deal and study about the virus. World Health Organization—Coronavirus Disease (COVID-19) Outbreak; Centers for Disease Control and Prevention—Coronavirus Disease 2019 (COVID-19); Chinese Centers for Disease Control—Tracking the Epidemic; Johns Hopkins Center for Health Security; COVID-19 Resource Center; Open Critical Care—Covid Guidelines Dashboard and Gavi, the Vaccine Alliance; The Global Fund to Fight AIDS, Tuberculosis and Malaria; The Bill & Melinda Gates Foundation; Africa Centre for Disease Control and Prevention and many more have been working to overcome from this pandemic (https://www.nejm.org/coronavirus).

> Covid *-19 has governments at all levels operating in a context of radical uncertainty. The regional and local impact of the* Covid-*19 crisis is highly heterogeneous, with significant implications for crisis management and policy responses. This paper takes an in-depth look at the territorial impact of the* Covid-*19 crisis in its different dimensions: health, economic, social, and fiscal. It provides examples of responses by national and subnational governments to help mitigate the territorial effects of the crisis and offers ten takeaways on managing* Covid-*19's territorial impact (OECD, 2020).*

All universities and research institutions have prioritized to address Covid-19 and to some extend since world already have many vaccines available and already begin to at the global scale; therefore, there is a hope of survival. Please have a look at the **Appendix 1 and 2** for the list of administered vaccines for Covid-19. However, what about those of our relatives who passed away, what about the survivors who lost the way of life and severely facing the turmoil condition due to inequality between haves and have nots.

9.2 The Sentiment

Well, still many of us are under Covid-19 radar, still getting sick, being hospitalized, and even having last breath. We all, let us say who has access to information (any kind), to the news, know about the dreadful nature of Covid-19, which has already taken thousands of lives. Many of us are in virus trap or on the way to its shadow. As we already know that Covid-19 is so dangerous that its entry into our body is like entry of poison, which has insofar no direct treatment. We all are noticing that the virus is super painful in various ways. First, there is fear of death, that is what is happening now, and we never know how long it will be continued; secondly, it hits directly to the human civilization, because it is a pandemic so, whoever got the virus—he or she is out from the inherited family system. It breaks the chain of love–emotion system, sensitivity, and immediately, forces us to think, "who is important." It shoots into the eternal part of the feelings and penetrates into the thinking pattern which separates us from us to I and makes each of us as individuals. It is the direct threat of our social system. We all know that death is the ultimate reality; however, none of "us" or "I" are prepared or will be prepared and ready to accept (except few who might have different issues in lives). Covid-19: now we can add new dictionary meaning as "pain on humanity." This pain has two folds: (1) The pain of victims, who are under control system—we mean who are identified as corona positive, have symbols and already in the hospital or self or forced isolation distance box (who may defeat the virus and come to normal life or virus will defeat and they will be distancing forever); (2) the pain of the loved ones and relatives of the victims. This twirled pain what human society has been facing now, cannot be expressed in any languages. How to feel the pain of a dying person, who wanted to tell/share some secrete of his or her life; or the pain of family members, who wanted to say goodbye to their loved ones in their last breath. Due to the risk of virus transformation, it is not permitted to be close to anyone. Even family members are not allowed to perform final rituals of loved ones. The pain of death of the family member and not having access to say goodbye creates a psychological pain cycle—which can create social phobia. And it is happening now. What will be the social impact of such tragedy? How will society overcome that pain and loss of family members? It is breaking our hearts, social system, society, and the essence of future direction of lives in a way that it is almost impossible to recover.

The globe is already entered toward the social devastation, and there is a chance of frustrations among us, could create the deviant behavior and it could lead to system failure, not only socially, economically, and culturally but also in political arena. The trust on power could be altered and there is a chance of increase of famine with epidemic. As we begin to notice, people are so freighted, they already began to strike back to the social and political norms "Necessity knows no laws" (see news related to lockdown). This pandemic is breaking world order toward the negativism, and it is hard to foresee the degree of social breakdown; however, the symbols show that, it will be the hardest harm in the human civilization. Still, we could minimize the havoc impact, if we disciplined ourselves as responsible citizens. It depends on how the leaderships of each countries work collectively, mutually, and wisely. It also depends on how we as responsible citizens provide positive environment to support the leadership. It is not a time of blame but is the time to contribute to save humanity.

This pandemic created complete uncertainty on human civilization and broke the chain of the social, political, economic, religious, and environmental system, which we have been practicing, from generation to generation. Now, we all are Horrified, Terrified, Having Phobia of uncertainty, wounded internally and externally and even not able to prepare to make sequential efforts to fight with this pandemic. The late response or preparedness occurred because we could not foresee the connection of world, were not able to think that "the pain neighbor is facing could be my pain too." If that reality was realized on time, it would not spread as it is now. Now there is no point of heating the wounded part. However, it is just in the beginning stage in many parts of the world, therefore, it is absolutely essential to take the protective measures as much as possible (the media has been publishing it every single hour—what to do—so it is not necessary to illustrate here—we all know—what to do). Most of us might be at home, still scared, frightened; however, what about the some of us who are in travel, who are trapped in different countries, or migrants, or the people in refugee camps due to inhumane greed of power. What if the pandemic penetrates in that already broken, dismantled, sunk in sorrow frame of society, where people are dying every day without food and water? What about some of us who are still living in slums, or the no nation situation? How will these come under the watch system? What will be the consequences of this phobic uncertainty? (Now we have various kind of vaccines to control the pandemic, though is it possible to forget those who left us due to this virus?). **Please see Annex 1 and 2.**

The blackish cloud of the pandemic has captured our entire planet and the thunderstorms of Covid, have penetrated, wounded our hearts in a way that we are scared and feared within internally, eternally, and also externally. The concern is so serious, that we already stopped regular communication, and whatever we are talking is all about the fear of Corona Virus (we do not have any agenda to discuss except virus and it created phobia).

The communication pattern itself is dangerous, because it is heating the emotional side, sensitive side, and the side of humanity, through which we are different to animal. How, an epidemic like Covid came, it is the subject of long debate and research. However, the heat of this virus might be the first in the human history, where, the world is so much connected due to advancement in communication system. Now all of us almost citizens of every country know that, ultimately, Corona may take our lives or if not at least provide maximum torture we can imagine. Many of us who have strong immune system, good health, good food, care, medicinal excess, and healthcare system will win against Corona, and who are already having health problems, have no infrastructures and resources to fight; have to say goodbye.

So, absolutely, all the governments need to join hands, to fight against this, all power should pour to resolve this crisis. This is the attack on humanity, so it is the responsibility of every citizen of the world to survive first and help to survive other, without any discrimination of race, gender, ethnicity, nationalities, social-cultural and economic strata, and nations geographical boundaries. It is the time to give up the greediness of self-centric me and only me motive and think that—I may be the next victim of virus. And also, it is the time to think that my meaning of life is only significant if we all are here in this planet. It is the time to feel and think the bottom line of life—what does it mean? It is the time to contribute to the society, whatever we have with the factual truth that I may be the next victim and may not see this beautiful world tomorrow. Therefore, this is the time to take a lesson and suppress/remove the erratic nature we have. It is time to remove cruel, egotistic, deceitful, imprudent, malicious, and other "me or I centric" nature we have. This is the time to nurture love, forgiveness, calmness, wisdom, reassurance and feel and realize the value of own life and lives of others.

It is the time to think that there is something more powerful than weapons, wealth, power, and pride. We need to appreciate the role of media and technology, which is giving us the way to see, feel and realize what is happening around and in the surroundings and the globe. A small case appeared in China; now it became a global challenge. In the beginning when China was suffering, the world leaderships did not realize that, we may be the next victim; did not realized that we all are connected, and we all are the same in one way or another. Instead of sympathizing, we simply reacted that, it is their problem, not mine. However, now, Corona forced us to realize the fact that we all humans are interconnected to each other. It clearly shows that we all are tightly connected—and it is the time to think and apply—the philosophy of Basudaiva Kutumbakam (all Earth is our house and all living beings are our relatives—http://ologyjournals.com/aaeoa/aaeoa_00019.pdf), or (live and let other live—http://ologyjournals.com/aaeoa/aaeoa_00020.pdf). Insofar, we all are maintaining social distances, however, in my opinion it is also the time for strengthening love, respect, and honor between us, not being in near but being dear and loved one. We request, urge to all my loved ones, (we mean all men and women of the world, one way or another I am connected with you and you are connected with me—we all are humans) to remain connected within our own niche first, and beyond our niche and to the globe (there are ample ways, we all already know them, regularly using them one way or another); and selflessly contribute to the society, whatever we have as much as possible, as wisely as possible, without any expectation. Contribution could be anything—"whatever we have."

9.3 The Factual Side

Now, the people of the world are divided into two thoughts: Covid-19 prevention and control, and proximity of hunger due to the prolonged lockdown. Especially the rich are more afraid of the disease because the hunger never touches them. But the oppressed-class people are not as worried about the disease as they are about hunger. The bare-foot–marching people seen in the highway shows the same, as there is no place to eat on the way due to the lockdown, neither have they had affordability. Even now, with the emphasis on prevention of Corona infection around the world, international organizations and nation-states do not seem to be as concerned about starvation as they seem to have been. People are dying of hunger three times a day more than Covid-19. This number is estimated to have increased further in the current lockdown. So, journalist Paula Froelich used "luxury" to the later part of the lockdown in an article of New York Post. It is because the well-off people and elites have been "enjoying" the lockdown, whereas the oppressed section of people are stranded and traumatized with the shock of life, livelihood, and hunger. We can judge it simply through the social media posting as well. In fact, the prolonged lockdown has caused massive fissures in our society—mainly between those who can afford to stay home and those who can't. Working-class and blue-collar people, many of whom live month to month, are destitute due to the prolonged lockdown. So, it became luxury to minimize the challenges of Covid-19 for the well-off or elites, whereas it is becoming fatal to hunger to the working-class people. It looks like a romanticism on poor. It seems it is the easy-going measure to the government as well.

The symbolic architecture of lockdown

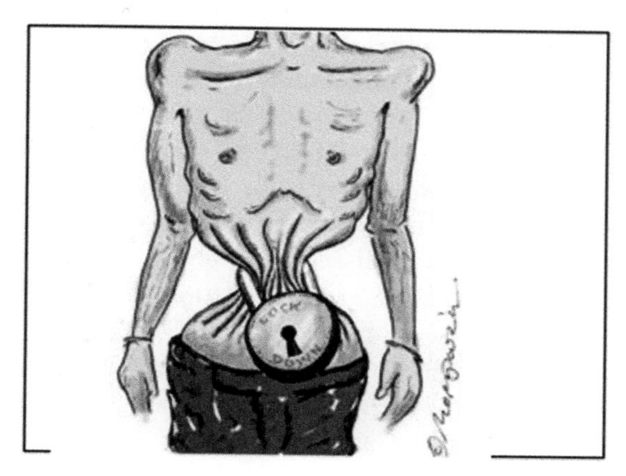

Nutrition is the one of the lead ingredients for the natural healing of the disease. The human body is designed to prevent itself from the external attack and to build the immune system needed for self-healing. Healthy body, psychosocial well-being, and nutritious food are the essence of the immune system. If someone is to stay in confinement for a long time, s/he will lose all these things resulting in the loss of immunity, which will lead to a sudden onset of any disease. So, people have to keep active themselves to gain the immunity. To keep human body active and productive s/he needs to work in a regular manner. Therefore, in order to fight the disease, it is very important to control hunger. The Global Nutrition Report 2020 published recently highlighted the link between inequities in the food system and health inequalities. While undernourished people have a weaker immune system and can be at greater risks of the virus, people with obesity and diabetes have been found to be more likely to die from Covid-19. The report warned that there is a real risk that, as nations strive to control the virus, the gains they have made in reducing hunger and malnutrition will be lost.

The death toll of Covid-19 each day, several times it crossed 7000, is shocking everyone. However, very few people are aware that hunger is the biggest cause of death even today. Ten million people die every year (more than 25,000 per day) due to hunger and hunger-related diseases. About 8% are the victims of hunger caused by high-profile earthquakes, floods, droughts, and wars. Over 2.5 million Indians die of hunger every year that counts over

7000 each day. A child dies of starvation every 10 seconds. As per UNICEF report (2018), approximately 3.1 million children die from undernutrition each year. That is nearly half of all deaths in children under the age of five. It can be estimated that this number has increased significantly in the current situation of lockdown. The WFP estimates that 32 million schools in 120 countries are closed due to the current halt. It is certain that a large number of children will be malnourished when their parents stop working. According to the ILO, about 2.7 billion workers (four out of five of the world's human resources) are in full or partial layoffs. It has made the poor even more vulnerable. Accordingly, famine has already started and further exacerbated as supply chain disruptions have disrupted food supplies.

The Harvard professor, Dr. Cornel West opined that economic viruses were already at work before Corona viruses. UN agencies had forecasted acute food insecurity for 2020 driven by conflicts in Yemen and Central Africa and the Middle East, an outbreak of desert locusts in East Africa, extreme weather events in the Caribbean, adverse climate in the Sahel and West Africa and socio-political crisis and high food prices in Latin America, even before the Covid-19 outbreak. Some 821 million people go to bed hungry every night all over the world, and the latest Global Report on Food Crisis reveals that 135 million people on earth are marching toward the brink of starvation. The WFP analysis shows that, due to the current pandemic, an additional 130 million people could be pushed to the brink of starvation, that is a total of 265 million people, by the end of 2020. About two billion people are overweight or obese and 30% of all deaths are linked to nutrition-related diseases. As millions of people faced being pushed to the brink of starvation, WFP warns of a "hunger pandemic." If we don't prepare and act now—to secure access, avoid funding shortfalls and disruptions to trade—we could be facing multiple famines within a short few months.

According to UNCTAD, the UN body dealing with trade, global trade is projected to fall by a record 27% in the second quarter of the year. According to IMF report (2020), for the first time since the Great Depression in the 1930s, both advanced economies and emerging market and developing economies are in recession and projected to contract sharply by 3% in an average in 2020. The ILO estimates that the global market will lose up to 25 million jobs due to the pandemic. The most vulnerable people in society have been the first and hardest hit. It will have a serious impact on labor-sending-countries as it will lead to a sharp decline in remittance income. Households dependent on remittance income will lose the ability to cope with hunger. A recent Oxfam report estimates that the Covid pandemic will add 500 million poor people worldwide. This will widen the proximity of the famine. However, all these figures are interim. The actual data may be even more alarming because its multidimensional impact has not been studied properly. In a country like Nepal that has had a poor healthcare system, informal and unorganized labor market, economy based on remittances and import tariffs, and weak governance capacity, the impact and panic of the Covid-19 is even harder. In countries where a large portion of income has to be spent on food, unemployment benefits, insurance, and other social protection provisions are not effective; famine is a natural consequence. The longevity of this crisis can be projected with the fact that the WHO advised the need of maintaining social distance by 2022.

According to the FAO's 2019 report on Nepal, about 2.5 million (8.7%) Nepalese are malnourished. It is still in the "Sensitive" (Index 20.8) category in the World Hunger Index. Nepal's own statistics (Nepal Demographic Health Survey 2016) shows that 4.6 million people are suffering from food insecurity, with 20% of households mildly food-insecure, 22% are moderately food-insecure, and 10% are severely food-insecure. Similarly, about one million (36%) children under the age of five are severely malnourished. Karnali and the hilly regions of the Far-west province are the centers of hunger. Every Nepali has to spend about 50% of their income on food stuffs, while about 22% of households have to spend up to 75% of their income on food. Despite being an agricultural country, it suffers from about 85,000 metric tons of food deficiency annually. The amazing fact is that 60% of the farmers themselves are suffering from food shortage. Remittance income is an effective coping strategy to meet food deficiency. According to a 2009 World Bank study, about half a million households, including those in India, were employed abroad. This number of migrant-workers going abroad, except India, is estimated to have reached 2.5 million. From this, the country received Rs. 784 billion in remittance income last year alone. Now this income seems to be declining drastically. The WB estimates remittance to decline by 14% resulting in a loss of approximately 145 billion NRs in 2020.

It is estimated that about one million people working abroad, except India, will be back immediately after the lockdown is released or eased. Likewise, according to the latest labor force survey report, out of 20.7 million working aged people only 7.1 million are in the employment. Out of the employed people, around 62% work in informal sector that is the most unsecured job. Now most of the jobs are halted and it seems around 50% jobs, especially informal ones, are hard to revive. About 500,000 youths are coming to the labor market each year. It creates harsh pressure to the government to create employment in a massive scale. This is sure that lack of employment weakens the coping strategy against hunger. In the Great Earthquake of 2015, 1.4 million people were in urgent need of food aid and about 1 million people were pushed below the poverty line. But the current impact seems to be even more serious. If present trends continue, food security is expected to increase among: households engaged in informal labor, precarious labor, service sector and daily wage work, as well as households with return migrants and income losses. Moreover, the transportation constraints are posing problems in all the four pillars of food security: availability, access, utilization, and stabilization.

According to the UN, from farm to fork, the food supply chain accounts for 29% of global greenhouse gas emissions as it has been dealing by the big corporate houses. About a third of all food produced goes to waste. It would preserve enough food to feed 2 billion people—more than twice the number of undernourished people across the globe. It would also increase household income, improve food availability, reduce food imports, and improve the balance of trade. However, for that it needs to restore the distribution chain intact, otherwise, the lockdown may further increase the food wastage as the farmers being not able to do the harvesting properly and bring production to the market.

The Zero Hunger Challenge, an international commitment envisioned for "world without hunger," has re-emphasized ending hunger by 2025 sustainably. Likewise, the United Nation's SDGs call on all countries to end hunger in all its forms by 2030. But the present pandemic made it harder to achieve these goals in accredited time. It does not seem possible to revive the economy and the livelihood of the people through the continuation of its ongoing plans, strategies, structures, and working style. For this, the nation-state needs to make focused plans and strategies with smart operational structures for each level of governments. The state of transition, a recent change in governance system with federal structure, has posed an additional challenge to our country, Nepal. It also created a situation to rethink on our political system and governance approach. The country's economy and the livelihood of its citizens cannot be revived in an equitable manner by following the existing policies as usual.

It needs focused protective intervention from the state which focuses on building local food regime—peasants command over food system—that *La Via Campesina*, the peasants' movement, has been campaigning since 1996. The Constitution of Nepal has included the Right to Food and Food Sovereignty as the fundamental rights while the Act (2018) has detailed various measures to its realization. The implicit spirit of the food sovereignty is to build the local food regime that regenerates human immunity, protects environment, respects cultural diversity, and rewards human pluralism. It strengthens the community resilience and also reduces carbon emission.

Now it is special time to reshape the economy in order to bring emancipatory shift in the life of the vulnerable people. For this, protective intervention, or protectionism to restate the local food system, local production and market mechanism, and the local economy would help develop the independent and strong local government which will strengthen the federal system of the country. Not only that, but it will also lead the nation-state toward Liberal Socialism through transformative democracy. Inclusive society from bottom to top is the basic part of protective liberalism, whereas building inclusive governance from the bottom is the hardcore strategy of the liberal socialism. Therefore, the "Coronomics" can be the foundation for rising the new economic policy including local food regime, homegrown approach to eradicate hunger, and promote self-sufficient local economy.

So, What:

The human civilization history has seen many pandemics; however, those were mostly area specific. The case of Covid-19, or any future communicable diseases is/will not be the same because we all citizen of the world are connected in a way that, if any small event occurs in one part of the world, it can spread to the globe, because we are connected through modernism (technology). Therefore, we should understand that our small negative action in the society could have global negative impact. In addition to Covid-19, the world is already facing various challenges: political, social, religious, etc., inequalities, ageing, AIDS, atomic energy, children, climate change, economic colonization, democracy fights, poverty, food insecurity, gender inequality, lack of access to healthcare, human rights, questions on international law and justice, migration, challenges on peace and security, population growth, refugees challenges, scarcity of water, misuse of technology, drug, growing individualistic approach among youth, deviant behavior to involve in temporary pleasures, corruption, wars, violation of social norms and values, and so on. In addition to that the inequality is growing and the discriminations on the basis of race, caste, color, country of origin, religion, migratory status, gender, and sexual orientations are still in practice. The economic inequality is widening and widening, and the Covid-19 is playing a favorable role to increase inequality. Scholarly world has an important role to overcome from these challenges.

If So,

It clearly shows that we all are tightly connected—and it is the time to think and apply—the philosophy of Basudaiva Kutumbakam (all Earth is our house and all living being are our relative—http://ologyjournals.com/aaeoa/aaeoa_00019.pdf), or (live and let other live—http://ologyjournals.com/aaeoa/aaeoa_00020.pdf). We should not use our energy to blame governments, authorities, or anyone, no one to be blamed, but we should offer the positivity, knowledge, and power of togetherness. At the bottom of the heart, we should think that we may not have time to share or express the thankfulness among us. So, let us not be greedy to express the gratitude and to acknowledge the contributions anyone has poured in our journey.

References

Bhandari S, Rankawat G, Singh A, Wadhwani D, Patel B. (2020), Evaluation of interleukin-6 and its association with the severity of disease in COVID-19 patients. Indian J Med Spec 2020;11:132-6

Bhandari, Medani P, (2019). Live and let other live- the harmony with nature /living beings-in reference to sustainable development (SD)- is contemporary world's economic and social phenomena is favorable for the sustainability of the planet in reference to India, Nepal, Bangladesh, and Pakistan? Adv Agr Environ Sci. (2019);2(1): 37−57. DOI: 10.30881/aaeoa.00020 http://ologyjournals.com/aaeoa/aaeoa_00020.pdf

Bhandari, Medani P. (2020), Bashudaibha Kutumbakkam, all living being are our relatives, with Reference to journalist role on Corona Crisis-1, https://saralpatrika.com/content/31756/2020-05-30, Issue 31756, Kathmandu, Nepal

Bhandari, Medani P. (2020), Bashudaibha Kutumbakkam, All living being are our relatives, with Reference to journalist role on Corona Crisis-2, Issue 273/743 https://www.enepalese.com/2020/05/273743.html,Virginia,USA

Bhandari, Medani P. (2020), Editorial- the Role of Scholarly World – Science Writing, Scientific Journal of Bielsko-Biala School of Finance and Law, ASEJ 2020; 24 (2): 4-4, https://asej.eu/resources/html/article/details?id=207187

Bhandari, Medani P. (2020), Second thoughts, In the COVID-19 Regime, What Role Does Intellectual Society Play?, The Society of Transnational Academic Researchers (STAR Scholars Network), USA, Bulletin 20/2, https://starscholars.org/in-the-Covid-19-regime-what-role-does-intellectual-society-play/

Bhandari, Medani P. (2020), The Phobia Corona (COVID 19) - What Can We Do, Scientific Journal of Bielsko-Biala School of Finance and Law, ASEJ 2020, 24 (1): 1-3, GICID: 01.3001.0014.0769, https://asej.eu/resources/html/article/details?id=202946

Bhandari, Medani P. (2020), The Re-Revolution of Humanities, With Reference to Corona Crisis, KhabarHub, issue- https://www.khabarhub.com/2020/11/163167/ Kathmandu, Nepal

Bhandari, Medani P. (2020), The Wariness of the World, A global concern, with reference of Covid-19, Annapurna Post, issue 155397, http://annapurnapost.com/news/155397, Kathmandu, Nepal- http://annapurnapost.com/news/155397?fbclid=IwAR2xaczZeEBd_TZgJ8zLm_GSiL3lBxSOTydQZCnu7WUbKlawB9ath9ebDrI

Bhandari, Medani. P. (2020). In the Covid-19 Regime – What Role Intellectual Society Can Play. International Journal of Science Annals, 3(2), 5–7. doi:10.26697/ijsa.2020.2.1 https://ijsa.culturehealth.org/en/arhiv https://ekrpoch.culturehealth.org/handle/lib/71

Brandily, B. (2020), A Poorly Understood Disease? The Unequal Distribution of Excess Mortality Due to COVID-19 Across French Municipalities.

Bk, Man Bahadur (2020), The Luxury Lockdown: Tackle to Covid-19, Fatal to Hunger, Spotlight Nepal, https://www.spotlightnepal.com/2020/05/21/luxury-lockdown-tackle-Covid-19-fatal-hunger/

Bk, Man Bahadur (2019), Eradicating Hunger: Rebuilding Food Regime, EKTA Books, Nepal

Bk, Man Bahadur (2008), Social Inclusion in Microfinance, Janautthan Pratisthan, Nepal

Chernick, H., D. Copeland and A. Reschovsky (2020), "THE FISCAL EFFECTS OF THE COVID-19 PANDEMIC ON CITIES: AN INITIAL ASSESSMENT", National Tax Journal, https://doi.org/10.17310/ntj.2020.3.04.

Chung, D. and H. Soh (2020), Korea's response to COVID-19: Early lessons in tackling the pandemic, https://blogs.worldbank.org/eastasiapacific/koreas-response-Covid-19-early-lessons-tackling-pandemic.

COE(2020), European Committee on Democracy and Governance and COVID-19, https://www.coe.int/en/web/good-governance/cddg-and-Covid#{%2264787140%22:[14]}.

Council of State Governments (2020), COVID-19: Fiscal Impact to States and Strategies for Recovery, https://web.csg.org/Covid19/wp-content/uploads/sites/10/2020/07/fiscal-impact.pdf.

Crutzen P.J. (2006) The "Anthropocene". In: Ehlers E., Krafft T. (eds) Earth System Science in the Anthropocene. Springer, Berlin, Heidelberg. https://doi.org/10.1007/3-540-26590-2_3

Cucinotta, D., & Vanelli, M. (2020). WHO Declares COVID-19 a Pandemic. *Acta bio-medica : Atenei Parmensis*, *91*(1), 157–160. https://doi.org/10.23750/abm.v91i1.9397 https://www.ncbi.nlm.nih.gov/pmc/articles/PMC7569573/

Google Web (2020), Corona Virus and Its Popularity, Google.com https://www.google.com/search?sxsrf=ALeKk039xNzt_ZgXs2ep_quLXcYT4BhX6w%3A1585462914102&ei=gj6AXuLyBa-U0PEP-K6K-AY&q=corona+epidemic&oq=corona+ep&gs_lcp=CgZwc3ktYWIQARgDMgIIADICCAAyAggAMgIIADICCAA6BAgjECc6BggAEAgQHjoKCAAQgwEQRhD7AToFCAAQkQI6CggAEIMBEBQQhwI6BAgAEApQ0f0BWPKpAmDV0wJoAXAAeACAAbcCiAGfB5IBBzAuNC4wLjGYAQCgAQGqAQdnd3Mtd2l6&sclient=psy-ab

Green MS; Swartz T; Mayshar E; Lev B; Leventhal A; Slater PE; Shemer Js (January 2002). "When is an epidemic an epidemic?". *Isr. Med. Assoc. J.* **4** (1): 3–6. PMID 11802306 https://sfamjournals.onlinelibrary.wiley.com/doi/pdf/10.1046/j.1365-2672.2001.01492.x

Leonard E. Egede, M.D., and Rebekah J. Walker, Ph.D. (2020), Structural Racism, Social Risk Factors, and Covid-19 — A Dangerous Convergence for Black Americans, New England Journal of Medicine, N Engl J Med 2020; 383:e77 DOI: 10.1056/NEJMp2023616 https://www.nejm.org/doi/full/10.1056/NEJMp2023616

LePan, Nicholas (2020). Visualizing the history of Pandemics- https://www.visualcapitalist.com/history-of-pandemics-deadliest/ https://www.worldometers.info/coronavirus/

McKinsey & Cie (2020), Navigating the post-COVID-19 era: A strategic framework for European recovery.

Muro, Whiton and Maxim (2020), "COVID-19 is hitting the nation's largest metros the hardest, making a "restart" of the economy more difficult", The Avenue, Brookings, https://www.brookings.edu/blog/the-avenue/2020/04/01/why-it-will-be-difficult-to-restart-the-economy-after-Covid-19/?utm_campaign=brookings-comm&utm_source=hs_email&utm_medium=email&utm_content=85726548.

OECD (2020), OECD Policy Responses to Coronavirus (COVID-19)- The territorial impact of COVID-19: Managing the crisis across levels of government (Updated 10 November 2020), This note was written by Dorothée Allain-Dupré, Isabelle Chatry, Antoine Kornprobst and Maria-Varinia Michalun, with substantial inputs from Charlotte Lafitte, Antti Moisio, Louise Phung, Kate Power, Yingyin Wu, and Isidora Zapata, from the OECD Centre on SMEs, Entrepreneurship, Regions and Cities. Comments and inputs received from Delegates of the OECD Regional Development Policy Committee are gratefully acknowledged- http://www.oecd.org/coronavirus/policy-responses/the-territorial-impact-of-Covid-19-managing-the-crisis-across-levels-of-government-d3e314e1/

Pardo, J., Shukla, A. M., Chamarthi, G., & Gupte, A. (2020). The journey of remdesivir: from Ebola to COVID-19. *Drugs in context*, *9*, 2020-4-14. https://doi.org/10.7573/dic.2020-4-14 https://www.ncbi.nlm.nih.gov/pmc/articles/PMC7250494/

Patel, J. A., Nielsen, F., Badiani, A. A., Assi, S., Unadkat, V. A., Patel, B., Ravindrane, R., & Wardle, H. (2020). Poverty, inequality and COVID-19: the forgotten vulnerable. *Public health*, *183*, 110–111. https://doi.org/10.1016/j.puhe.2020.05.006- https://www.ncbi.nlm.nih.gov/pmc/articles/PMC7221360/

Pooladi M, Entezari M, Hashemi M, Bahonar A, Hushmandi K, Raei M. (2020), Investigating the Efficient Management of Different Countries in the COVID-19 Pandemic. J Mar Med. 2020; 2 (1) :18-25URL: http://jmarmed.ir/article-1-79-en.html

Potter C.W. (2008), A history of influenza, sfamjournals.onlinelibrary.wiley First published: 07 July 2008 https://doi.org/10.1046/j.1365-2672.2001.01492.x https://sfamjournals.onlinelibrary.wiley.com/doi/full/10.1046/j.1365-2672.2001.01492.x

Sah, Ranjit, Nayanum Pokhrel, Zareena Fathah, Akihiko Ozaki, Divya Bhandari, Yasuhiro Kotera, Niranjan Prasad Shah, Shailendra Sigde, Kranti Suresh Vora, SenthilKumar Natesan, Shailesh Kumar Patel Ruchi Tiwari, Yashpal Singh Malik Mohd. Iqbal Yatoo, Alfonso J Rodriguez-Morales and Kuldeep Dhama (2020), SARS-CoV-2 / COVID-19: Salient Facts and Strategies to Combat Ongoing Pandemic. J Pure Appl Microbiol. 2020;14(3):. https://www.researchgate.net/profile/Alfonso_Rodriguez-Morales/publication/343699472_SARS-CoV-2_COVID-19_Salient_Facts_and_Strategies_to_Combat_Ongoing_Pandemic/links/5f43c852299bf13404ec1cca/SARS-CoV-2-COVID-19-Salient-Facts-and-Strategies-to-Combat-Ongoing-Pandemic.pdf

Silwal, A., Bhuwan Chalise, & Deepa Limbu. (2020). Risk Perception and Public Effort to Cope Covid-19 in Nepal. *JOURNAL OF CREATIVE WRITING | ISSN 2410-6259*, *4*(2), 81 - 110. Retrieved from https://journals.discinternational.org/index.php/jocw/article/view/64

Skjesol, G. and J. Tritter (2020), The COVID-19 pandemic in Norway: The dominance of social implications in framing the policy response.

UN (2020), COVID-19: Embracing digital government during the pandemic and beyond, https://www.un.org/development/desa/dpad/publication/un-desa-policy-brief-61-Covid-19-embracing-digital-government-during-the-pandemic-and-beyond/.

Watts, Sheldon (1999), Epidemics and History: Disease, Power and Imperialism, Yale University Press; 1st edition (November 10, 1999)

WHO (2020), Advice on the use of masks in the context of COVID-19, https://www.who.int/publications/i/item/advice-on-the-use-of-masks-in-the-community-during-home-care-and-in-healthcare-settings-in-the-context-of-the-novel-coronavirus-(2019-ncov)-outbreak.

WHO (2020), Considerations for mass gatherings in the context of COVID-19: annex: considerations in adjusting public health and social measures in the context of COVID-19, https://www.who.int/publications/i/item/considerations-for-mass-gatherings-in-the-context-of-Covid-19-annex-considerations-in-adjusting-public-health-and-social-measures-in-the-context-of-Covid-19.

WHO (2020), Covid strategy update (April 2020), https://www.who.int/publications/i/item/Covid-19-strategy-update---14-april-2020.

WHO (2020), Laboratory testing strategy recommendations for COVID-19, https://apps.who.int/iris/bitstream/handle/10665/331509/WHO-COVID-19-lab_testing-2020.1-eng.pdf.

Worldometers (2020); A Timeline of Historical Pandemics, https://www.worldometers.info/coronavirus/https://www.worldometers.info/coronavirus/

Annex 1

Authorized/approved vaccines: Source: COVID-19 vaccine tracker; Posted 01 March 2021 | By Jeff https://www.raps.org/news-and-articles/news-articles/2020/3/Covid-19-vaccine-tracker

Name	Vaccine Type	Primary Developer	Country of Origin	Authorization/ Approval
Comirnaty (BNT162b2)	mRNA-based vaccine	Pfizer, BioN-Tech; Fosun Pharma	Multinational	Albania, Andorra, Argentina, Aruba, Australia, Bahrain, Canada, Chile, Colombia, Costa Rica, Ecuador, EU, Faroe Islands, Greenland, Iceland, Iraq, Israel, Japan, Jordan, Kuwait, Liechtenstein, Malaysia, Mexico, Monaco, New Zealand, North Macedonia, Norway, Oman, Panama, Philippines, Qatar, Saint Vincent and the Grenadines, Saudi Arabia, Serbia, Singapore, Switzerland, UAE, UK, US, Vatican City, WHO
Moderna COVID-19 Vaccine (mRNA-1273)	mRNA-based vaccine	Moderna, BARDA, NIAID	US	Canada, EU, Faroe Islands, Greenland, Iceland, Israel, Liechtenstein, Norway, Qatar, Saint Vincent and the Grenadines, Singapore, Switzerland,United Kingdom, United States

Continued

Annex 1 (*Continued*)

Name	Vaccine Type	Primary Developer	Country of Origin	Authorization/ Approval
COVID-19 Vaccine AstraZeneca (AZD1222); also known as Covishield	Adenovirus vaccine	BARDA, OWS	UK	Argentina, Bahrain, Bangladesh, Barbados, Brazil, Chile, Dominican Republic, Ecuador, El Salvador, Egypt, EU, Guyana, Hungary, India, Iraq, Maldives, Mauritius, Mexico, Morocco, Myanmar, Nepal, Nigeria, Pakistan, Philippines, Saint Vincent and the Grenadines, South Africa, South Korea, Sri Lanka, Taiwan, Thailand, UK, Vietnam
Sputnik V	Recombinant adenovirus vaccine (rAd26 and rAd5)	Gamaleya Research Institute, Acellena Contract Drug Research and Development	Russia	Algeria, Argentina, Armenia, Bahrain, Belarus, Bolivia, Egypt, Gabon, Ghana, Guatemala, Guinea, Guyana, Honduras, Hungary, Iran, Kazakhstan, Kyrgyzstan, Lebanon, Mexico, Mongolia, Montenegro, Myanmar, Nicaragua, Pakistan, Palestine, Paraguay, Republika Srpska, Russia, Saint Vincent and the Grenadines, San Marino, Serbia, Syria, Tunisia, Turkmenistan, United Arab Emirates, Uzbekistan, Venezuela

Annex 1 (*Continued*)

Name	Vaccine Type	Primary Developer	Country of Origin	Authorization/ Approval
CoronaVac	Inactivated vaccine (formalin with alum adjuvant)	Sinovac	China	Azerbaijan, Bolivia, Brazil, Cambodia, China, Chile, Colombia, Hong Kong, Indonesia, Laos, Mexico, Thailand, Turkey, Philippines, Uruguay
BBIBP-CorV	Inactivated vaccine	Beijing Institute of Biological Prod-ucts; China National Pharmaceuti-cal Group (Sinopharm)	China	Argentina, Bahrain, Cambodia, China, Egypt, Hungary, Iraq, Jordan, Laos, Macau, Morocco, Nepal, Pakistan, Peru, Senegal, Serbia, Seychelles, UAE, Zimbabwe
JNJ-78436735 (formerly Ad26.COV2.S)	Non-replicating viral vector	Janssen Vaccines (Johnson & Johnson)	The Netherlands, US	Bahrain, Saint Vincent and the Grenadines, US
EpiVacCorona	Peptide vaccine	Federal Budgetary Research Institution State Research Center of Virology and Biotechnol-ogy	Russia	Russia, Turkmenistan
Convidicea (Ad5-nCoV)	Recombinant vaccine (adenovirus type 5 vector)	CanSino Biologics	China	Mexico, China (military use), Pakistan
Covaxin	Inactivated vaccine	Bharat Biotech, ICMR	India	India

Continued

Annex 1 (*Continued*)

Name	Vaccine Type	Primary Developer	Country of Origin	Authorization/ Approval
No name announced	Inactivated vaccine	Wuhan Institute of Biological Products; China National Pharmaceutical Group (Sinopharm)	China	China
CoviVac	Inactivated vaccine	Chumakov Federal Scientific Center for Research and Development of Immune and Biological Products	Russia	Russia

Source: COVID-19 vaccine tracker; Posted 01 March 2021 | By Jeff https: //www.raps.org/news-and-articles/news-articles/2020/3/Covid-19-vaccine-tracker

Annex 2

Vaccine candidates in development: Source: COVID-19 vaccine tracker; Posted 01 March 2021 | By Jeff https://www.raps.org/news-and-articles/news-articles/2020/3/Covid-19-vaccine-tracker

Candidate	Mechanism	Sponsor	Trial Phase	Institution
NVX-CoV2373	Nanoparticle vaccine	Novavax	Phase 3	Novavax
ZF2001	Recombinant vaccine	Anhui Zhifei Longcom Biopharmaceutical, Institute of Microbiology of the Chinese Academy of Sciences	Phase 3	Various
ZyCoV-D	DNA vaccine (plasmid)	Zydus Cadila	Phase 3	Zydus Cadila
CVnCoV	mRNA-based vaccine	CureVac; GSK	Phase 2b/3	CureVac

Continued

Annex 2 (*Continued*)

Candidate	Mechanism	Sponsor	Trial Phase	Institution
Bacillus Calmette-Guerin (BCG) vaccine	Live-attenuated vaccine	University of Melbourne and Murdoch Children's Research Institute; Radboud University Medical Center; Faustman Lab at Massachusetts General Hospital	Phase 2/3	University of Melbourne and Murdoch Children's Research Institute; Radboud University Medical Center; Faustman Lab at Massachusetts General Hospital
INO-4800	DNA vaccine (plasmid)	Inovio Pharmaceuticals	Phase 2/3	Center for Pharmaceutical Research, Kansas City. Mo.; University of Pennsylvania, Philadelphia
VIR-7831	Plant-based adjuvant vaccine	Medicago; GSK; Dynavax	Phase 2/3	Medicago
No name announced	Adenovirus-based vaccine	ImmunityBio; NantKwest	Phase 2/3	
UB-612	Multitope peptide-based vaccine	COVAXX	Phase 2/3	United Biomedical Inc. (UBI)
NVX-CoV2373	Nanoparticle vaccine	Novavax	Phase 3	Novavax
Abdala (CIGB 66)	Protein subunit vaccine	Center for Genetic Engineering and Biotechnology	Phase 2	Center for Genetic Engineering and Biotechnology

Continued

Annex 2 (*Continued*)

Candidate	Mechanism	Sponsor	Trial Phase	Institution
BNT162	mRNA-based vaccine	Pfizer, BioNTech	Phase 1/2/3	Multiple study sites in Europe, North America and China
AdCLD-CoV19	Adenovirus-based vaccine	Cellid; LG Chem	Phase 1/2a	Korea University Guro Hospital
Nanocovax	Recombinant vaccine (Spike protein)	Nanogen Biopharmaceutical	Phase 1/2	Military Medical Academy (Vietnam)
EuCorVac-19	nanoparticle vaccine	EuBiologics	Phase 1/2	Eunpyeong St. Mary's Hospital
Mambisa (CIGB 669)	Protein subunit vaccine	Center for Genetic Engineering and Biotechnology	Phase 1/2	Center for Genetic Engineering and Biotech-nology
IIBR-100	Recombinant vesicular stomatitis virus (rVSV) vaccine	Israel Institute for Biological Research	Phase 1/2	Hadassah Medical Center; Sheba Medical Center Hospital
No name announced	SF9 cell vaccine candidate	West China Hospital, Sichuan University	Phase 1/2	West China Hospital, Sichuan University
Soberana 1 and 2	Monovalent/ conjugate vaccine	Finlay Institute of Vaccines	Phase 1/2	Finlay Institute of Vaccines
VLA2001	Inactivated vaccine	Valneva; National Institute for Health Research (NIHR)	Phase 1/2	Multiple NIHR testing sites
No name announced	Adjuvanted protein subunit vaccine	CEPI	Phase 1/2	

Annex 2 (*Continued*)

Candidate	Mechanism	Sponsor	Trial Phase	Institution
AG0301-COVID19	DNA vaccine	AnGes, Inc.	Phase 1/2	AnGes, Inc.; Japan Agency for Medical Research and Development
GX-19N	DNA vaccine	Genexine	Phase 1/2	
ARCT-021 (LUNAR-COV19)	Self-replicating RNA vaccine	Arcturus Therapeutics and Duke-NUS Medical School	Phase 1/2	Duke-NUS Medical School, Singapore
No name announced	Protein subunit vaccine	Sanofi; GlaxoSmithKline	Phase 1/2	Various
No name announced	Inactivated vaccine	Chinese Academy of Medical Sciences, Institute of Medical Biology	Phase 1/2	West China Second University Hospital, Yunnan Center for Disease Control and Prevention
HDT-301 (HGCO19)	RNA vaccine	University of Washington; National Institutes of Health Rocky Mountain Laboratories; HDT Bio Corp; Gennova Biopharmaceuticals	Phase 1/2	
AV-COVID-19	Dendritic cell vaccine	Aivita Biomedical, Inc.	Phase 1b/2	Rumah Sakit Umum Pusat Dr Kariadi
PTX-COVID19-B	mRNA-based vaccine	Providence Therapeutics; Canadian government	Phase 1	
COVI-VAC	Intranasal vaccine	Codagenix; Serum Institute of India	Phase 1	
CORVax12	DNA vaccine (plasmid)	OncoSec; Providence Cancer Institute	Phase 1	Providence Portland Medical Center

Continued

Annex 2 (*Continued*)

Candidate	Mechanism	Sponsor	Trial Phase	Institution
MVA-SARS-2-S	Modified vaccinia virus Ankara (MVA) vector vaccine candidate	Universitätsklinikum Hamburg-Eppendorf; German Center for Infection Research; Philipps University Marburg Medical Center; Ludwig-Maximilians - University of Munich	Phase 1	University Medical Center Hamburg-Eppendorf
COH04S1	Modified vaccinia virus Ankara (MVA) vector vaccine candidate	City of Hope Medical Center; National Cancer Institute	Phase 1	City of Hope Medical Center
pVAC	Multi-peptide vaccine candidate	University Hospital Tuebingen	Phase 1	University Hospital Tuebingen
AdimrSC-2f	Protein subunit vaccine	Adimmune	Phase 1	Adimmune
bacTRL-Spike	Monovalent oral vaccine (bifidobacteria)	Symvivo	Phase 1	Symvivo Corporation
COVAX-19	Monovalent recombinant protein vaccine	Vaxine Pty Ltd.	Phase 1	Royal Adelaide Hospital
DelNS1-2019-nCoV-RBD-OPT1	Replicating viral vector	Xiamen University, Beijing Wantai Biological Pharmacy	Phase 1	Jiangsu Provincial Centre For Disease Control and Prevention
GRAd-COV2	Adenovirus-based vaccine	ReiThera; Leuko-care; Univercells	Phase 1	Lazzaro Spallanzani National Institute for Infectious Diseases
UQ-CSL V451	Protein subunit vaccine	CSL; The University of Queensland	Phase 1	
SCB-2019	Protein subunit vaccine	GlaxoSmithKline, Sanofi, Clover Biopharmaceuticals, Dynavax and Xiamen Innovax; CEPI	Phase 1	Linear Clinical Research (Australia)

Annex 2 (*Continued*)

Candidate	Mechanism	Sponsor	Trial Phase	Institution
VXA-CoV2-1	Recombinant vaccine (adenovirus type 5 vector)	Vaxart	Phase 1	Vaxart
AdCOVID	Intranasal vaccine	Altimmune	Phase 1	University of Alabama at Birmingham
AAVCOVID	Gene-based vaccine	Massachusetts Eye and Ear; Massachusetts General Hospital; University of Pennsylvania	Pre-clinical	
ChAd-SARS-CoV-2-S	Adenovirus-based vaccine	Washington University School of Medicine in St. Louis	Pre-clinical	Washington University School of Medicine in St. Louis
HaloVax	Self-assembling vaccine	Voltron Therapeutics, Inc.; Hoth Therapeutics, Inc.	Pre-clinical	MGH Vaccine and Immunother-apy Center
LineaDNA	DNA vaccine	Takis Biotech	Pre-clinical	Takis Biotech
MRT5500	mRNA-based vaccine	Sanofi, Translate Bio	Pre-clinical	
No name announced	Ii-Key peptide COVID-19 vaccine	Generex Biotechnology	Pre-clinical	Generex
No name announced	Protein subunit vaccine	University of Saskatchewan Vaccine and Infectious Disease Organization-International Vaccine Centre	Pre-clinical	University of Saskatchewan Vaccine and Infectious Disease Organization-International Vaccine Centre
No name announced	mRNA-based vaccine	Chulalongkorn University's Center of Excellence in Vaccine Research and Development	Pre-clinical	

Continued

Annex 2 (*Continued*)

Candidate	Mechanism	Sponsor	Trial Phase	Institution
No name announced	gp96-based vaccine	Heat Biologics	Pre-clinical	University of Miami Miller School of Medicine
No name announced	Inactivated vaccine	Shenzhen Kangtai Biological Products	Pre-clinical	
PittCoVacc	Recombinant protein subunit vaccine (delivered through microneedle array)	UPMC/University of Pittsburgh School of Medicine	Pre-clinical	University of Pittsburgh
T-COVIDTM	Intranasal vaccine	Altimmune	Pre-clinical	
LNP-nCoVsaRNA	Self-amplifying RNA vaccine	Imperial College London	No longer being studied	Imperial College London
V590	Recombinant vaccine (vesicular stomatitis virus)	Merck; IAVI	No longer being studied	
V591	Measles vector vaccine	University of Pittsburgh's Center for Vaccine Research	No longer being studied	University of Pitts-burgh; Themis Bio-sciences; Insti-tut Pasteur

Source: COVID-19 vaccine tracker; Posted 01 March 2021 | By Jeff https://www.raps.org/news-and-articles/news-articles/2020/3/Covid-19-vaccine-tracker

10

Family Labor, Gender and Labor-saving Farm Technology Use Among Smallholder Farmers in Terms of Transformation

Prem B. Bhandari[1] **and Medani P. Bhandari**[2]

[1]Managing Director of South Asia Research Consult, Inc. Michigan, USA
[2]Professor and Advisor of Gandaki University, Pokhara, Nepal, Prof. Akamai University, USA and Sumy State University, Ukraine

Abstract

Discourse on agricultural transformation—from subsistence-based to market-based commercial agriculture—has received much attention to address the challenges of food security worldwide. Family labor is the common source of labor input to perform various gender-specific activities in subsistence farming. In many developing countries including Nepal, the availability of family labor may be an important obstacle in the transformation of agriculture toward the market-based labor-saving inputs and mechanization-dependent agriculture. Although negative consequences of green revolution technologies are widespread, these inputs nevertheless are important for increasing agricultural production. This chapter unveils this reality by using the uniquely detailed data from a rural food insecure agrarian setting of Nepal by examining the relationships between family labor availability and use of modern labor-saving technologies in agriculture among smallholder farmers. We use the labor demand framework to examine the relationships. Results from multi-nominal logistic regression revealed that the availability of family labor, both males and females, discouraged the use of such technologies in crop production net of household- and neighborhood-level factors. Moreover, the results further provide evidence that both the availability of male and female laborers are equally important in hindering the use of labor-saving farm technologies. These findings provide important insights in leveraging problems of food insecurity through smallholder agricultural transformation in developing countries.

Keywords: inequality, labor, labor-saving, technology, rural, Nepal, South Asia

10.1 Introduction

Food security, at the individual, household, national, regional, and global levels [is achieved] when all people, at all times, have physical, social, and economic access to sufficient, safe, and nutritious food to meet their dietary needs and food preferences for a healthy and active life (FAO 2001).

Gender-based **inequalities** all along the food production chain "from farm to plate" impede the attainment of food and nutritional security (World Bank, FAO and IFAD, 2009). Maximizing the impact of agricultural development on food security entails enhancing women's roles as agricultural producers as well as the primary caretakers of their families. Food security is a primary goal of sustainable agricultural development and a cornerstone for economic and social development (World Bank, FAO and IFAD, 2009:11). Still, majority of Nepal living depends on subsistence farming system and women are the main labor force in agriculture, however, their contributions are rarely acknowledged.

National food security requires both the production and the ability to import food from global markets to meet a nation's consumption needs.

Household food security is year-round access to an adequate supply of nutritious and safe food to meet the nutritional needs of all household members (men and women, boys and girls).

Nutritional security requires that household members have access not only to food, but also to health care, a hygienic environment, and knowledge of personal hygiene. Food security is necessary but not sufficient for ensuring nutrition security. (International Fund for Agricultural Development [IFAD]) (World Bank, FAO and IFAD, 2009:12)

As shown in the Figure 10.1, food security is essentially built on three pillars: food availability, food access, and food utilization. However, the people below the poverty line have no food available all the year, no access of nutritional food, although they thrive for the maximum utilization of available food. Even women and children especially girls get the second choice in their plate.

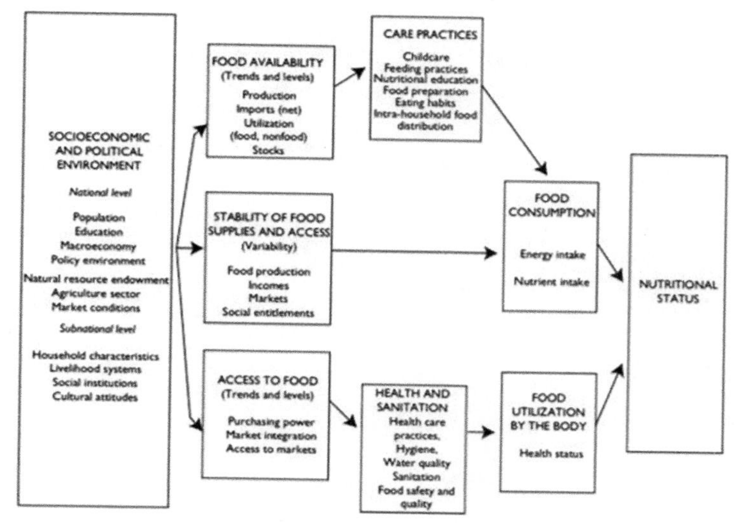

Figure 10.1 Elements in Achieving Food and Nutrition Security
Source: IFAD, FAO, and WFP 2000; as in World Bank, FAO and IFAD, 2009:12.

Because (as noted in World Bank, FAO and IFAD, 2009:5)

- Gender asymmetries in access to and control over assets
- Gender asymmetries in participation and power in land, labor, financial, and product markets
- Gender-differentiated distribution of risks and gains along value chains
- Gender asymmetries in market information, extension services, skills, and training
- Gender asymmetries in participation and leadership in rural organizations
- Gender asymmetries in rights, empowerment, and political voice
- Gender asymmetries in household composition and labor availability (dependency ratios, migration, and disability)
- Physical and agroecological risks and their gender differentiated impacts and vulnerability.

World Bank, ILO, and IFAD (2009:19) accept the fact that—"Women play a triple role in agricultural households: productive, reproductive, and social. The productive role, performed by both men and women, focuses on economic activities; the reproductive role, almost exclusively done by women, includes childbearing and rearing; household maintenance, including

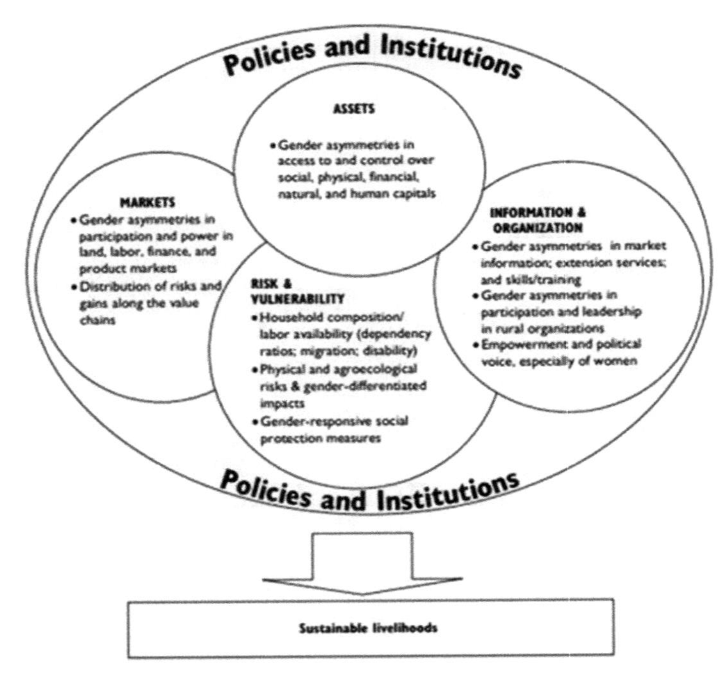

Figure 10.2 Sustainable Livelihoods through a Gender Lens
Source: World Bank, ILO, and IFAD (2009:5).

cooking, fetching water, and fuelwood; and the social role or community building, often dominated by women, which includes arranging funerals, weddings, and social events" (19). "Gender issues must be addressed in development. First, gender dimension is crucial for economic reasons and from the efficiency point of view. This is especially true in the agriculture sector, where gender inequalities in access to and control over resources are persistent, undermining a sustainable and inclusive development of the sector. Second, equity or distributional issues are related to gender differences in outcomes. Gender differences, arising from the socially constructed relationship between men and women, affect the distribution of resources between them and cause many disparities in development outcomes. Third, gender roles and relations affect food security and household welfare, critical indicators of human development. Last, but not least, gender equality is a basic human right, one that has value in and of itself" (World Bank, ILO and IFAD 2009:2). These statements of World Bank, ILO and IFAD totally fits with the Nepal case, along with the other developing world's countries.

Nepal's women relying in subsistence farming not only bear triple load to handle the family. They do everything to survive and save the family system, but their contribution is rarely acknowledged in the family as well as in the society, because society is still governing through male centric traditional system.

There are hundreds of books and papers which evaluate the rural livelihood of subsistence farmers and the role of women agriculture activities (Bhandari, Banskota and Sharma, 1999; Berdegue and Escobar, 2001; Curry and Tempelman, 2006; Deere and Leon, 2003; Doss, 2005; FAO, 2007; GENRD, 2006; World Bank, FAO and IFAD, 2009; Aikens, 1992; Bar-Yosef and Meadow, 1995; Bellwood, 1995; Byrd, 1994; Harlan, 1971; Harris, 1996; Higham, 1995; Higham and Lu, 1998; McGuire and Schiffer, 1983; Tzounis et al., 2017 and Jayaraman et al., 2015a; Bhandari, 2006, Bhandari, 2013). The findings of research have the same message that "In many parts of the world—for example, sub-Saharan Africa (SSA) and South Asia—despite women being the main farmers or producers, their roles are largely unrecognized. .... The poor, especially women, face obstacles in making their voices heard even in democratic systems and in increasing accountability and governance reforms in many areas (World Bank 2007a). For instance, recent studies stress that women's representation and gender integration into national plans and agricultural sector strategies remain a challenge (World Bank 2005b). .... Women face considerable gender-related constraints and vulnerabilities compared to men because of existing structures in households and societies. Property grabbing from women and children is common" (World Bank, ILO and IFAD 2009). There is no doubt of the statements, however, the world is still not being able to tackle the issue and make aware that each stakeholder in the farming process hold same stake, whoever are the part of it. This chapter also reveals the similar fact with the example of Nepal, where 66% people directly engaged in farming, combinedly with livestock.

10.2 The Global Scenario

The United Nation's World Food Programme (WFP) reports that 110 out of 210 countries—primarily poor countries with subsistence agriculture—are facing food security problems and this number is expected to continue growing (FAO et al., 2013). Despite the report of significant decline in the number of undernourished worldwide, still about 842 million people are estimated to have been in chronic hunger in the period 2011–2013. About

12% of the global population or one in eight persons are estimated to be not receiving enough food regularly to run active life. The vast majority of these undernourished people (827 million) live in developing countries. South Asia alone hosts 295 million (35% of the total) of them. Nepal is one of the most food insecure countries in the world with about 25% of the population below poverty with a ranking of 157 among 187 countries(UNDP, 2011; Joshi et al., 2012).Of the Nepal's 75 districts, 38 districts are characterized as food insecure (Ministry of Agriculture and Cooperatives, 2012). Subsistence nature of agriculture with low level of agricultural production and productivity associated with low labor productivity is considered one of the main reasons behind food insecurity (World Bank, 2013; FAO et al., 2013).

World agriculture has made a dramatic shift away from traditional farming systems toward increasingly mechanized, commercial farming systems during the second half of the 20th century (Mamdani, 1972; Self, 2008). This shift to commercial farming in peasant economies has many socio-economic, environmental, and political implications. Some scholars argued negative consequences such as unequal distribution of economic benefits (Griffin, 1974; Jacoby, 1972), unemployment effects (Griffin, 1974; Jacoby, 1972), environmental effects (Biswas, 1994; Pimentel and Pimentel, 1991), and possible peasant revolutions (Scott, 1977; Paige, 1975; Skocpol, 1982). Contrarily, others advocate for the important role agriculture plays in reducing world hunger and food insecurity (World Bank, 2008; FAO et al., 2013; APP, 1995). Proponents of the technological revolution in agriculture—including agricultural modernization—have greatly emphasized the positive aspects of transitioning away from traditional, subsistence farming to mechanized, commercial farming. These positive aspects include increases in food production and productivity, declines in food prices, and overall socio-economic development (e.g., Hazell and Ramaswamy, 1991; Mellor, 1976).

In developing countries, farm sizes are small (1.2 and 1.8 hectares in Asia and Sub-Saharan Africa, respectively) and family labor is commonly used to perform almost all types of agricultural operations such as land preparation, water management, fertilizer application, harvesting and post-harvest processing and storage (World Bank, 2008; World Bank, 2013). South Asia is not an exception. The average size of operational holdings (actual area cultivated) is only 0.5 hectares in Bangladesh, 0.8 hectares in Nepal and Sri Lanka, 1.4 hectares in India, and 3.0 hectares in Pakistan (Thapa, 2009). Moreover, a large majority of farmers are small holders. For example, in Nepal, 93% of operational holdings are operated by small farmers (<2 hectares) covering 69% of the cultivated area (Thapa, 2009). In Bangladesh, farms less than 2 hectares account for 96% of operational holdings with a share of 69% of cultivated area.

It is reported that, in South Asia, the labor productivity is much lower in agriculture sector compared to other sectors. For instance, in Nepal, the productivity of agricultural labor is Nepali Rupees 700 (approximately US $ 7) per person compared to the labor productivity of NRs 2817 (approximately US $ 28) per person in non-agriculture sector (ADB 7762-NEP, 2011). Thus, enhancing agricultural productivity (hence labor productivity) requires improvement in the use of modern farm technologies, through investment in areas such as of irrigation, farm roads, land improvement, agricultural mechanization, and the use of fertilizers and pesticides (Joshi et al., 2012).

Previous studies have primarily examined various economic factors such as prices, land size, and incomes contributing to technology use (Feder and O'Mara, 1981). Other researchers focused on micro-level explanations including household demographic and socio-economic characteristics on the use of farm technologies (Schutjer and Van der Veen, 1977; Feder and Umali, 1993; Rauniyar and Goode, 1992). This study, however, contributes to the existing literature by examining the relationships between labor availability and the use of modern labor-saving technologies among smallholder farmers in a rural agricultural setting. By examining this link, this study offers important insights in leveraging the transformation of smallholder agriculture in developing countries.

10.3 Background and Research Questions

Agriculture sector remains the major source of income and employment for the majority of Nepalese. It absorbs about 60% of the labor force for employment but has very low labor productivity (Upreti et al. 2008; ADB 7762-NEP, 2011). Farm sizes are very small, which declined to 0.7 hectares in 2010 from 1.1 hectares in 1995. In 2010, 52% of the holdings operated less than 0.5 hectares of land (93% holdings less than 2 hectares in 2009).

Agricultural productivity or yield (production per unit of land) in Nepal has remained stagnant or in some years declined during the last three decades. There is a wide gap in potential and actual agricultural productivity (ADB 7762-NEP, 2011). One of the main reasons for low agricultural yield is the low use of modern farm inputs and technologies (APP, 1995; ADB 7762-NEP, 2011; World Bank, 2008). In 2010, only 54% of the arable land was provided with irrigation. Most land is irrigable during rainy season only. Use of fertilizers is low, at 31 kg/hectare, one of the lowest among the neighboring countries in 1990 (APP, 1995). In fact, it is reported that the use of chemical fertilizers has actually declined to 19.6 kg/ha in 2000 (Leclerc and Hall, 2007). The level of mechanization is also low (APP, 1995; ADB 7762-NEP, 2011; Pariyar et al., 2001). Therefore, modernization of agriculture by providing farmers with new technologies is essential to reduce ever increasing food insecurity in Nepal (APP, 1995). With this view in mind, the Nepalese government formulated and implemented a 20-year Agricultural Perspective Plan (APP) in 1995 with a strong focus on developing agriculture sector by encouraging farmers to use green revolution technologies such as mechanization, irrigation, fertilizers, and high-yielding varieties of seeds. More recently, Nepal's Government has planned to invest Rs 65.77 billion in the agriculture sector over a period of three years (2013–14 to 2015–16) to boost productivity and spur economic growth particularly through improving land and labor productivity (The Kathmandu Post, 2014).

Inadequate and untimely supply of quality inputs has been considered a major impediment behind low use of modern inputs in Nepal (APP, 1995; Parajuli, 2007; ADB 7762-NEP, 2011). Moreover, studies also reported macroeconomic factors such as the demand and supply of fertilizers (ESCAP/FAO/UNIDO, 1997), fertilizer policy issues (Joshi, 1998; Tamrakar, 1998), and fertilizer trade liberalization issues (Basnyat, 1999). These studies primarily focused on issues of fertilizer acquisition, pricing mechanisms, and the distribution systems in the country. Studies of factors affecting modern inputs use at the micro-level are limited, however. In 2003, a study conducted by the Ministry of Agriculture and Cooperatives (2003) examined factors such as the price of fertilizer, prices of major agricultural outputs, wealth of household, size of cultivated land, and irrigation as some of the important determinants of fertilizer use. Regarding agricultural mechanization, very little research has examined the impact of mechanization on crop production, employment, and income (Pudasaini, 1979) and the use of mechanization in the Nepalese agriculture (Salokhe and Ramalingham, 1998; Shrestha, 1990).

Interestingly, however, it is reported that 75% farmers were well aware of the modern inputs and their value even in early 1970s (Parajuli, 2007). Despite the fact, their use in Nepali agriculture up till now is still very low. It is further reported that farmers were hesitant to take risks due to the high cost of farm machinery, fuel, fertilizers, and pesticides. While aforementioned findings may be equally relevant, there is a paucity of studies that examine the role household-level labor availability may have on the use of various labor-saving modern inputs in crop production. Because the modern agricultural technologies such as mechanization, fertilizers, and pesticides are labor-saving in nature (Boserup, 1965), we argue that none- or low-use of these inputs may be associated with the availability of household labor in a context where family labor is the major source of farm labor. If cheap labor is already available to carry out farm activities, it is expected that the household might be reluctant to use labor-saving modern inputs. With this background, this study attempts to answer: *(i) to what extent does the availability of family labor influence the use of technologies in crop production, net of socioeconomic and neighborhood contextual factors?* Moreover, some of the agricultural operations in rural agrarian countries are gender specific (Acharya and Bennet, 1981; Agarwal, 1992; Sachs, 1996; Boserup, 1990; Kazinga and Wahha, 2013). Therefore, it is likely that the use of technology may replace gender-specific labor requirements in some specific sorts of operations and, thus, the presence of gender-specific labor in a household is expected to influence the use of labor-saving technology in farming. Therefore, this paper also attempts to answer *(ii) does the extent to which labor availability and technology use differ by type of labor—males and females—net of socioeconomic and neighborhood contextual factors?*

10.4 The Setting

The Western Chitwan valley, situated in the southern plain of central Nepal, is the study setting. Before the 1950s, the valley was covered with dense forests and was infamous for malarial infestation. With U.S. assistance, however, the Nepalese government initiated a rehabilitation program in the valley during the 1950s by clearing the forest. Since then, the area has witnessed a rapid inflow of migrants attracted by the free distribution of land for agricultural purposes at the beginning of the settlement, and by the subsequent growth of modern amenities and services in recent decades. Currently, the valley is inhabited mostly by in-migrants. Chitwan's central location and relatively well-developed transportation network have been the catalytic forces for transforming it into a hub for business and tourism. This has resulted in a rapid proliferation of government services, businesses, and wage labor opportunities in the district (Shivakoti et al., 1999).

Population in the valley is an admixture of Indo-Aryan and Tibeto-Mongoloid origins. The household economy is primarily subsistence-based farming. A large majority of farmers practice crop–livestock integrated mixed farming systems (Bhandari, 2004, Bhandari, 2013; Bhandari and Ghimire, 2013). Land is generally used to produce food. Animals are kept for milk, meat, eggs, draft power and manure. To a large extent, the labor needed for performing farm and other household activities comes from within the household. More recently, however, agriculture is experiencing modernization and the family mode of agricultural production has been rapidly changing throughout Nepal (Ministry of Agriculture and Cooperatives, 2003; Pariyar et al., 2001).

10.5 Theoretical Background

Everett M. Rogers (1960) offered the theory of diffusion of new ideas and subsequent adoption behaviors of farmers. According to Rogers, diffusion and adoption of new ideas takes place through five different stages: awareness, interest, evaluation, trial, and final adoption of a new technology. He also pointed out other factors affecting the rate of adoption. For Rogers, if a new idea is affordable, simple, divisible (can be tried in a small amount), visible (outputs can be seen), and compatible to the farmer's condition, the rate of adoption is faster. Although many other factors have been studied to explain modern technology use in agriculture (Feder and O'Mara, 1981; Rauniyar and Goode, 1992; Schutjer and Van der Veen, 1977), Godoy et al., 1998 concluded that there is no single micro-level theory to explain technology use by farm households and therefore, pointed towards a need to develop a theory of adoption.

In this study, we utilize the household labor demand framework which is derived from the "new home economics," that originates from Gary S. Becker (1991) to assess the relationship between family labor availability and the use of labor-saving technologies in agriculture. In many developing countries, a household is both a producer as well as a consumer and farm households are the primary units of decision making regarding farming practices (Becker, 1991; Ellis, 1993; Feder and Umali, 1993). The use of technologies—particularly those designed to perform labor intensive jobs—replace labor (Agarwal, 1983; Binswanger, 1978; Schutjer and Van der Veen, 1977; Boserup, 1965; Mamdani, 1972). Therefore, we expect that the availability of family labor may have important implications in the decision to use such labor-saving modern farm technologies.

Modern farm technologies are broadly grouped as: mechanical (tractors, pumpsets, and farm implements) and bio-chemical (chemical fertilizers and pesticides) technologies (Bartsch, 1977). Biologically, the effects of these two technology packages on agricultural production differ. While the use of mechanical technology increases labor productivity and agricultural production by improving the physical condition of soil and by timely completion of agronomic operations, the use of bio-chemical technologies increases production by directly affecting plant physiology. Therefore, the factors contributing to the use of these two technological packages may differ (Schutjer and Van der Veen, 1977). More importantly, some of the agricultural operations are gender-specific (Acharya and Bennet, 1981; Agarwal, 1992; Sachs, 1996; Boserup, 1990). Boserup, 1990 indicated that in Africa, plowing of fields is primarily done by males and hoeing or weeding is done by females. This situation is not an exception to the Nepalese agriculture. Moreover, application of farmyard manure, weeding, and thinning out of disease and insect-infested plants are primarily carried out by women. It is likely that use of technology may replace either male or female labor depending upon the nature of agricultural operations performed. Therefore, the presence of gender-specific labor in a household may affect the use of labor-saving technologies differently. Below, we discuss the mechanisms the household labor availability may influence the use of labor-saving modern mechanical and bio-chemical technologies in a poor rural agrarian setting.

10.5.1 Linkages Between Labor Availability and the Use of Mechanical Technologies

In Nepal, land preparation for crop cultivation is generally performed by using human and animal labor. Men are primarily responsible for plowing land. If there is a shortage of male labor in a household, alternatives are either to hire bullocks and a man or to hire a tractor (in *Terai*, the flat plain area). The use of tractors and power tillers for plowing land is gradually increasing. It is reported that the use of a tractor requires only one-fifth the labor that was needed to plow land compared to using a bullock (Agarwal, 1992; Bartsch, 1977). Since, a shift from human and bullock labor to a tractor replaces male labor, it is hypothesized that a household with more working-age males per unit of cultivated land is expected to be less likely to use a tractor. Farmers also use farm implements such as corn shellers, threshers, sprayers, and chaff cutters (Pariyar et al., 2001). Corn shellers are used for loosening grains from corn and sprayers are used for spraying chemicals such as pesticides and herbicides. A chaff cutter is used for cutting straw or fodder into small pieces. Although male labor is also used, females typically loosen corn grains. Similarly, a chaff cutter saves men's time compared to women. The use of a sprayer generally increases male labor and saves female labor by reducing their time for weeding or removing diseased plants from the field. Altogether, these farm implements replace the need for human labor (Binswanger, 1978; Tunisia et al., 1990).

Use of rainfall and canal water is the common method used in irrigating crop fields in Asia. Nepal's agriculture is no exception. In the Chitwan Valley, irrigation is provided by canal water during the monsoon season. However, in uplands, a pumpset is used. During dry seasons, canals are generally dry and pumpset is the only source for regular supply of water. These days deep tube wells are also in practice. Evidence is limited whether the use of a pumpset is a labor-saving or a labor-using technology. However, there are findings that traditional methods such as the use of the Persian wheel (an animal powered wheel with pots) and *charsa* (use of bullocks for lifting water from the well), commonly used methods in India, are labor-intensive as compared to pumpset irrigation (Bartsch, 1977). Billings and Singh (Agarwal, 1983) in India reported that the substitution of a pumpset for Persian wheels reduced human labor requirement to one-fourth of the previous level. Bartsch further reported that manual labor is greatly reduced when a pumpset is used as compared to gravity flow. It is therefore hypothesized that: (a) *availability of working-age family members per unit of cultivated land reduces the likelihood of using labor-saving mechanical technologies; and (b)altogether, availability of working-age males per unit of cultivated land will have much stronger effect compared to females to reduce the likelihood of using labor-saving mechanical technologies.*

10.5.2 Linkages Between Labor Availability and the Uses of Bio-chemical Technologies

Bio-chemical technologies refer to chemical fertilizers and pesticides (insecticides and herbicides). In placecountry-regionNepal, farmyard manure (FYM) or compost is the commonly used material to replenish soil nutrients. Recently, the use of chemical fertilizer is also increasing. In Swaziland, the use of chemical fertilizer is considered to be a labor-intensive technology, where it is frequently used as basal-dose and top-dressing (Rauniyar and Goode, 1992). Arnon, 1987 also reported that the application of fertilizers may increase labor demand due to the need for more frequent and intensive weeding. In India, Bartsch, 1977 indicated similar findings. In Nepal, anecdotal evidence suggests that the application of FYM demands a much higher level of human labor as compared to the use of chemical fertilizers. Labor is required to raise animals, prepare compost, carry out and apply the compost to the field. It requires a significant amount of labor as compared to buying, storing, and application of chemical fertilizer. FYM is primarily applied by women, although men and children also perform this task. Chemical fertilizer is applied primarily by men.

Similarly, manual weeding of unwanted plants is a common practice in Nepal and the task of weeding is performed by women. Although the application of pesticides is minimal in placecountry-regionNepal, their use tends to replace female labor. Rani and Malaviya, 1992 reported that one acre of land required 12.42 days for weeding by women in India. When herbicides were applied, the time required decreased to 0.42 days per acre. Therefore, it is hypothesized that (c) *availability of family labor in a household reduces the likelihood of using chemical fertilizers and pesticides; and (d) altogether, availability of working-age females per unit of cultivated land will have much stronger effect compared to males to reduce the likelihood of using chemical fertilizers and pesticides in agriculture.*

10.6 Data

This study used the Chitwan Valley Family Study (CVFS) household- and neighborhood-level data collected in 1996.The data was collected as part of the Population and Environment Study (PopEnv).[1] The CVFS was primarily designed to examine the influence of rapidly changing social contexts on demographic processes including timing of marriage, childbearing, and contraceptive use. The focus of the Population and Environment Study was to investigate the reciprocal relationships between marriage, childbearing, migration and other demographic variables, and environmental outcomes such as changes in land use, flora diversity, and water quality and vice versa. The data was collected at three different levels—neighborhood, household, and individual. The data were collected from households in 151 neighborhoods scattered throughout the valley. A neighborhood was defined as a geographic cluster of five to fifteen households. These neighborhoods were chosen as an equal probability, systematic sample of neighborhoods in western Chitwan, and the characteristics of this sample closely resemble the characteristics of the entire Chitwan Valley population (Barber et al., 1997). Of particular interest, the access to non-family community services came from this neighborhood-level data. Next, the household-level information was

[1]1Both the Chitwan Valley Family Study and the Population and Environment Study were supported by the National Institute of Child Health and Human Development (NICHD). W.G. Axinn, Professor of Sociology, University of StateplaceMichigan is the Principal Investigator.

collected through household census and household agriculture and consumption surveys in 1996. This study utilized data from 1225 farm households within the neighborhoods. The census collected information on age, sex, marital status, and individual relationships within the household. The agriculture and consumption survey collected information on household resources and assets, consumption, and agricultural practices. Of particular interest, the survey collected information on the use of various farm technologies such as tractors, chemical fertilizers, pesticides, and farm implements in crop production, the size of cultivated land, land ownership, and livestock holdings. The data was collected through paper-pencil based face-to-face interviews with 99% response rate. Individual-level measures, age, and education of the household head come from the individual-level data.

10.7 Measures

*Outcome meas*ures. There are two outcome measures—use of mechanical technologies and bio-chemical technologies. *Mechanical technology* included the use of a tractor, pumpset, and farm implements. Tractor use was measured by asking "Did your household use a tractor to plough the land for planting ….. crop?" Similarly, the ownership of a pumpset and farm implements such as a thresher, chaff cutter, sprayer, corn sheller, and other implements was measured as a dichotomy. The responses are coded "1" if a household used a technology and "0" otherwise. A three category summated index was created: (a) a household used none of them; (b) a household used any one of them; and (c) a household used any two or more of them. *Bio-chemical technology* included the use of chemical fertilizers and pesticides. Use of chemical fertilizers and pesticides was measured by asking whether a household used any chemical fertilizers and pesticides in crop production in the past 3 years. The responses were coded "1" if used and "0" otherwise. A three category summated index was created: (a) a household used none of them; (b) a household used any one of them; and (c) a household used both of them.

Explanatory measures. Presence of working-age labor per unit of cultivated land is the major explanatory measure. Data on the number of working-age men and women 15 to 64 years of age living in a household at the time of survey was collected in 1996. As used by Rauniyar and Goode (1992) in their study of Swaziland, a household level measure of family labor availability, total, men, and women per hectare of cultivated land was created Because majority of farmers have small land size, the availability of

Table 10.1 Descriptive Statistics of Measures, 1996 (N = 1225).

Measures	Descriptive Statistics			
	Mean	Std. dev.	Minimum	Maximum
Technology use				
Package I: Bio-chemical technology use				
Fertilizer (used = 1)	0.83	0.38	0.00	1.00
Pesticides/herbicides (used = 1)	0.23	0.42	0.00	1.00
Index				
Used both	0.21	0.41	0.00	1.00
Used any one	0.63	0.48	0.00	1.00
Package II: Mechanical technology use				
Tractor (used = 1)	0.77	0.42	0.00	1.00
Pumpset (own = 1)	0.04	0.19	0.00	1.00
Improved farm implements (own = 1)	0.14	0.35	0.00	1.00
Index				
Used any two or more	0.14	0.35	0.00	1.00
Used any one	0.66	0.48	0.00	1.00
Household labor availability				
Number of working age females/household	1.67	0.99	0.00	8.00
Number of working age males/household	1.72	0.96	0.00	10.00
Number of working age males and females/household	3.39	1.66	1.00	15.00
Household size	5.76	2.54	1.00	26.00
Household-level controls				
Age of head of the household (years)	41.78	12.52	15.00	80.00
Migration of individual from household (yes = 1)	0.25	0.43	0.00	1.00
Total cultivated land (*kattha)*	25.04	23.44	1.00	200.00
Land fragmentation (number of parcels)	2.12	1.23	1.00	6.00
Irrigated land (percent)	58.14	41.46	0.00	100.00
Type (quality) of cultivated land				
Khet only (yes = 1)	0.31	0.46	0.00	1.00
Bari only (yes = 1)	0.22	0.41	0.00	1.00
Khet and *Bari* both (yes = 1)	0.47	0.50	0.00	1.00
Land ownership: Full-owners (yes = 1)	0.72	0.45	0.00	1.00
Part-owners (yes = 1)	0.20	0.40	0.00	1.00
Sharecroppers (yes = 1)	0.08	0.27	0.00	1.00
Livestock ownership (yes = 1)	0.90	0.30	0.00	1.00

Continued

Table 10.1 (*Continued*)

Measures	Descriptive Statistics			
	Mean	Std. dev.	Minimum	Maximum
Education of head of the household (years)	4.18	4.53	0.00	16.00
Exposure to media (yes = 1)	0.54	0.50	0.00	1.00
Ethnicity: Bahun/Chhetri	0.49	0.50	0.00	1.00
Dalit	0.11	0.32	0.00	1.00
Hill *Janajati* (Indigenous)	0.16	0.37	0.00	1.00
Newar	0.06	0.24	0.00	1.00
Terai *Janajati* (Indigenous)	0.18	0.39	0.00	1.00
Neighborhood-level controls				
Number of services within a 10-minute walk	0.77	0.70	0.00	3.00
Presence of Small Farmer Group (yes = 1)	0.20	0.40	0.00	1.00
Proximity to urban center				
Strata 1 (close to urban center)	0.23	0.42	0.00	1.00
Strata 2 (between strata 1 and 3)	0.33	0.47	0.00	1.00
Strata 3 (farthest from the urban center)	0.44	0.50	0.00	1.00

1 hectare = 1.5 *bigha* = 30 *kattha*

family labor per unit of land is an appropriate factor in the decision to use labor-saving technologies. Therefore, labor availability is adjusted for land size.

Controls. The models of relationships between family labor availability and labor-saving technology use also included a series of controls known to influence these relationships. The controls included: (i) age of the elderly person or the household head; (b) migration of family member(s) (coded as "1" if any member is away from home for work reason, and "0" otherwise); (iii) quality of cultivated land as (a) cultivated only *khet* land, (b) cultivated both *khet* and *bari* land, and (c) cultivate only *bari* land (percent of irrigated land was also used in the models of bio-chemical technology use); (iv) land ownerships; (v) land fragmentation (number of land parcels); (vi) livestock ownership; (vii) education of the household head or the elderly person; (viii) ownership of a radio and/or television; (ix) caste/ethnicity (grouped as Brahmin and Chhetri, Dalit, Newar, Hill *Janajati*, and Terai *Janajati*); (x) access to community services (such as banks, cooperatives, markets, and transportation); (xi) presence of Small Farmers Development Program (SFDP), and (xii) proximity to the largest urban center of Narayangarh.

10.8 Analytic Strategy

First, descriptive statistics of all the measures used in the analysis are presented (Table10.1). Second, bivariate relationships were examined (results not shown). Finally, as both the outcome measures, the use of mechanical and bio-chemical technologies have more than two nominal categories, multinomial logistic regression models were estimated to examine the relationships between farm technology use and family labor availability adjusting for all other factors (Hosmer and Lemeshow, 2000) According to Hosmer and Lemeshow (2000), the multinomial logit equation is:

$$g_1(x) = \ln\left[\frac{\Pr(y = j)/x}{\Pr(y = J)/x}\right] = \alpha + \beta_1 x_1 + \beta_2 x_2 + ... + \beta_n x_n$$

Where, $g_1(x)$ is the logit function, $\Pr(y = j)$ is the probability of the *i*th category of the dependent variable, α is the intercept, βs are the regression (slope) coefficients, and *x*s are the covariates. Models are estimated separately for mechanical and bio-chemical technologies and are presented as unstandardized β-coefficients and odds ratios (in parentheses). For simplicity, results are interpreted as odds ratios which are "the odds of having an event occurring versus not occurring, per unit change in an explanatory variable, other thing being equal" (Liao, 1994:16). Results for the association between labor availability and mechanical technology use and bio-chemical technology use (total: model 1a and 1b; male: model 2a and 2b; and female: model 3a and 3b), net of controls are provided in Tables 10.2 and 10.3, respectively. Within each group, the results in the first model (e.g., model 1a) is the relationships between labor availability and any one technology use versus no use and the results in the second model (e.g., model 1b) is the relationship between labor availability and two or more technology use versuss no use.

10.9 Results and Discussion

Seventy-seven percent of the households reported that they used a tractor for plowing of crop fields, 14% reported they owned improved farm implements, and only 4% owned a pumpset (Table 10.1). Of the total, 20% households used none of these three technologies, 66% households used any one of them and 14% of them used any two or more of them. Similarly, 83% households reported using chemical fertilizers and 23% reported using pesticides/herbicides. Altogether, 16% households used none of these two chemicals, 83% of them used any one of them and 21% used both of them.

Table 10.2 Multinomial logistic regression models of the relationships between household labor availability and mechanical technology use (N = 1225).

Measures	Total Models		Gender Disaggregated Models			
	Total Labor		Male Labor		Female Labor	
	Used any one Input vs. None (Model 1a)	Used Both Inputs vs. None (Model 1b)	Used any one Input vs. None (Model 2a)	Used both Inputs vs. None (Model 2b)	Used any one Input vs. None (Model 3a)	Used both Inputs vs. None (Model 3b)
Household labor availability						
Number of working-age labor/*hectare*	–0.054 (0.948)***	–0.209 (0.812)***	–0.081 (0.922)***	–0.301 (0.740)***	–0.105 (0.900)***	–0.357 (0.700)***
Household-level controls						
Age of head of the household (years)	–0.007 (0.993)	0.018 (1.019)+	–0.007 (0.993)	0.017 (1.017)	–0.007 (0.993)	0.020 (1.020)+
Migration of individual from household (yes = 1)	0.461 (1.586)*	0.546 (1.727)+	0.442 (1.556)*	0.491 (1.635)+	0.448 (1.566)*	0.486 (1.625)+
Quality of land (Ref = *Bari* only)						
Khet only	0.176 (1.192)	1.299 (3.667)**	0.224 (1.251)	1.493 (4.450)***	0.170 (1.186)	1.361 (3.900)**
Khet and *Bari* only	–0.063 (0.939)	0.671 (1.956)	0.021 (1.021)	0.932 (2.538)*	–0.073 (0.930)	0.755 (2.128)+
Land ownership (Ref = Sharecroppers)						
Full owners (yes = 1)	0.235 (1.264)	1.899 (6.676)*	0.176 (1.192)	1.828 (6.220)*	0.274 (1.315)	1.918 (6.809)*
Part-owners (yes = 1)	0.124 (1.132)	1.097 (2.996)	0.108 (1.114)	1.097 (2.994)	0.158 (1.171)	1.117 (3.057)
Fragmentation of holding (no. of land parcels)	0.211 (1.234)*	0.488 (1.628)***	0.244 (1.276)*	0.554 (1.741)***	0.212 (1.236)*	0.500 (1.648)***

Table 10.2 *(Continued)*

Measures	Total Models		Gender Disaggregated Models			
	Total Labor		Male Labor		Female Labor	
	Used any one Input vs. None (Model 1a)	Used Both Inputs vs. None (Model 1b)	Used any one Input vs. None (Model 2a)	Used both Inputs vs. None (Model 2b)	Used any one Input vs. None (Model 3a)	Used both Inputs vs. None (Model 3b)
Livestock ownership (yes = 1)	0.150 (1.162)	0.075 (1.078)	0.240 (1.272)	0.247 (1.281)	0.122 (1.130)	0.176 (1.192)
Education of head of the household (years)	0.044 (1.045)+	0.101 (1.106)***	0.048 (1.049)*	0.106 (1.112)***	0.042 (1.043)+	0.100 (1.105)**
Exposure to media (yes = 1)	–0.078 (0.925)	0.731 (2.076)**	–0.078 (0.925)	0.751 (2.120)**	–0.073 (0.930)	0.748 (2.112)**
Ethnicity (Ref = Bahun/Chhetri)						
Dalit	–0.483 (0.617)+	–1.758 (0.172)**	–0.493 (0.611)*	–1.772 (0.170)**	–0.544 (0.580)*	–1.888 (0.151)**
Hill *Janajati* (Indigenous)	–0.012 (0.988)	–0.234 (0.791)	–0.006 (0.994)	–0.191 (0.826)	–0.031 (0.969)	–0.273 (0.761)
Newar	–0.050 (0.951)	0.033 (1.033)	–0.057 (0.945)	0.008 (1.008)	–0.071 (0.932)	0.058 (1.059)
Terai *Janajati* (Indigenous)	–0.384 (0.681)+	–0.703 (0.495)+	–0.430 (0.650)+	–0.798 (0.450)*	–0.405 (0.667)+	–0.763 (0.466)*
Neighborhood-level controls						
No. of services within a 10-minute walk	0.257 (1.293)*	0.073 (1.075)	0.253 (1.288)*	0.055 (1.056)	0.255 (1.291)*	0.064 (1.066)
Presence of Small Farmer Group (yes = 1)	–0.739 (0.477)**	–1.361 (0.256)***	–0.733 (0.481)**	–1.343 (0.261)***	–0.743 (0.476)**	–1.380 (0.252)***

(Continued)

Table 10.2 *(Continued)*

Measures	Total Models		Gender Disaggregated Models			
	Total Labor		Male Labor		Female Labor	
	Used any one Input vs. None (Model 1a)	Used Both Inputs vs. None (Model 1b)	Used any one Input vs. None (Model 2a)	Used both Inputs vs. None (Model 2b)	Used any one Input vs. None (Model 3a)	Used both Inputs vs. None (Model 3b)
Proximity to urban center (Ref = strata 1)						
Strata 2 (between strata 1 and 3)	–0.381 (0.683)+	0.183 (1.201)	–0.384 (0.681)+	0.202 (1.224)	–0.361 (0.697)+	0.252 (1.286)
Strata 3 (farthest from the urban center)	0.172 (1.188)	1.622 (5.065)***	0.176 (1.192)	1.639 (5.149)***	0.199 (1.220)	1.734 (5.665)***
Intercept	1.162	–4.928***	0.886	–5.685***	1.155*	–5.359***
Chi-Square	341.146***		326.079***		337.689***	
–2 Log likelihood	1804.761		1819.828		1808.218	
Degrees of freedom	38		38		38	
McFadden Pseudo R-square	0.159		0.152		0.157	

t-statistic*** = p<.001; ** = p<.01; * = p<.05; + = <.10 1 hectare = 1.5 *bigha* = 30 *kattha* Figures in parentheses are odds ratios.

Table 10.3 Multinomial logistic regression models of the relationships between household labor availability and bio-chemical technology use (N = 1225).

Measures	Total Models		Gender Disaggregated Models			
	Total Labor		Male Labor		Female Labor	
	Used any one Input vs. None (Model 1a)	Used Both Inputs vs. None (Model 1b)	Used any one Input vs. None (Model 2a)	Used both Inputs vs. None (Model 2b)	Used any one Input vs. None (Model 3a)	Used both Inputs vs. None (Model 3b)
Household labor availability						
Number of working-age labor/*hectare*	–0.026 (0.975)*	–0.052 (0.949)**	–0.046 (0.955)*	–0.084 (0.919)**	–0.052 (0.949)*	–0.108 (0.898)***
Household-level controls						
Age of head of the household (years)	0.004 (1.004)	0.013 (1.014)	0.004 (1.004)	0.013 (1.013)	0.004 (1.004)	0.013 (1.013)
Migration of individual from household (yes = 1)	–0.086 (0.918)	–0.075 (0.928)	–0.083 (0.920)	–0.076 (0.926)	–0.090 (0.914)	–0.081 (0.922)
Irrigated land (percent)	0.002 (1.002)	0.004 (1.004)	0.002 (1.002)	0.004 (1.004)	0.002 (1.002)	0.004 (1.004)
Land ownership (Ref = Sharecroppers)						
Full owners (yes = 1)	0.293 (1.340)	0.850 (2.340)*	0.268 (1.307)	0.814 (2.257)+	0.318 (1.374)	0.884 (2.421)*
Part-owners (yes = 1)	–0.056 (0.945)	0.189 (1.208)	–0.066 (0.936)	0.182 (1.199)	–0.038 (0.963)	0.211 (1.234)
Fragmentation of holding (no. of land parcels)	0.449 (1.567)***	0.493 (1.638)***	0.466 (1.593)***	0.524 (1.688)***	0.446 (1.563)***	0.487 (1.627)***
Livestock ownership (yes = 1)	0.102 (1.108)	–0.010 (0.990)	0.138 (1.148)	0.069 (1.072)	0.080 (1.083)	–0.040 (0.960)

(Continued)

Table 10.3 *(Continued)*

Measures	Total Models		Gender Disaggregated Models			
	Total Labor		Male Labor		Female Labor	
	Used any one Input vs. None (Model 1a)	Used Both Inputs vs. None (Model 1b)	Used any one Input vs. None (Model 2a)	Used both Inputs vs. None (Model 2b)	Used any one Input vs. None (Model 3a)	Used both Inputs vs. None (Model 3b)
Education of head of the household (years)	0.051 (1.052)+	0.106 (1.111)***	0.052 (1.054)+	0.108 (1.115)***	0.049 (1.050)+	0.103 (1.108)***
Ownership of radio and television (yes = 1)	0.130 (1.139)	0.335 (1.399)	0.127 (1.135)	0.335 (1.397)	0.132 (1.141)	0.338 (1.402)
Ethnicity (Ref=Bahun/Chhetri)						
Dalit	–0.801 (0.449)**	–0.777 (0.460)*	–0.798 (0.450)**	–0.785 (0.456)*	–0.832 (0.435)**	–0.819 (0.441)*
Hill *Janajati* (Indigenous)	–0.200 (0.818)	–0.025 (0.975)	–0.202 (0.817)	–0.026 (0.975)	–0.211 (0.809)	–0.042 (0.959)
Newar	0.011 (1.011)	–0.582 (0.559)	0.011 (1.011)	–0.581 (0.559)	–0.005 (0.995)	–0.597 (0.550)
Terai *Janajati* (Indigenous)	–1.437 (0.238)***	–1.480 (0.228)***	–1.450 (0.235)***	–1.509 (0.221)***	–1.449 (0.235)***	–1.494 (0.225)***
Neighborhood-level controls						
No. of services within a 10-minute walk	–0.160 (0.852)	–0.316 (0.729)+	–0.158 (0.853)	–0.317 (0.729)+	–0.160 (0.852)	–0.315 (0.730)+
Presence of Small Farmer Group (yes = 1)	–0.516 (0.597)+	–0.642 (0.526)+	–0.515 (0.597)	–0.639 (0.528)+	–0.520 (0.595)+	–0.648 (0.523)+
Proximity to urban center (Ref = strata 1)						
Strata 2 (between strata 1 and 3)	–0.860 (0.423)***	–0.443 (0.642)	–0.861 (0.423)***	–0.441 (0.643)	–0.853 (0.426)***	–0.431 (0.650)

Table 10.3 *(Continued)*

Measures	Total Models		Gender Disaggregated Models			
	Total Labor		Male Labor		Female Labor	
	Used any one Input vs. None (Model 1a)	Used Both Inputs vs. None (Model 1b)	Used any one Input vs. None (Model 2a)	Used both Inputs vs. None (Model 2b)	Used any one Input vs. None (Model 3a)	Used both Inputs vs. None (Model 3b)
Strata 3 (farthest from the urban center)	0.552 (1.737)+	0.739 (2.095)*	0.554 (1.740)+	0.746 (2.108)*	0.563 (1.756)+	0.758 (2.133)*
Intercept	0.928	−1.377+	0.850	−1.575*	0.950	−1.338+
Chi-Square	226.869***		224.617***		227.953***	
−2 Log likelihood	1989.992		1992.244		1988.908	
Degrees of freedom	36		36		36	
McFadden Pseudo R-square	0.102		0.101		0.103	

t-statistic *** = $p<.001$; ** = $p<.01$; * = $p<.05$; + = $<.10$ 1 hectare = 1.5 *bigha* = 30 *kattha* Figures in parentheses are odds ratios.

A household, on average, consisted of about six individuals (mean = 5.76) (an average of 5.38 for Nepal and 5.79 for the central Terai in 2001). On average, a household had about 3.39 working-age individuals: 1.67 men and 1.72 women (4.06 working age persons per hectare of cultivated land). A typical household head was about 42 years old. One in every four households had at least one member away from home for work reasons. A typical household had 25.04 *kattha* (0.83 hectare; 1 hectare = 30 *kattha*) of cultivated land. About 58% of the total cultivated land was irrigated and a large majority of the households reported that most of their land was irrigated during the monsoon season only. About 72% households were full owners, about one-fifth (20%) of them were part-owners and 8% of them were sharecroppers. The average number of parcels per household was 2.12. Ninety percent of the households reported that they kept animals (also a proxy of bullock ownership) such as cattle, buffalo, sheep, and goats. On average, a typical head of the household had slightly over 4 (4.18) years of schooling. Slightly over one-half (54%) of the households owned either a radio or a television or both. One-half of the households belonged to Brahmin/Chhetri, 18% belonged to the Terai *Janajati*, 16% belonged to the Hill *Janajati*, 11% were from *Dalit* and only 6% of them were Newar. Less than one service (mean = 0.77) was available within a 10-minute walk from the neighborhood. About 20% of the households belonged to a neighborhood where at least one member of the SFDP was present. About 23% of the households were in the area close to the urban center (strata 1), 44% of them were farthest from the urban center (strata 3) and the rest (33%) of them were in between these two areas. Below, the results of multivariate analysis are described.

10.9.1 Labor Availability and the Uses of Mechanical Technologies

The associations between family labor availability and the use of mechanical technologies (tractor, pumpset, and farm implements)are provided in Table 10.2. The results from the first set of models for total labor availability (model 1a and 1b) reveal that the increase in family labor availability per unit of cultivated land is negatively and statistically significantly associated with the use of mechanical technologies. For example, net of household- and community-level controls, a one person increase in total family labor per hectare of cultivated land reduced the odds of using any one item of mechanical technology by about 5% (odds ratio = 0.948; $p<.001$; model 1a) and two or more items of mechanical technologies by 19% (odds ratio =

0.812; $p<.001$, model 1b). Moreover, when the results are compared between users of any one mechanical technology versus nonusers (model 1a) and users of two or more mechanical technologies versus none (model 1b), the magnitude of the associations was higher for two or more units. This finding is consistent with the hypothesis that increased family labor availability may be negatively associated with the likelihood of using labor-saving mechanical technologies in farming.

The relationships between gender disaggregated family labor availability and the use or nonuse of mechanical technologies in farming were further examined. The associations between the presence of working-age males (models 2a and 2b; Table 10.2) and females (models 3a and 3b; Table 10.2) per hectare of cultivated land and the use of mechanical technologies reveal that, adjusting for all other factors, a one person increase in the availability of working-age male or female per hectare of cultivated land significantly reduced the odds of using either one or both items of mechanical technologies. For example, a one person increase in male laborer per hectare of cultivated land decreased the odds of using any one input (vs. using none) by 8% (odds ratio = 0.922; $p<.001$; model 2a) and both inputs (vs. using none) by 26% (odds ratio = 0.740; $p<.001$; model 2b). Similar were the results for female labor availability (models 3a and 3b). However, interestingly, contrary to the expectation, the magnitudes of the associations for females were slightly stronger than those of males in both models.

10.9.2 Labor Availability and the Uses of Bio-chemical Technologies

Associations between family labor availability and bio-chemical technology use net of household- and neighborhood-level controls are provided in Table 10.3 (models 1a and 1b). Results revealed that increases in working-age family labor per hectare of cultivated land significantly decreased the likelihood of using bio-chemical inputs in crop production. For example, a one person increase in working-age family labor per hectare of cultivated land significantly decreased the odds of using any one item of bio-chemical input, either chemical fertilizer or pesticide, by about 3% (odds ratio = 0.975; $p<.05$, model 1a), net of all other factors. Similarly, a one person increase in family labor per hectare of cultivated land decreased the odds of using both items of bio-chemical inputs by over 5% (odds ratio = 0.949; $p<.01$), net of all other factors.

Table 10.3 also presents the results of the associations between the presence of working-age male (models 2a and 2b) and female (3a and 3b) family members per hectare of cultivated land and the use of one or more units of bio-chemical inputs. Adjusting for all other factors, a one person increase in the availability of working-age members—either male or female—per hectare of cultivated land significantly reduced the odds of using any one or both items of bio-chemical inputs. For example, a one person increase in male laborer per hectare of cultivated land decreased the odds of using any one input by 5% (odds ratio = 0.955; $p<.05$; model 2a) and both inputs by 8% (odds ratio= 0.919; $p<.01$; model 2b). Similar were the results for female labor availability, with slightly stronger associations with female laborers than males. Interestingly, the magnitude of the associations between labor-saving technology use and female labor availability per unit of land is marginally but consistently greater across all models than the magnitude of the associations for male labor availability suggesting the significance of the availability of women labor force in the decision to use labor-saving technologies in agriculture.

10.9.3 Other Relationships

The findings also reveal the importance of other household- and neighborhood-level factors in the decision to use of modern technologies. The findings in the expected direction of these theoretically important measures suggest internal validity thus providing confidence in our results. As expected, education was positively associated with the use of modern technologies. Similarly, access to communication or a proxy measure for wealth or income—ownership of a radio and/or a television—positively influenced the use of mechanical technologies suggesting their important roles in technology use decisions. Migration of individuals was also positively associated with the use of mechanical technologies. Land ownership was significantly associated with the use of both technologies. Full land owners were more likely than sharecroppers to use them. This evidence is important in the context where land ownership has always been an issue for the development of Nepalese agriculture (NPC, 2003). In placecountry-regionNepal, dual land ownership prevails and emphasis is provided to abolish this system. The use of mechanical technologies also differed by quality of land. Those who cultivated *khet* land were more likely to use two or more items of mechanical technologies than those who cultivated only *bari* land. Although availability of irrigated land was positively associated with the use of bio-chemical inputs, the association was not statistically significant. The number of parcels cultivated by a farm household was found to increase the use of both technologies in crop production. This result is surprising, however. It could be due to the difficulty in transporting and applying farmyard manure in the distant fields as reported in Ethiopia (Gebeyehu, 1995). By caste/ethnicity, as expected, the findings revealed that the Terai *Janajati* and *Dalit* households were relatively disadvantaged in terms of using both bio-chemical and mechanical technologies compared to the Brahmin/Chhetri.

Despite the belief that no or low use of modern inputs is primarily due to their inadequate and untimely supply (APP, 1995; ANZDEC Limited, 2002; NPC, 2003), the results revealed, at least in the valley, that the associations between the use of farm technologies and the access to services (such as banks, cooperatives, and bus services), the presence of the SFD Program, and rural–urban location of farm households, however, were not clear. While the increased access to services increased the use of mechanical inputs, which is expected but decreased the likelihood of using chemical fertilizers and pesticides, which is in contrary to expectation. Rural–urban location of farm households also has a mixed effect on the use of various farm inputs. Households living in remote areas were more likely to use both of these farm inputs compared to those who are living in the vicinity of urban areas. It could be because of the fact that the households near the urban center may have other alternative income sources than farming and agriculture may not have received attention from the farmers.

10.10 Conclusion and Implications

Food insecurity is a global challenge. Whereas women play a triple role in agricultural households: productive, reproductive, and social (World Bank, ILO, and IFAD2009:19). This situation applies mostly to undernourished people live in developing countries and are mostly the subsistence-based smallholder farmers. Although controversies abound about the roles of green revolution technologies worldwide, their roles cannot be underestimated in increasing food production and therefore, in reducing world hunger and food insecurity. It is well recognized that many farmers in Asia and Africa are smallholders. Low use of production enhancing modern technologies by them and associated market access have been the major challenges in increasing agricultural production and thus, in alleviating the problem of food insecurity in those countries. Our results revealed that one of the reasons behind low use of modern inputs is due to the availability of family labor and their use among smallholder farmers. Previously, however, this empirical support was limited. This study contributes to the existing literature by examining these relationships between household labor availability and the use of modern labor-saving mechanical and bio-chemical technologies among smallholder farmers in a rural subsistence agricultural setting that is experiencing rapid commercialization more recently.

The findings provide evidence that the availability of working-age family members per unit of cultivated land discourage the use of both—mechanical and bio-chemical labor-saving technologies in agriculture. This could be the reasons behind low labor productivity in agriculture in Asia (World Bank, 2013; Ministry of Agriculture and Cooperatives, 2012). In addition, households having larger number of livestock may be more likely to reduce the use of chemical fertilizers because FYM can be a substitute of chemical fertilizer. Thus, the relationship between household labor availability and farm mechanization, for example, using tractor and other machines seems more salient as mechanization can be a substitute of labor availability mainly male. Moreover, from a gender perspective, the presence of both working-age men and women labor force per unit of land is equally important in the decision to use both of these technologies. Interestingly, the magnitude of the associations between labor-saving technology use and female labor availability per unit of land is marginally but consistently greater across all models than the magnitude of the associations for male labor availability. This is an important observation in a context where women's role in the economy is still neglected. Although the actual mechanism is not clear, this could be because women spend more time in household work including farming than men (FAO, 2000; Kumar and Hotchkiss, 1988; NESAC, 1998) and replace men's work wherever possible, for example, digging of crop fields, manual threshing and loosening of corn grains instead of using machines (corn sheller), etc. For example, FAO (2000) reported that women spend 10.8 hours per day in agriculture compared to 7.5 hours per day for men. This study provides important insights on the role of family labor availability on technology use which might be important for leveraging persistent food insecurity problem facing rural agrarian settings of developing countries.

This evidence is salient in the present context, where the country is experiencing unprecedented levels of out-migration, shortage of male labor, and increasing dependence on remittances. This shortage of labor due to out-migration may have been the main reasons behind increasing use of technologies by farmers. Additionally, both the large gender gap in out-migration and the low status of women in rural agricultural settings may also have important consequences in rural agriculture. Feminization of Nepali agriculture is another recent phenomenon. Due to unbalanced male out-migration, women are increasingly overburdened and are performing not only their traditional activities, but also the activities that were previously performed only by males (CBS, 2011; Maharjan et al., 2012; Gartaula et al., 2010). Given the gendered nature of farming operations, important consequences on women including changes in their roles, their time allocations, and health status can be expected, requiring further understanding.

Moreover, the existing agricultural development policies in Nepal basically focus on ensuring distribution of agricultural inputs while neglecting the role the availability of family labor that may play in agricultural modernization (ANZDEC Ltd., 2000). For example, thus far, the Ministry of Agriculture and Cooperatives has emphasized the distribution of inputs and their prices with the assumption that assured supply of inputs would encourage farmers to use them. This is reflected in the national policy documents. Obviously, the availability of inputs may be a constraint in the Hills and the high Hills and other remote districts of the country where the distribution of inputs is obstructed by rugged geographic terrain and transportation difficulties. However, such problems are not prominent in the Terai, particularly in the Chitwan Valley. Therefore, in a country where the family is the major source of labor and almost all activities including plowing, irrigating, weeding, and roughing of infested plants are performed by household labor, the provision of modern inputs may not be the primary solution to increasing their use.

The transition from subsistence, family-based farming to commercial farming is not without cost. Experience from the green revolution has already raised genuine concerns about its unintended negative consequences beyond increased production such as unequal distribution of economic benefits, unemployment, adverse health effects, and possible peasant revolutions (Griffin, 1974; Jacoby, 1972; Scott, 1977; Paige, 1975; Skocpol, 1982); and health and environmental effects (Pimentel and Pimentel, 1991). Therefore, it is crucial to gain a better understanding of the environmental and health effects caused by the use of chemical fertilizers and pesticides, along with the potential unemployment effects on both men and women.

Finally, we acknowledge various limitations of this study. First, despite the uniqueness and richness of the data used here, it is cross-sectional and was collected in one point in time in 1996. Therefore, these findings are rather associations than cause-effect relationships. Second, the data is collected from only one part of a district in the Terai plain. Therefore, findings will have to be used rather cautiously. For example, the findings related to mechanical technology use may not be appropriate for policy purposes for the Hill and Mountain districts of Nepal, where large machines (e.g., large tractors) cannot be used due to the topography. Third, the findings revealed a strong negative association between female labor pool in a household and the use of mechanical inputs. A further study is needed to explore mechanisms and changes in gender roles at this critical juncture when Nepali agriculture is rapidly being feminized.

Ethical Consideration

The data is publicly available through the Inter-University Consortium for Political and Social Research (ICPSR) at the University of Michigan at http://www.icpsr.umich.edu/icpsrweb/ICPSR/studies/4538?archive=ICPSR&q=nepal.

The author is certified with the human subjects protection "Program for Education and Evaluation in Responsible Research and Scholarship" at the University of Michigan. Thus, an independent ethical approval for the data used in this paper is not required.

Acknowledgement

This research was supported by a number of grants from the National Institute of Child Health and Human Development (NICHD) (Grant # R01-HD032912, Grant # R01-HD033551, and Grant # R01 HD033551–13) and a NICHD center grant to the Population Studies Center at the University of Michigan (R24 HD041028). I thank William G. Axinn (PI) and Dr. Dirgha Ghimire for providing access to the data. I also thank Drs. Shannon Stokes and Leif Jensen, my mentors at the Pennsylvania State University for their guidance and supervision. In addition, the research staff at the Institute for Social and Environmental Research-Nepal for their contributions to the research reported here and Cathy Sun for assisting with data management. Last but not least, I owe a special debt of gratitude to the respondents who continuously welcome to their homes and share their invaluable experiences, opinions, thoughts and have devoted countless hours responding to our survey questionnaires. All errors and omissions remain the responsibility of the author.

References

Acharya, M., and Bennet, L. (1981),The status of women in Nepal: The rural women of Nepal. Kathmandu, Nepal: Tribhuvan University, Centre for Economic Development and Administration.

ADB 7762-NEP. (2011), Assessment report. Technical assistance for the preparation of the agricultural development strategy. Kathmandu, Nepal: Asian Development Bank.

Agarwal, B. (1983), Mechanization in Indian agriculture: An analytical study based on the Punjab. Delhi School of Economics Monograph in Economics No. 6.Allied Publishers Private Limited, New Delhi.

Agarwal, B. (1992), Impact of HYV rice technology on rural women. In: Punia, R. K., Editor, 1997.*Women in agriculture: Their status and role.* Northern Book Center, New Delhi.

Aikens, M. C. (1992), Hunting, fishing, and gathering in Pacific Northeast Asia: Pleistocene continuities and holocene developments. In Aikens, M. C., and Rhee, S. N (eds.), Pacific Northeast Asia in Prehistory: Hunter–Fisher–Gatherers, Farmers and Sociopolitical Elites, Washington State University Press, Pullman, pp. 99–104.

Antonis Tzounis,Nikolaos Katsoulas,Thomas Bartzanas and Constantinos Kittas, (2017), Internet of Things in agriculture, recent advances and future challenges, Biosystems Engineering, Volume 164, December 2017, Pages 31–48, https://fardapaper.ir/mohavaha/uploads/2018/07/Fardapaper-Internet-of-Things-in-agriculture-recent-advances-and-future-challenges.pdf

ANZDEC Limited. (2002), Nepal agriculture sector performance review. Main report prepared for the Ministry of Agriculture and Co-operatives, Nepal and Asian Development Bank. Volume 1.ANZDEC Limited, New Zealand.

APP.(1995), *Nepal Agriculture Perspective Plan.* Agricultural Projects Services Center, Kathmandu and John Mellor Associates, Inc., Washington, D.C.

Arnon, I. (1987), *Modernization of Agriculture in Developing Countries.* John Wiley and Sons, Chichester, New York, Brisbane, Toronto, Singapore.

Barber, J. S, Shivakoti, G., Axinn, W. G. and Gajurel, K. (1997), Sampling strategies for rural settings: A detailed example from the Chitwan Valley Family Study, Nepal. *Nepal Population Journal*, 6(5), pp. 193–203.

Bartsch, W. H. (1977), *Employment and Technology Choice in Asian Agriculture.* Praeger Publishers, New York, London.

Bar-Yosef, O., and Meadow, R. H. (1995), The origins of agriculture in the near east. In Price, D. T. and Gebauer, A. B. (eds.), Last Hunters First Farmers, School of American Research Press, Santa Fe, pp. 39–94.

Basnyat, B. B. (1999), Meeting the fertilizer challenge: A study on Nepal's fertilizer trade liberalization. *Research Report Series.* No 41. October 1999. Winrock. International, Policy Analysis in Agriculture and Related Resource Management, Kathmandu, Nepal.

Becker, Gary S. (1991), *A Treatise on the Family: Enlarged Edition.* Harvard University Press, Cambridge, Massachusetts.

Bellwood, P. (1995), Early agriculture, language history and the archaeological record in China and Southeast Asia. In Yeung, C. T., and Li, W. B. (eds.), Archaeology in Southeast Asia. The University Museum and Art Gallery, Hong Kong, pp. 11–22.

Berdegue, Julio, and Germán Escobar. (2001), "Agricultural Knowledge and Information Systems and Poverty Reduction." AKIS/ART Discussion Paper, World Bank, Rural Development Department, Washington, DC.

Bhandari, Medani P.; Banskota, K. and Sharma, Bikash (1999), Roads and Agricultural development in Eastern Development Region, A Correlation analysis, Winrock international, Kathmandu, Nepal

Bhandari, P. (2004), Relative deprivation and migration in an agricultural setting of Nepal. *Population and Environment*, 25(5), pp. 475–499.

Bhandari, P. and Ghimire, D. (2013), Rural agricultural change and fertility transition in Nepal. *Rural Sociology*,78(2), pp. 229–252.

Bhandari, P. B. (2006), Technology use in agriculture and occupational mobility of farm households in Nepal: Demographic and socioeconomic correlates (Unpublished doctoral dissertation). The Pennsylvania State University, Department of Agricultural Economics and Rural Sociology, University Park.

Bhandari, P. B. (2013), Rural livelihood change? Household capital, community resources and livelihood transition. *Rural Studies*, 32, pp. 126–136.

Biswas, M. R. (1994), Agriculture and Environment: A Review, 1972–1992.*Ambio,* 23 (3), pp. 192–197.

Boserup, E. (1965),*The Conditions of Agricultural Growth: The Economics of Agrarian Change under Population Pressure.* Chicago and George Allen and Unwin Ltd: Aldine Publishing Company.

Boserup, E. (1990), *Economic and Demographic Relationships in Development: Essays Selected and Introduced by T. Paul Schultz.* Baltimore, London: The Johns Hopkins University Press.

Byrd, B. F. (1994), Public and private, domestic and corporate: The emergence of the Southwest Asian village. American Antiquity 59: 639–666.

Central Bureau of Statistics. (1993), *Population Census – 1991*. National Planning Commission Secretariat, Central Bureau of Statistics, Kathmandu, Nepal.

Central Bureau of Statistics.(2011), *National Population and Housing Census 2011*.National Report. National Planning Commission Secretariat, Government of Nepal, Kathmandu, Nepal.

Curry, John, and Diana Tempelman. (2006), "Improving the Use of Gender and Population Factors in Agricultural Statistics: A Review of FAO's Support to Member Countries in Gender Statistics." Food and Agriculture Organization (FAO), Rome.

Deere, Carmen Diana, and Magdalena Leon. (2003), "The Gender Asset Gap: Land in Latin America." World Development 31: 925–47.

Doss, Cheryl. (2005), "The Effects of Intrahousehold Property Ownership on Expenditure Patterns in Ghana." Journal of African Economies 15: 149–80.

Ellis, F. (1993), *Peasant Economics: Farm Households and Agrarian Development*. Cambridge University Press, Cambridge.

ESCAP/FAO/UNIDO. (1997), *Supply, Marketing, Distribution and Use of Fertilizer in Nepal*. ESCAP/FAO/UNIDO, Fertilizer Advisory, Development and Information Network for Asia and the Pacific, United Nations. Thailand, Bangkok.

FAO, WFP, IFAD. (2013), *The State of Food Insecurity in the World 2013: The Multiple Dimensions of Food Security*. FAO, Rome.

Feder, G. and O'Mara, G. T.(1981), Farm size and the diffusion of green revolution technology. *Economic Development and Cultural Change*,30. pp. 59–76.

Feder, G. and Umali, D. L. (1993), Adoption of agricultural innovations in developing countries: A survey. *Technological Forecasting and Social Change*,43, pp. 215–239.

Food and Agriculture Organization (FAO). (2007), "Progress Report on the Implementation of the FAO Gender and Development Plan of Action." FAO, Rome. Gender and Rural Development Thematic Group

Gartaula, H. N., Anke, N. and Visser, L. (2010), Feminization of agriculture as an effect of male outmigration: Unexpected outcomes from Jhapa District, Eastern Nepal. *The International Journal of Interdisciplinary Social Sciences*, 5(2), pp. 565–577.

Gebeyehu, Y. (1995), Population pressure, agricultural land fragmentation and land use: A case study of Dale and Shashemene Weredas, Southern

Ethiopia. Pp. 43–64in D. Aredo and M. Demeke, Editors, Ethiopian agriculture: *Problems of Transformation.* Proceedings of the Fourth Annual Conference on the Ethiopean Economy. Addis Ababa.

GENRD (2006), "FY06 Gender Portfolio Review." World Bank, Washington, DC.

Godoy, R., Franks, J. R. and Claudio, M. A. (1998), Adoption of modern agricultural technologies by lowland indigenous groups in Bolivia: The role of households, villages, ethnicity, and markets. *Human Ecology,* 26(3), pp. 351–369.

Harlan, J. R. (1971), Agricultural origins: Centers and noncenters. Science 174: 468–474.

Harris, D. R. (1996), Introduction. In Harris, D. R. (ed.), The Origins and Spread of Agriculture and Pastoralism in Eurasia, University College of London Press, London, pp. 1–11.

Hazell, P. B. R. and Ramaswamy, C. (1991), *The Green Revolution Reconsidered: The Impact of High-yielding Rice Varieties in South India.* Baltimore and London: The Johns Hopkins University Press.

Higham, C. (1995), The Transition to Rice Cultivation in Southeast Asia. In Price, D. T., and Gebauer, A. B. (eds.), Last Hunters First Farmers, School of American Research Press, Santa Fe, pp. 127–156.

Higham, C., and Lu, T. L. (1998), The origins and dispersal of rice cultivation, Antiquity 72 (278): 867–877.

Hosmer, D. W. and Lemeshow, S. (2000), *Applied Logistic Regression.* A Wiley-Interscience Publication, John Wiley & Sons, Inc., New York, Chichester, Weinheim, Brisbane, Singapore, Toronto.

Jacoby, E. H. (1972), Effects of 'green revolution' in South and South-East Asia. *Modern Asian Studies,* 6(1), pp. 63–69.

Jayaraman, P. P., Palmer, D., Zaslavsky, A., & Georgakopoulos, D. (2015a), Do-it-yourself digital agriculture applications with semantically enhanced IoT platform. In 2015 IEEE tenth international conference on intelligent sensors, sensor networks and information processing (ISSNIP) (pp. 1e6). https://doi.org/10.1109/ISSNIP.2015.7106951.

Jayaraman, P., Palmer, D., Zaslavsky, A., & Salehi, A. (2015b), Addressing information processing needs of digital agriculture with OpenIoT platform. And open-source Retrieved from http://link.springer.com/chapter/10.1007/978-3-319-16546-2_11.

Joshi, B.H. (1998), *Fertilizer Policy and Institutions.* Nepal Agriculture Perspective Plan, Implementation Status and Future Actions. John Mellor Associates, Inc. and Agricultural Projects Services Centre, Kathmandu, Nepal.

Joshi, K. D., Conroy, C. and Witcombe, J. R., (2012), Agriculture, seed, and innovation in Nepal: Industry and policy issues for the future. International Food Policy Research Institute (IFPRI). Project Paper, December 2012. Washington, D.C.

Kumar, S. K., and Hotchkiss, D. (1988), *Consequences of Deforestation for Women's Time Allocation, Agricultural Production, and Nutrition in Hill Areas of Nepal.* Research Report No. 69.International Food Policy Research Institute, Washington, DC.

Leclerc, Gregoire and Hall, C. A. S. (Editors).(2007), *Making World Development Work: Scientific alternatives to Neoclassical Economic Theory.* New Mexico: University of New Mexico Press.

Maharjan, A., Bauer, S. and Knerr, B., (2012), International migration, remittances, and subsistence farming: Evidence from Nepal. *International Migration*, doi:10.1111/j.1468-2435.2012.00767

Mamdani, M. (1972), *The Myth of Population Control: Family, Caste, and Class in an Indian Village.* Monthly Review Press, New York and London.

McGuire, R. H., and Schiffer, M. B. (1983), A Theory of Architectural Design. Journal of Anthropological Archaeology 2: 277–303.

Mellor, J. W. (1976), *The New Economics of Growth.* Ithaca, N.Y. :Cornell University Press.

Ministry of Agriculture and Cooperatives, (2003), Nepal fertilizer use study. A study funded by the UK Department for International Development (DFID), Kathmandu, Nepal.

Ministry of Agriculture and Cooperatives. (2012), *Agriculture Atlas of Nepal.* Ministry of Agriculture and Cooperative, His Majesty's Government of Nepal, Kathmandu.

National Planning Commission. (2003), *The Tenth Plan: Poverty Reduction Strategy Paper 2002-2007.*His Majesty's Government of Nepal, National Planning Commission, Kathmandu, Nepal.

NESAC. (1998), N*epal Human Development Report 1998.*Nepal South Asia Centre, Kathmandu, Nepal.

Paige, J. M. (1975), *Agrarian Revolution: Social Movements and Export Agriculture in the Underdeveloped World.* New York: Free Press.

Pariyar, M., Shrestha, K. B. and Dhakal, N. (2001), Baseline study on agricultural mechanization needs in Nepal. Facilitation Unit, Rice-Wheat Consortium for the Indo-Gangetic Plains, National Agricultural Science Centre (NASC) Complex, DPS Marg, Pusa Campus, New Delhi 110 012, India.

Pimentel, D. and Pimentel, M. (1991), Comment: Adverse environmental consequences of the green revolution. Pp. 239–332 in K. Davis and M. Bernstam, Editors, *Resources, Environment and Population.* The Population Council of Oxford University Press, New York.

Rani, S. and Malaviya, A. (1992), Farm mechanization and women in paddy cultivation. Pp. 261–67 in R. K. Punia, Editor, *Women in Agriculture: Their Status and Role.* Northern Book Center, New Delhi.

Rauniyar, G. P. and Goode, F. M. (1992), *Technology Adoption on Small Farms. World Development*, 20(2), pp. 275–282.

Rogers, E. M. (1960), *Social Change in Rural Society: A Textbook in Rural Sociology.* New York :Appleton-Century-Crofts, Inc.

Sachs, C. (1996), *Gendered Fields: Rural Women, Agriculture, and Environment.* Colorado: Westview Press, Inc.

Salokhe, V. M. and Ramalingham, N.(1998), Agricultural Mechanization in the South and South-East Asia. Presented at the Plenary Session of the International Conference of the Philippines Society of Agricultural Engineers, April 21–24, Los Banos, Philippines.

Scott, J. C. (1977), *The Moral Economy of the Peasant: Rebellion and Subsistence in Southeast Asia.* New Haven: Yale University Press.

Self, S. (2008), Developing countries and fertility: Role of agricultural technology. *International Journal of Development Studies,* 7 (1). pp. 62–75.

Shivakoti, G. P., Axinn, W. G., Bhandari, P., and Chhetri, N. (1999), The impact of community context on land use in an agricultural society. *Population and Environment,* 20, pp. 191–213.

Shrestha, N. R. (1990), *Landlessness and Migration in Nepal.* Westview Press, Boulder, CO.

Skocpol, T. (1982), Review: What makes peasants revolutionary? *Comparative Politics*, 14 (3), pp. 351–375.

Tamrakar, A. M. (1998), Fertilizer policy in Nepal-1998 under changed perspective. Draft Final Report. Consultant' Report, Fertilizer Advisory, Development and Information Network for Asia and the Pacific, United Nations. Thailand, Bangkok.

Thapa, G. (2009), Smallholder Farming in Transforming Economies of Asia and the Pacific: Challenges and Opportunities. A Discussion Paper prepared for the side event organized during the Thirty third session of IFAD's Governing Council, 18 February 2009. (Retrieved on June 14, 2016 https://www.ifad.org/documents/10180/a194177c-54b7-43d0-a82e-9bad6 4b76e82).

The Kathmandu Post. (2014), Agriculture Development Strategy. The Kathmandu Post dated June 20, 2014, retrieved on August 15, 2014 (Available online at http://www.ekantipur.com/the-kathmandu-post/2013/06/20/money/agriculture-development-strategy/250264.html).

UNDP.(2011), *Human Development Report 2011 Sustainability and Equity: A Better Future for All.* UNDP, Washington, D.C.

World Bank, FAO, and IFAD. (2009), Gender in Agriculture Sourcebook. World Bank: Washington, DC. http://www.fao.org/3/aj288e/aj288e.pdf

World Bank. (1998), *Poverty in Nepal at the Turn of the Twenty-first Century.* Report No. 18639-NEP, Poverty Reduction and Economic Management Unit, South Asia Region.

World Bank. (2008), *World Development Report 2008: Agriculture for Development.* World Bank Publications, Washington, D.C.

World Bank. (2013), *World Development Report 2013: Jobs.* World Bank Publications, Washington, D.C.

11

Income Inequality and Poverty Status of Households Around National Parks in Nigeria

Daniel Etim Jacob[1,*] **and Imaobong Ufot Nelson**[2]

[1]Forestry and Wildlife Department, University of Uyo, Nigeria
[2]Biodiversity Preservation Center, Uyo, Nigeria
E-mail: danieljacob@uniuyo.edu.ng
*Correspondence Author

Abstract

Support zone communities of Nigeria National Parks had been benefiting from the various goods and services provided by the forest, before the change in forest status coupled with its new resource governance approach that restricts their access to the park resources, thus impacting negatively on their livelihood. This study therefore evaluates income inequality, poverty incidence, and its determinants among rural households in the area. Data for the study was obtained using household data collected through questionnaire administered randomly among 1009 households across the three national parks. The data obtained were analyzed using descriptive statistics, logit regression, Foster–Greer–Thorbecke (FGT) poverty measure, and Gini coefficient. The result obtained indicates that majority of the household heads were male (92.57%), between the age class of 21 and 40 years (44.90%), had non-formal education (38.16%), were farmers (65.21%), owned land (95.44%), with a household size of 1 to 5 (36.67%), and an annual income range of ₦401,000 to ₦600,000 (24.58%). The total sampled households in the three national parks had a 0.358 level of income inequality with a mean per-capita income of ₦304,323.39. Poverty indices for the study area showed that poverty line was ₦202,882.30, incidence 57.58%, depth 163.20%, and severity 22.59%. Also, education level of household head, household size and age of household head where significant factors ($p<0.05$) of poverty among rural households, while income inequality was location specific. The study recommends more investment and development be focused on improving, localising, and building upon on the existing livelihood strategies on ground with more emphasises on improving vocational trainings and agricultural activities for the people as it would serve as a catalyst in alleviating their poverty and also enhancing their overall wellbeing.

Keywords: Poverty Incidence, National Parks, Support zone communities, per-capita income, Nigeria.

11.1 Introduction

In most developing countries (Nigeria inclusive), the governance strategy applied in protected area governance is marred with conflicts (Jacob, 2008; Andrew-Essien and Bisong, 2009; Jacob and Ogogo, 2011; Jacob et al., 2013) and debates (Mosetlhi, 2012; Jacob et al., 2018). The frequent arrest and punishment of trespassers into park territories for livelihood activities have come under serious criticism. Murphree (1991) argues that conservation cannot be attainable when the local people are paying the costs. The parks restrict local people's access to their livelihood assets (e.g., natural, physical, financial, human and social capitals). These restrictions undermine the capacity of buffer or support zone communities to combine their different capitals in the pursuit of different livelihood strategies (Enuoh, 2014). This scenario culminates in livelihood outcomes of poverty in the buffer zone communities as the parks undermine households' capacity for asset combination in the pursuit of sustainable livelihood strategies and outcomes.

Given the growing consensus that protected areas should be part of the solution to poor people's problems, and not create new ones (Abbot et al., 2001), integrated conservation and development projects (ICDPs) were introduced in the 1980s, with the assumption that development activities in the support zone communities by the park will in some way affect the attitudes and behavior of the people, so that they become more supportive of conservation measures that regulate resource use in the park (Abbot et al., 2001; Jacob, 2017; Bhandari, 2014; Bhandari, 2018; Bhandari, 2020). However, after about a decade of ICDP activities in the tropics, project reviewers and researchers alleged that the strategy was a mix of success and failure

(Abbot et al., 2001; Enuoh, 2014). Oates (1999) and Terborgh (1999) out rightly reported ICDP projects as a failure because they did not enhance the actualization of biodiversity conservation objectives. These authors called for a return to authoritarian protection or fortress conservation, as the only strategy of effective biodiversity conservation in the tropics (Enuoh, 2014). Their prescriptions ignore the problem of colonial nationalization of the forest territories of local communities, property rights struggles, and ignored community demands for the payment of compensation by parks.

However, while a growing number of conservation stakeholders were skeptical of ICDPs, the United Nation (UN) Millennium Ecosystems Assessment Report and the UN Millennium Development Goals (MDGs) placed emphasis on global sustainable development strategies that tackled biodiversity conservation and poverty alleviation simultaneously (Enuoh, 2014). MDG 1 sought to eradicate poverty globally, while MDG 7 sought to ensure environmental sustainability. On grounds of the above alleged failure of ICDPs, Agrawal and Redford (2006) argued that biodiversity conservation and poverty alleviation cannot be achieved together. Sanderson (2005) stressed that poverty alleviation and biodiversity conservation can only take place in the most difficult settings, places of extreme ecological vulnerability, very low population densities, and no state presence.

Similarly, Barrett et al. (2005) argued that the common assumption that poverty reduction and environmental sustainability goals are inherently complementary does not appear to stand up well to empirical scrutiny. However, the reality that surrounds tropical biodiversity conservation is that wherever parks and protected areas exist, abject poverty (with people living on less than one dollar per day) also exists among the surrounding communities. Wildlife raiding of agricultural crops is more intense in park buffer zone communities than in the nonbuffer zone communities (Jacob et al., 2013; Enuoh, 2014; Jacob et al., 2015). This implies that biodiversity conservation in parks and protected areas impoverishes or impact negatively on rural agricultural production, return on agricultural investments, food scarcity, and livelihood outcomes in buffer zone communities.

Poverty is multi-dimensional in nature and the relationship between poverty and biodiversity conservation in tropical parks and protected areas appear unclear and less understood by some conservationists, conservation organizations, researchers, and policy makers, hence the calls for more research that will establish the relationship between biodiversity conservation and rural poverty (Wilkie et al., 2006). In response to the call for more research to be conducted to establish the links between biodiversity conservation and poverty in order to inform conservation interventions, the paper seeks to ascertain income inequality and determinant of poverty among support zone communities of national parks in Nigeria.

11.2 Materials and Methods

11.2.1 Study Area

Nigeria is located in the western part of Africa between latitudes 4°16'N and 13°52'N; and between longitudes 24°9'E and 14°37'E (Figure 11.1). It occupies a total land area of 923,768 km^2 with a 2014 population estimate of about 167,912,561 million people (82,098,000 females and 85,814,560 males) with a population growth of 3.2% (Oyedele, 2014). By virtue of its geographical extent, Nigeria spans different climatic and ecological zones. The variable climatic conditions and physical features have consequently endowed Nigeria with a very rich biodiversity.

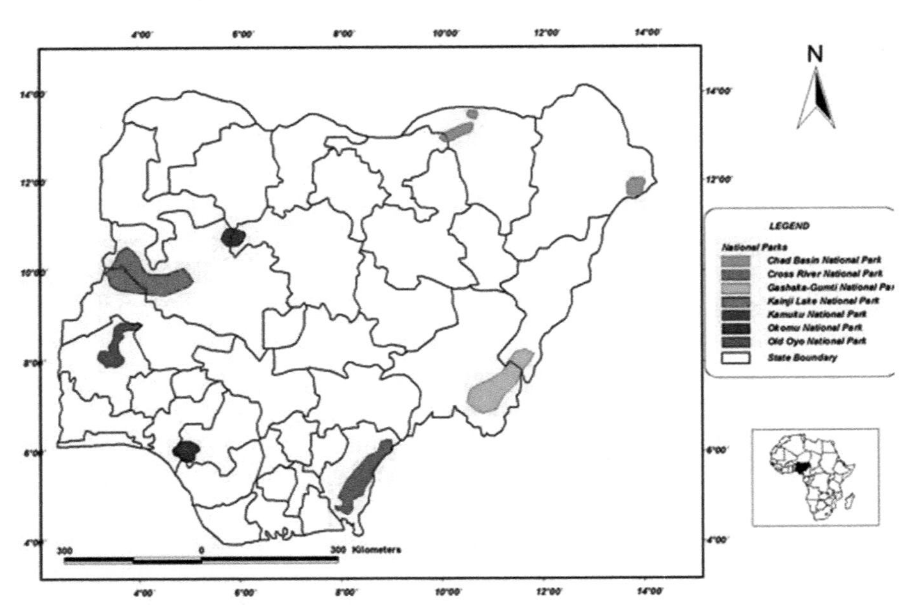

Figure 11.1 Map of Nigeria showing location of national parks.
Source: Ogunjinmi et al. (2012).

Table 11.1 Nigerian National Parks, their ecological zones and coverage in 1995 and 2007.

National Park	State of Location	Ecological Zone	Area (km^2) 1995[a]	Area (km^2) 2007[b]
Chad Basin	Borno	Northern Guinea/ Sudan Sahel savanna	2258	2429.43
Kainji Lake	Niger, Kwara	Northern Guinea/ Sudan Sahel savanna	5382	3710.37
Kamuku	Kaduna	Northern Guinea/ Sudan Sahel savanna	1121	695.36
Gashaka-Gumti	Adamawa	Northern Guinea/ Sudan Sahel savanna	6731	6989.15
Old Oyo	Oyo	Southern Guinea	2512	1665.14
Cross River	Cross River	High Forest	4000	2368.27
Okomu	Edo	High Forest	181	67.59

Source: [a] FORMECU, 1995; [b] Mohammed et al., 2013.

11.2.2 Site Selection

To ensure effective representation and selection of the National Parks in the country, the National Parks were stratified into ecological zones and from each zone, the National Park with the smallest area based on Mohammed et al. (2013) area coverage analysis of 2007 (Table 11.1) were selected for the study. The reason for the selection of the National Park with the smallest area in each zone is based on the observation of Wells et al. (1992) that parks with relatively small areas have a more probability of being degraded or destroyed than those with large area coverage due to poverty. From the above mention criteria for selection, the selected National Parks were Kamuku (Northern guinea/Sudan Sahel savanna), Old Oyo (Southern Guinea) and Okomu National Park (High forest).

11.2.3 Sampling Design and Data Collection

The target population of the study consists of people who were located in villages within 3 km from the boundary of each of the National Parks (Table 11.2). Thirty percent (30%) of the villages from each park were purposively selected based on proximity to access road and from them, 20% of the household in each village were randomly selected to ensure effective comparison, variation, and representativeness of the households in the geographical sub-units (villages). This is in accordance with the observation of Angelsen et al. (2011), Udeagha (2015) and Jacob et al. (2016).

Table 11.2 Sampling unit selection design.

National Park	Village Sampling Frame	30% Sampling Size (Villages)	Mean Household Sampling Frame/ Village	20% Household Sampling Size/ Village	Total Household Sample Per Park	Total questionnaires returned per park
Kamuku	27	9	271*	54	486	463
Old Oyo	23	7	282*	56	392	369
Okomu	12	4	248*	50	200	177
Total	55	18	801	160	1078	1009

*2006 household population census.

Data collection exercise took place between August 2015 and June 2016, involving household questionnaire surveys, informant interviews and on-site data collection and inspection. A semi-structured questionnaire was randomly administered to the household heads or their representatives to gather factual data and perceptions on the study variables. Both closed and open-ended questions were used. The questionnaire was designed and used in accordance with guidelines for questionnaire design in measuring livelihood and environmental dependence (Rubin and Babbie, 2008; Angelsen et al., 2011).

11.3 Data Collection

The study utilised both secondary and primary data. Secondary data collection sources included: official documents as well as relevant literature and research reports specific to the area of study and phenomenon under investigation. Structured questionnaire were used in gathering of primary data using 20% randomly selected households from the household in the study area in other to ensure variation and representativeness (Udeagha, 2015; Heubach, 2012; Angelsen et al., 2011; Rubin and Babbie, 2008; Babbie, 2005) and was also supplemented by other ethnographic and participatory methods of data collection such as; field observation, visual photography, transact walk, In-Depth Interviews (IDIs), and Focus Group Discussions (FGDs) with relevant community stakeholders who have ample knowledge of the subject matter. This was done with the assistance of community leaders and their council.

11.4 Analytical Techniques

i. Poverty indices

The Foster–Greer–Thorbecke (FGT) model was used to measure poverty status among the rural households in the study area using Equation (11.1) as described by Jacob et al. (2016).

The FGT poverty index is:

$$P_{\alpha}(y,z) = \frac{1}{2}\sum_{i=1}^{q}\left[\frac{z-y_i}{z}\right]^{\alpha} \tag{11.1}$$

Where:

P = Foster, Greer, and Thorbecke index
z = the poverty line for the household
y_i = household income
n = total number of households in population
q = the number of poor households
$[\frac{z-y_i}{z}]$ = proportion of shortfall in income below the poverty line
α = non-negative poverty aversion parameter (0, 1 or 2).

The analysis of the poverty status of the households were decomposed into the three indicators i.e., prevalence of poverty (P_0), poverty depth (P_1), and severity of poverty (P_2).

A poverty line was constructed to categorize the respondents into poor and non-poor groups using the two-third mean per-capita income as the benchmark as described by Ruben and van den Berg (2001), Yunez-Naude and Taylor (2001), Adewunmi et al. (2011) and Fonta and Ayuk (2013). Households whose mean per-capita income fell below the poverty line where regarded as being poor, while those whose per-capita income where above the benchmark were regarded as non-poor.

$$\text{Per-capita income (PCI)} = \text{Income/Household size} \tag{11.2}$$

$$\text{Total Per-capita income (TPCI)} = \sum \text{PCI} \tag{11.3}$$

$$\text{Mean TPCI (MTPCI)} = \text{TPCI/Total number of households} \tag{11.4}$$

$$\text{Poverty line (PL)} = 2/3\times\ \text{MTPCI} \tag{11.5}$$

The incidence of poverty or headcount index was calculated using Equation (11.6) below where $\alpha = 0$ in FGT. This measures the proportion of the population that is poor or fall below the poverty line.

$$P_o = \left[\frac{1}{n}\right] q = \left[\frac{q}{n}\right] \tag{11.6}$$

Where: n = total number of households in population, q = the number of poor households' proportion of shortfall in income below the poverty line.

Poverty depth or poverty gap index is the measure of the extent to which individuals fall below the poverty line as a proportion of the poverty line. This was calculated using the formula where $\alpha = 1$, hence the expression becomes;

$$P_1 = \frac{1}{2} \sum_{i=1}^{q} \left[\frac{z - y_i}{z}\right] \tag{11.7}$$

The Equation (11.7) reflects both incidence and depth of poverty or the proportion of the poverty line that the average poor will require to attain to the poverty line.

Poverty severity index is the measure of the squares of the poverty gap relative to the poverty line. This was calculated using the formula where α = 2, hence the expression becomes;

$$P_2 = \frac{1}{n} \sum_{i=1}^{q} \left[\frac{z - y_i}{z}\right]^2 \tag{11.8}$$

If α = 2, the index measures the severity of poverty which is the mean of square proportion of the poverty gap. When multiplied by 100, it gives the percentage by which a poor household's per capita expenditure should increase to push them out of poverty.

ii. Estimation of total income inequality

The Gini coefficient for total income is used to compute the income inequality among the study area as recommended by Cheong (1999). The Gini coefficient for total income was calculated as

$$\frac{\sum_{i=1}^{n} \sum_{i=1}^{n} |TIi_t - TI_t}{2\pi^2 \mu} \tag{11.9}$$

Where; μ = Mean household income, n = Total population, TI_i= Share on individual i of total household income, and TI_j = share of individual j of total household income.

iii. Determinant of poverty status

To identify determinants of poverty status of households among the parks support zone households, a logit regression was carried out. The model was chosen because of the dichotomous dependent variables and because the technique has no restrictive distribution assumptions (Jacob et al., 2016).

The logistic (logit) probability function is given as

$$P_i = 1/1 + e - zi = f(Zi) \tag{11.10}$$

Where Pi is the probability that a household i (i = 1, 2 ... n) will be poor. Index Zi is a random variable which predicts the probability of a household being poor or non-poor. The probability Pi in Equation (11.9) is further transformed to give Equation (11.10).

$$Pi = ezi/1 + ezi \tag{11.11}$$

Therefore, for the ith observation, a household will be

$$\text{Zi} = \text{In Pi}/1 - \text{Pi} = \beta\text{o} + \Sigma\beta\text{oX} \tag{11.12}$$

Therefore, ln (P/1–P) = 1, if the household is poor while ln (P/1-P) = 0, if otherwise, i.e., non-poor. Implicitly, the model is empirically estimated as

$$\text{Y} = \beta\text{o} + \beta\text{1X1} + \beta\text{2X2} + \beta\text{3X3} + \beta\text{4X4}\ldots\ldots\ldots\ldots\ldots\beta\text{9X9} \tag{11.13}$$

where:

Y = Poverty status of farm households sampled (1= if poor, 0 otherwise)
X1 = Age of household head (years), X2 = Land (hectares), X3 =Highest educational level (years of formal schooling), X4 = Household size (number), X5 = Gender (male = 1, female = 0), X6 = Occupation (Off-farm = 1, farm = 0), X7 = No of adult in household, X8 = Cattle equivalent (number), ξ = errors term

11.5 Results and Discussions

11.5.1 Demographic Characteristics of Sampled Households

The result in Table 11.2 indicates the socio-economic characteristics of the respondents (N = 1009). Households headed by male (311.33±158.94, N = 934) were significantly ($t = 3.23$, $p<0.05$, $df = 2$) different from households headed by the female (14.67±25.40, N = 75). This is an indication that

Table 11.3 Definition of dependent and independent variables included in the econometric model and expected signs (a *priori* Statement).

Variables	Type of Variable	Expected Signs	
Age	Continuous	+	−
Land	Continuous	+	−
Education	Continuous	+	−
Household size	Continuous	+	+
Gender (Female)	Binary/dummy	+	+
Occupation (off-farm)	Binary/dummy	+	−
Adult equivalent	Continuous	+	−
Cattle equivalent	Continuous	+	−

majority (92.57%) of the households had an elderly male to dictate the affairs in each family. This is in accordance with the observation of Olorunsanya and Omotesho (2011), Olawuyi and Adetunji (2013), and Jacob (2017) that majority of the rural households in Nigeria are headed by the male. The male dominance in the study area still subscribes to the patriarchal view that men provide for the family and have the power and authority to control the general affairs of the household unit, including decision-making (Silver et al., 2015).

Table 11.2 also indicates that there existed significant ($F = 3.53$, $p<0.10$) variation among the age classes. Majority (44.90%, M = 151.00±79.30, N = 453) of the sampled respondents were within the age class of 21 to 40 years, followed by those in the age class of 41 to 60 years (27.45%, M = 92.33±28.02, N = 277) and those greater than 60 years (15.56%, M = 52.33±25, N = 157), while those belonging to less than 20 years were the least (12.09%, M = 40.67±26.54, N = 122). The result implies that majority of the respondents are in their prime, hence, they are in their economically active and productive age (Jacob et al., 2013; Jacob et al., 2015; Silver et al., 2015; Nelson et al., 2017; Nelson et al., 2018).

Level of education did not vary significantly ($F = 1.54$, $p>0.05$) among the households. However, majority of the household heads in the study area had non-formal education (38.16%, M = 128.33±108.25, N = 385), followed by secondary (27.65%, M = 93±26.89, 279), primary (24.28%, M = 81.67±12.66, N = 245) and the least was tertiary education (9.91%, M = 33.33±8.74, N = 33.33±8.74). In general, it could be said that more than 61.84% of the household heads in the study area were literate and had acquired various forms of formal education with an average number of years spent in school being 6.696 years. This schooling years falls under post primary level of education. This schooling year is higher than 4.89 years reported for most of rural households in Uganda (Uganda

Bureau of Statistics, 2002; Balikoowa, 2008). The high literary rate in the study area agrees with Olawuyi and Adetunji (2013), Jacob et al. (2013), Silver et al. (2015) and Oluwatusin and Sekumade (2016) that majority of the households in the rural areas in the country have had formal education, which according to Jacob et al. (2013) has the potential for making up some of the deficiency in non-formal education and positively influencing the adoption of innovation. With their level of education, the respondents possess the ability to participate effectively in resource management decisions of the park to ensure sustainable conservation of the park resources while also meeting the needs of their households (Emelue et al., 2014).

Occupationally, there was significant ($F = 5.70$, $p<0.01$) difference between the households in the study area. Farming was their main occupation in the study area (65.20%, M = 219.33±142.59, N = 658). This is followed by trading (21.07%, M = 70.67±49.10, N = 212) and studentship (3.17%, M = 10.67±3.51, N = 32), while Nurse/Traditional birth attendant (0.39%, M = 1.33±1.52, N = 4) was the least occupation practiced by the sampled respondents. The high rate of farming household in the study area is in accordance with the observations of Chianu et al. (2004), Tumusiime (2006), Balikoowa (2008) and Olayide et al. (2009) that agriculture is the dominant livelihood activities of rural communities.

Ownership of land by households were significantly ($t = 2.69$, $p<0.10$, df =2) different from those households who did not own land (= 321.00±171.13, N = 963, vs. = 15.33±26.59, N = 46). This implies that majority (95.44%) of the households had possession of land in the study area. This agrees with the observation of Balikoowa (2008) that land possession is usually location specific, hence majority of the people living in the rural area are more likely to own land than those in the urban areas.

Number of land owned by a household also varied significantly ($F = 7.41$, $p<0.05$) in the study area. Majority of the respondents (78.29%, M = 251.33±136.88, N = 732) owned between 1 and 2 parcels of land, followed by those with 3–4 (4.56%, M = 60.00±31.48, N = 175), while those who owned 5 parcels of land and above (2.99%, M = 9.67±3.06, N = 28) were the least. The possession of more than one parcel of land in the study area indicates land fragmentation in the study area. This could be attributed to the practice of inheritance whereby the father apportions land among all his male children (Balikoowa, 2008). Where the family size is large, each male child is bound to inherit just a small portion of the land and may have to purchase more land to add to his inheritance so as to increase his own land holding.

The size of a household in the study area did not significantly ($F = 2.95$, $p>0.05$) differ among the parks. However, majority of the households in the study area had a family size of less than five members (36.67%, M = 123.33±42.00, N = 370), followed by those with 6 to 10 members (31.42%, M = 105.67±38.50, N = 317) and the households with more than 15 members were the least abundant (12.48%, M = 42.00±30.51, N = 126). The result agrees with the observation of Olorunsanya and Omotesho (2011), Javed and Asif (2011) and Oluwatusin and Sekumade (2016) who reported that rural areas are characterized by large family sizes ranging between 1 to 20 members per household. This could probably be as a result of the polygamous nature of most male-headed households in the study area (Olorunsanya and Omotesho, 2011).

Also, among the household, there was no significant ($F = 0.94$, $p>0.05$) difference between the various income classes in the study area. The distribution of annual income in the study area indicates that most (24.58%, M = 82.67±24.84, N = 248) of the households earn between ₦401,000.00 and ₦600,000.00 while those who earn between ₦801,000.00 and ₦1,000,000.00 were the least (10.80%, M = 36.33±11.59, N = 109). However, only 15.07% (M = 50.67±38.53, N = 152) of the households in the study area were able to earn an income of more than a million naira (>₦1,000,000.00). The variations in income in the study area could be attributed to the different livelihood activities of the households. The contribution of each source of income to total household income in the sample communities showed significant difference ($p<0.01$) among the parks although agriculture provided the bulk of household income with a mean of more than 63% across the three National Parks. This has been found in other studies on rural households (Chianu et al., 2004; Tumusiime, 2006; Manyong et al., 2006; Balikoowa, 2008; Badmus et al., 2009; Olayide et al., 2009; Akpan et al., 2015; Akpan and Udoh, 2016; Akpan et al., 2016).

11.5.2 Income Inequality Among the Study Area

Gini coefficient gives the overall picture of the level of inequality and well-being of the people in a community. The Gini coefficients of the sampled households in the three National Parks are presented in Figure 11.2. The result showed that households in Okomu National Park had the highest (0.3685) income inequality followed by households in Kamuku National Park (0.3604) and the least was households in Old Oyo National Park (0.3482). In general, the income inequality in the study area was 0.3583.

Table 11.4 Demographic characteristics of sampled respondents.

S/N	Variables		Total F	%	Mean±SD	Significant Level
1.	Gender	Male-headed	934	92.57	311.33±158.94a	3.23**
		Female-headed	75	7.43	14.67±25.40b	
		Total	1009	100		
2.	Age (years)	≤20	122	12.09	40.67±26.54a	3.53*
		21–40	453	44.9	151.00±79.30b	
		41–60	277	27.45	92.33±28.02a	
		>60	157	15.56	52.33±25.32ab	
		Total	1009	100		
3.	Educational Status	Non-formal	385	38.16	128.33±108.25	1.45ns
		Primary	245	24.28	81.67±12.66	
		Secondary	279	27.65	93±26.89	
		Tertiary	100	9.91	33.33±8.74	
		Total	1009	100		
4.	Main occupation	Farming	658	65.21	219.33±142.59a	5.70**
		Trading	212	21.07	70.67±49.10b	
		Tailor	13	1.27	4.33±1.53b	
		Civil servant	8	0.78	2.67±3.06b	
		Teaching	7	0.68	2.33±1.53b	
		Student	32	3.17	10.67±3.51b	
		Nurse/Birth attendant	4	0.39	1.33±1.52b	
		Artisan	75	7.43	25.00±33.45b	
		Total	1009	100		
5.	Land ownership	Yes	963	95.44	321.00±171.13a	2.69*
		No	46	4.56	15.33±26.59	
		Total	1009	100		
6.	Number of parcels of land owned	≤2	732	78.29	251.33±136.88a	7.41**
		3–4	175	18.72	60.00±31.48b	
		5 and above	28	2.99	9.67±3.06b	
		Total	935	100		

(Continued)

Table 11.4 (*Continued*)

S/N	Variables		Total F	%	Mean±SD	Significant Level
7.	Household size	≤5	370	36.67	123.33±42.00	2.95ns
		6–10	317	31.42	105.67±38.50	
		11–15	196	19.43	65.33±37.63	
		>15	126	12.48	42.00±30.51	
		Total	1009	100		
8.	Annual income of house-hold head (₦0,000)	≤200	154	15.26	51.33±30.66	0.94ns
		201–400	166	16.45	55.33±34.00	
		401–600	248	24.58	82.67±24.84	
		601–800	180	17.84	60.00±10.54	
		801–1000	109	10.8	36.33±11.59	
		>1000	152	15.07	50.67±38.53	
		Total	1009	100		

SD = Standard deviation, ns = Not significant, ** = Significant at 5% (p>0.05), * = Significant at 1% (p>0.01)
Mean with similar alphabet means they are not significantly different.

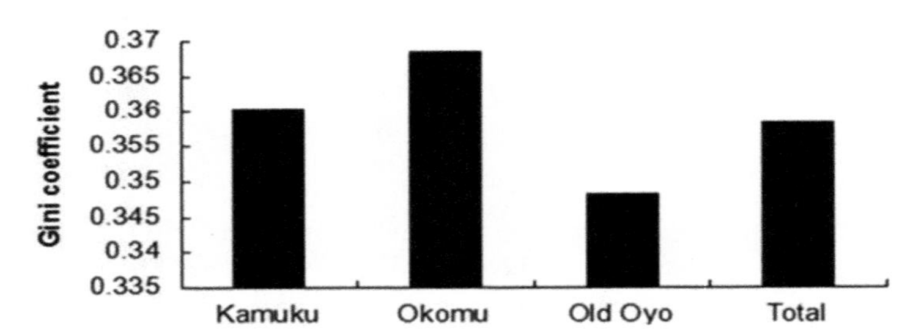

Figure 11.2 Income inequality in the study area.

The use of Gini coefficient in the study to evaluate income inequality among the households helps to gives the overall picture of the level of inequality and wellbeing of the people in the community (Cheong, 1999; Balikoowa, 2008; Jacob et al., 2016). The Gini coefficient of the sampled households in Okomu (0.369) and Kamuku (0.360) was lower than 0.411 and 0.384 recorded for Edo and Kaduna State respectively in 2012. However, the 0.348 Gini coefficient of households in Old Oyo was higher than 0.328 also recorded for households in Oyo State in 2012. This result implies that income inequality is geographical and location specific, thus even within the same State or region, there is likely to exist variations in income distribution among the households.

This scenario depicts that policies toward poverty alleviation should be highly localized for it to achieve better results that would serve as a pathway out of poverty for the rural poor households (Jacob et al., 2016). However, the introduction of alien livelihood strategies/activities without localizing and improving on the already existing once should be discourage as it would take time be adapted by the households therefore making them more vulnerable to poverty (Adewunmi et al., 2011; Igbalajobi et al., 2013). In general, the total sampled households in the three National Parks had a 0.358 level of income inequality. This is a reduction from the 0.506 reported for the country in 1996/97 (World Bank, 2008), 0.447 in 2011 (NBS, 2012), and 0.441 (Ohwotemu, 2012). The result (0.358) is also lower than the level of income inequality among rural communities in Nigeria (NBS, 2011; Jacob et al., 2016). The significant reduction in inequality among rural households in the study area could be attributed to location and climate which could have large effects on income levels and income distribution, through their effects on transport costs, disease burdens, and agricultural productivity among others. It could also the attributed to the effort of government to reduce poverty in the country. The reduction in income inequality in rural area is a laudable because inequality is an agent that can harm social cohesion and may exacerbate conflict. There is a general consensus in literatures that high levels of income inequality can, if unchecked, ferment internal conflict as a result of disparity in regional development (Adegoke, 2013).

11.5.3 Income Disposition and Per-capita Income Among the Parks

The results in Table 11.3 show the households income disposition and per-capita income among the various households around the National Parks in Nigeria. Household per-capita income among the parks ranged from ₦3,150,000.00 (8,750 USD) in Okomu to ₦7,500.00 (20.83 USD) in Old Oyo. Total per-capita income among the respondents in the parks ranged from ₦200,974,938.00 (558,263.71 USD) in Kamuku to and ₦35,044,651.00 (97,346.25 USD) in Old Oyo with a grand total of ₦296,934,459.70 (824,817.94 USD) and a mean of ₦304,323.39±383,507.45 (845.34 USD) annually. Among the parks, respondents in Okomu had the highest annual income (₦3,150,000.00; 8,750 USD) while Old Oyo had the least annual income range (₦90,000.00; 250 USD). Total income among the respondents in the parks ranged between ₦339,093,514.00 (941,926.42 USD) and ₦109,804,605.00 (305,012.79 USD) with a total of ₦681,405,750.00

(1,892,793.75 USD) and a mean of ₦675,327.80±501,234.20 (1,875.91 USD) annually. The high income of households in Kamuku National could due to their intensive livestock and farming activities. Cereal crops farming (maize, groundnut, millet, sorghum, rice, wheat, and soya beans) is a major agricultural activity in the northern part of the country. This could also be attributed to the people's culture and favorable climatic condition (Aregheore, 2009), hence, majority of the households obtained their income growing these crops and rearing livestock. This is in line with the observation of Adegboye (2016) that majority of the rural dwellers in the northern states of Nigeria are farmers.

Household size among the parks ranged between 20 and 1, with Kamuku having the highest household size (20), followed by Old Oyo (14), and lastly by Okomu (11). Total household size among the parks indicated that Kamuku also had the highest (3898) followed by Old Oyo (1205) and Okomu had the least with 474 members, while the total for all the respondents in the three parks were 5577 members with a mean household size of 5.46±3.86

Table 11.5 Income disposition and per-capita income among the parks.

Variables	Kamuku	Okomu	Old Oyo	Total
No. of Household	463	177	369	1009
Highest income/year (₦)	3150000.00	2124400.00	2085000.00	3150000.00
Lowest income/year (₦)	97500.00	97000.00	90000.00	90000.00
Total income/year (₦)	339093514.00	109804605.00	232507711.00	681405750.00
Mean income/year (₦)	732383.40 (480171.00)	620365.00 (628957.80)	630102.20 (449349.80)	675327.80 (501234.20)
Highest household size	20	11	14	20
Lowest household size	1	1	1	1
Total household size	3898	474	1205	5577
Mean household size	8.42 (3.18)	2.69 (2.07)	3.27 (2.59)	5.46 (3.86)
Highest per-capita income (₦)	2124400.00	3150000.00	1240000	3150000.00
Lowest per-capita income (₦)	9692.31	20333.33	7500.00	7500.00
Total per-capita income (₦)	200974938.00	60914870.70	35044651.00	296934459.70
Mean per-capita income (₦)	434071.14 (387107.20)	373646.16 (518033.67)	94971.95 (118406.90)	304323.39 (383507.45)

*values in parenthesis are standard deviation

members among the parks. These results corroborate with the studies of the household sizes in rural areas which is attributed to the polygamous of the made-headed households (Ogogo et al., 2010; Udeagha et al., 2013; Jacob et al., 2015; Jacob et al., 2016; Daniel et al., 2016).

11.5.4 Poverty Status of the Rural Households

The poverty line of the study area ranged from ₦25958.00 to ₦289380.80 with households in Kamuku National Park having the highest poverty line, while households in Okomu National Park had the least poverty line. This is an indication that households in Kamuku had more income than the households in support zones of the two other National Parks. The establishment of this poverty line helps in ascertain at what level of income a household or individual concerned in the study area encounters unacceptable difficulties in satisfying the basic needs of life. However, the poverty line varies in relation to the general level of development in an area. What is considered as poverty for one these areas may be wealth for another.

FGT poverty index was used to depict the extent of poverty among the households in the study area. The poverty aversion parameters employed shows that across the parks, Okomu had the highest incidence of poverty (0.5876), while Kamuku had the least incidence (0.4298). However, poverty incidence across the households in the three parks was 0.5758. This is an indication that 57.58% of the sampled households were living below the poverty line of ₦202882.30 (563.56 USD). This proportion invariably agreed with the proportion of poor households recorded as 63.8% (NBS, 2005) and 63.3% for Nigeria as estimated by NBS (2011).

In terms of poverty depth (poverty gap index), which is the measure that incorporates the extent to which a poor person's expenditure level falls below the poverty line, Kamuku had the highest index of 0.4898 followed by Old Oyo (0.3607), while Okomu had the least poverty depth of 0.2636. These indices obtained for the households in the park support zone communities are rather worrisome as it is higher than 0.2582 reported for rural areas in Nigeria (NBS, 2005). This implies that an average poor household in the study area will require about 63.20% of poverty line to come out of poverty. The result is in accordance with the report of JICA (2011) and NBS (2005) that the poverty gap in Nigeria tends to be deeper in the rural areas than the urban areas.

Poverty severity was highest in Okomu, followed by Kamuku, and Old Oyo had the least (0.1172) severity index among the parks. However, the households across the three parks had a poverty severity index of

Table 11.6 Poverty status of support zone communities of Nigeria National Parks.

Variables	Kamuku	Okomu	Old Oyo	Total
Poverty Incidence (P_0)	0.4298	0.5876	0.4959	**0.5758**
Poverty depth (P_1)	0.4898	0.2636	0.3607	**0.6320**
Poverty severity(P_2)	0.1308	0.1937	0.1172	**0.2259**
Poverty line	₦289380.80 803.83 USD	₦25958.00 72.10 USD	₦63314.63 175.87 USD	**₦202882.30** **563.56 USD**

0.2259. This value is higher than 0.1406 reported for rural areas in Nigeria (NBS, 2005; JICA, 2011) but lower than 23.34% observed by Jacob et al., 2015 for rural households in Ikot Ondo, Akwa Ibom State, and those reported by Adewunmi et al. (2011) and Igbalajobi et al. (2013) among rural households in some communities in Ogun and Osun States, Nigeria. Generally, the finding depicts the existence of poverty among the rural households in the study area and it behoves on the government to proffered adequate measures to alleviate poverty in these locations.

11.5.5 Poverty Status According to Gender of Household Head

The poverty status of households in the study area indicates that incidence of poverty was higher among the female-headed households (0.5867) than the male-headed households with an incidence of 0.4240 (Figure 11.3). However, poverty depth and severity was higher in male-headed households then the female-headed households. These results contradict the report of NBS (2003) and JICA (2011) who observed that male-headed households were more likely to be living in poverty than female-headed ones because their relative incidence of poverty varied increasingly from 29.2% to 58.2%, while that of the female-headed households also varied increasingly from 26.9% to 43.5% from 1998 to 2004. The change in gender poverty incidence could be attributed to the fact that most of the females who headed their households were widows and few had off-farm employment to support their income. The low-level of off-farm income among the female-headed households according to JICA (2011) is from inequality in education, legislations, occupational culture bias on gender, social attitude of women themselves, and so on. This is in accordance with the report of Okojie (2002) who observed that male-headed households are less likely to be poor than female-headed households.

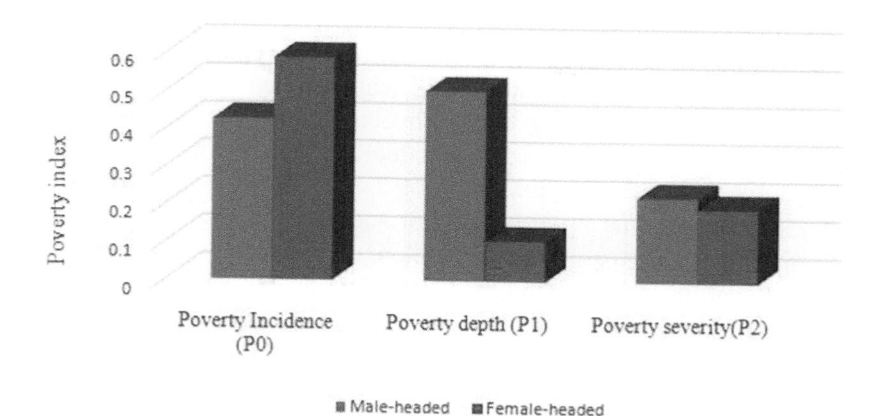

Figure 11.3 Gender poverty status among National Parks Communities.

11.5.6 Determinant of Poverty Status in Households of Support Zone Communities

The result of the logit regression of the determinants of poverty status among the households in the park support zone communities is presented in Table 11.5. The results showed that the X^2 statistics that tested the null hypothesis of all estimated coefficients taken together was equal to zero. The value of the X^2 statistics for the model is 22.64 and is significant at 1% confidence level. However, the value of the coefficient of determinant is low for most empirical studies. In this case, the pseudo R^2 is 0.0525, that is, 5.25% of the poverty status of the households is explained by the selected explanatory variables. The value of this pseudo R^2 suggests a reasonable efficiency of the model.

The estimated coefficients (intercepts included) of three of the estimated variables are statistically significant at 5%. These variables include age of the household head, education of the household head, and household size. Age of the household head was inversely (-0.0158665) related to poverty status and statistically significant at 5 per cent level. The implication of the negative or inverse relationship of age of the household head with incidence of poverty can be attributed to the fact that at the early stage of life, there is always greater energy which would probably help the household head to increase his output and income. Also, as the household head advance towards productive age, he become more experienced and finds it very easy to take risk and diversify his livelihood strategies and *vice versa.* This conforms with the findings of Anyanwu (2013), Ermias et al. (2014) and Udeagha (2015). However, as the household head gets older, his energy begins to depreciate, and his output and income also decline which increases the chances of the household falling into poverty. Igbalajobi et al. (2013) and Jacob et al. (2016) also reported that a decrease in the productive years of the household head significantly increases the probability of the household being poor because elderly persons decline in their strength and productivity as they get older as well as having increased health problems. This is in accordance with the findings of Ahmed et al. (2008), Ogwumike and Akinnibosun (2013) and Jacob et al. (2016) that age of the household head is very important for reducing the poverty status of the household.

The coefficient (–0.0446319) for level of education of the household head was significant ($p < 0.05$) and had the expected sign of being inversely related to the probability of the household being poor. The result is consistent with the findings of Olaniyan and Bankole (2005). It is worth mentioning that the chances of the household escaping poverty increases consistently as the educational level of the household head increases. The acquisition of education helps an individual to overcome the multi-dimensional poverty prevalence. Also, the education of the household head is beneficial for other family members (Majeed and Malik, 2014). This finding also agrees with Okojie (2002), who in his assessment of the relationship between education and poverty incidence in Nigeria using household data of 1980, 1985, 1992, and 1996 also found that all levels of education (primary, secondary, and tertiary) are significant in reducing the probability of a household being poor.

More so, household size was significant and inversely related (-0.1057377, $p<0.01$) to the probability of the household being poor. This implies that as the household size increases, the incidence of poverty decreases, thus not conforming to a *prior* statement. In other words, a larger household would have more labour to spread across various livelihood strategies and such household could derive more income from their livelihood activities (Fonta and Ayuk, 2013; Ermias et al., 2014). If the family size increases by one person then it increases the probability of the household not being poor by up to 10%. Generally, greater livelihood diversification could lessen poverty incidence in rural communities (Ellis, 2000; Illukpitiya and Yanagida, 2008; Udeagha, 2015; Jacob et al., 2016).

Genders was positive and had not statistically significant ($P>0.05$) relationship with incidence of poverty. This indicates that an increase in number of female household head, would increase the level of poverty incidence. However, the non-significance of the variable (gender) indicates that gender

Table 11.7 Determinants of poverty incidence among park support zone households.

Variables	Coefficient	Std. Err.	z	P>z
Age	–0.0158665	0.0072644	–2.18	0.029**
Land	0.0071873	0.0360189	0.20	0.842
Education	–0.0446319	0.0185695	2.40	0.016**
Household size	–0.1057377	0.0316884	–3.34	0.001***
Gender	0.000174	0.0027279	0.06	0.949
Occupation	–0.004965	0.0134245	–0.37	0.711
Adult equivalent	0.0063406	0.0105404	0.60	0.547
Cattle equivalent	0.0006833	0.0031514	0.22	0.828
Cons	1.196899	0.3658819	3.27	0.001***
LR chi^2	22.64			
Log likelihood	–204.42193			
Prob > chi^2	0.0000			
Pseudo R^2	0.0525			

Note: ***, ** and * denote statistical significance at 1, 5, 10 per cent levels, respectively.

is not a major determining factor of poverty incidence in the study area. Therefore, these conforms to the observation that poverty is a gender-neutral condition as female and male experience poverty in a distinctive way (Bastos et al., 2009). The result therefore counters the "feminization of poverty" which states that women are much more deprived and face severe hardships in pulling themselves out of poverty as compared to their male counterparts. Therefore, it should not be expected that female-headed household will increase the likelihood of the household being poor (Majeed and Malik, 2014).

Other variables such as land, adult equivalent and cattle equivalent were positively correlated, while occupation was inversely related to poverty incidence even though they were not significant (p>0.05). This implies that an increase in a single unit of any of these variables while all others being constant would either increase or reduce the level of poverty in a household. This finding is supported by Akinleye (2006) and Olaniyan and Bankole (2005).

11.6 Conclusion and Recommendation

This study examined the levels of income inequality and incidence of poverty among households in the support zone communities of Nigeria National Parks. The above findings depict the presence of income inequality among the various park households. The socioeconomic characteristics of the household were the major determinant of poverty incidence in the communities. Age, education and household size were inversely related to the poverty status of the households; and that a one per cent increase in these variables will reduce the probability of a household being poor by more than 1%, 4%, and 10% respectively. Therefore, it becomes very important for policy makers to come up with policies that would enable the rural household to irrevocably have a better standard of living.

Also, since most of the households inherently have access to land asset and low off-farm employment opportunities, efforts could be made to improve on this. Investments should be focused on improving agricultural practices/activities, encouraging gains in vocational education and trainings, diversification, improving upon and localising livelihood strategies and other forms of skill acquisition in order to spread/avoid the risk of being poor.

Access to credit facilities should be improved upon by establishing micro credit institutions and loan schemes that are tailored to benefit the rural people within the park vicinity. The same should be done elsewhere across the country with similar problems as it would further contribute to poverty reduction in rural areas, thus creating a clear pathway out of poverty.

Policies and programs should also be put in place that will help ameliorate the negative effects the park have on the households as it will go a long way in enhancing households' levels of productivity and incomes. However, such policies should be highly localised to enable it to achieve better outcome. Additionally, any poverty alleviation programme initiated in the area should effectively serve as safety nets for the households by decentralisation of its implementation in order to enhance resource governance which will reduce poverty.

References

Abbot, J. I.O.,Thomas, D. H. L., Gardner, A. A., Neba, S.E., and Khen, M.W. (2001). Understanding the links between conservation and development in the Bamenda Highlands, Cameroon. *World Development*, 29 (7): 1115–1136.

Adegboye, M. A. (2016). Socio-economic Status Categories of Rural Dwellers in Northern Nigeria. *Advances in Research*, 7(2), 1–10

Adejoke, Y. O. (2013). Disparity in Income Distinction in Nigeria: A Lorenz curve and Gini index Approach. *Universal Journal of Management and Social Sciences* 3(7):16–28.

Adewunmi, O. I., Adesimi, B., Ezekiel, O. A. (2011). Non-farm Activities and Poverty among Rural Farm Households in Yewa division of Ogun State, Nigeria," *Journal of Social. Science*, 26(3): 217–224

Agrawal, A., and Redford, K. (2006). Poverty, Development, and Biodiversity Conservation: Shooting in the dark? Working Paper No. 26, Wildlife Conservation Society.

Ahmed, N., Allison, E. H., Muir, J. F. (2008). Using the sustainable livelihoods framework to identify constraints and opportunities to the development of freshwater prawn farming in Southwest Bangladesh. *Journal of the World Aquaculture Society*, 39(5): 598–611

Akinleye, S. O. (2004). Characteristics and determinants of poverty among fish farmers in Lagos State. *Bowen Journal of Agriculture,* 3 (2) 2006: 141–150

Akpan, S. B., and Udoh, E. J. (2016). Farmers' decision to participate in government agricultural programmes in a volatile political environment: A case study of famers in the south-south region of Nigeria. *Russian Journal of Agriculture and Socio-Economic Sciences, 5*(53), 135–148.

Akpan, S. B., I. V. Patrick and A. Amama (2016). Level of income inequality and determinants of poverty incidence among youth farmers in Akwa Ibom State, Nigeria. *Journal of Sustainable Development* 9 (5): 162–174

Akpan, S. B., Patrick, I. V., James, S. U., and Agom, D. I. (2015). Determinants of decision and participation of rural youth in agricultural production: a case study of youth in southern region of Nigeria. *Russian Journal of Agriculture and Socio-Economic Sciences, 7*(43): 35–48. http://dx.doi.org/10.18551/rjoas.2015-07.05

Andrew-Essien, E. and Bisong, F. (2009). Conflicts, Conservation and Natural Resource use in Protected Area Systems: An Analysis of Recurrent Issues. *European Journal of Scientific Research* 25(1): 118–129

Angelsen, A., Larsen, H. O., Lund, J. F., Smith-Hall, C. and Wunder, S. (2011). Measuring Livelihood and Environmental Dependence: Method for research and fieldwork. Earthscan. p. 264.

Anyanwu, J. C. (2013). Marital Status Household Size and Poverty in Nigeria: Evidence from the 2009/2010 Survey Data. Working Paper Series N0 180 African Development Bank, Tunis Tunisia.

Aregheore, E. M. (2009). "Country Pasture/Forage Resource Profiles: Benin." Food and Agriculture Organization of the United Nations. http://www.fao.org/WAICENT/FAOINFO/agricULT/AGP/AGPC/doc/Counprof/PDFfiles/Benin.pdf

Babbie E. (2005). *The Basics of Social research.* 3rd ed., Toronto, Canada: Thomson and Wads worth, pp. 50–217.

Badmus, M. A., Aderinto, A. and Fagbola, O. (2009). Impact of Fadama II development project on market participation of women in Akinyele Local Government of Oyo State. Proceedings of the eighteenth Annual Congress of the Nigerian Rural Sociological Association held at Federal University of Technology Akure, Ondo state between 16th and 18th December 2009;53–59.

Balikoowa, K. (2008). Impacts of Bwindi Impenetrable National Park on local people's livelihoods. Unpublished MSc research report. Norwegian University of Life Sciences, Norway

Barrett, C.B., Lee, D.R., and McPeak, J. G. (2005). Institutional Arrangements for Rural Poverty Reduction and Resource Conservation. *World Development*, 33(2):193–197.

Bastos, A., Casaca F. S., Nunes, F. and Pereirinha, J. (2009).Women and Poverty: A Gender-Sensitive Approach. *Journal of Socio-Economics*, 38(5): 764–778.

Bhandari, Medani P. (2014). "Is Tourism Always Beneficial? A Case Study from Masai Mara National Reserve, Narok, Kenya". Pacific Journal of Science and Technology. 15(1):458–483.

Bhandari, Medani P. (2020) Second Edition- Green Web-II: Standards and Perspectives from the IUCN, Policy Development in Environment Conservation Domain with reference to India, Pakistan, Nepal, and Bangladesh, River Publishers, Denmark / the Netherlands. ISBN: 9788770221924 e-ISBN: 9788770221917

Bhandari, Medani P. (2018) Green Web-II: Standards and Perspectives from the IUCN, Published, sold and distributed by: River Publishers, Denmark / the Netherlands ISBN: 978-87-70220-12-5 (Hardback) 978-87-70220-11-8 (eBook).

Cheong, K. (1999) *A note on the interpretation and application of the Gini coefficient*, Working Paper No 99-1R, Manoa: University of Hawaii Chianu *et al.* (2004),

Chianu, J.N., H. Tsujii and P. Kormawa (2004). Agriculture in the savannas of northern Nigeria: pressures, transformations and damage and coping strategies. *Outlook on Agriculture*, 33: 247–253.

Daniel, K.S., Y., Udeagha, A.G. Umazi and Jacob, D. E. (2016). Socio-Cultural Importance of Sacred Forests Conservation in South Southern Nigeria. *African Journal of Sustainable Development* 6 (2): 151–162

Ellis F. (2000). The Determinants of Rural Livelihood Diversification in Developing Countries. *Journal of Agricultural Economics,* 51(2):289–302

Emelue, G.U., Jacob, D. E. and O. S. Godwin (2014). Assessment of Indigenous Wildlife Conservation Practices in Ika North East Local Government Area of Delta State, Nigeria. *Nigerian Journal of Agriculture, Food and Environment* 10(2):11–17.

Enuoh, O. O. O. (2014). Parks and Communities: Assessing the Social Impacts of the Creation of Cross River National Park, Nigeria, using the Sustainable Livelihoods Approach. Research on Humanities and Social Sciences, 4(21): 140–154.

Ermias M, Ewmetu, Z and Teketay D. (2014). Non-Timber Forest Products and Household Incomes in Bonga Forest area, Southwestern Ethiopia. *Journal of Forestry Research*, 25(1): 215–223.

Fonta, W.M and Ayuk, E. T. (2013). Measuring the Role of Forest Income in Mitigating Poverty and Equality: Evidence from South-Eastern Nigeria. Forests, Trees and Livelihoods, DOI:10.1080/14728028.2013.785783.Taylor&Fracis, pp. 1–21.

FORMECU (1995) An Assessment of Land use and vegetation changes in Nigeria 1978 - 1995. Geomatis, Ontario, Canada.

Heubach K. (2012). Socio-Economic Importance of Non-timber Forest Products for Rural Livelihood in West Africa Savannah Ecosystems: Current Status and Future Trends. Unpublished Ph.D Thesis, Johann Wolfgang Goethe University, Frankfurt, pp. 1–153.

Igbalajobi, O., Fatuase,A .I and Ajibefun, I. (2013). Determinants of Poverty Incidence among Rural Farmers in Ondo State, Nigeria. *American Journal of Rural Development*, 1 (5): 131–137.

Illukpitiya, P and Yanagida, J. F. (2008). Role of Income Diversification in Protecting Natural Forests: Evidence from Rural Households in Forest Margins of Sri Lanka. *Agroforestry systems,* 74:51–62.

Jacob, D. E. (2008). Buffer zone management in Cross River National park: Oban division. Undergraduate Research, University of Uyo, Uyo. pp. 66.

Jacob, D. E. (2017). Resource Governance Structure in selected Nigeria National Parks and Impact on Rural Livelihood. Ph.D thesis submitted to Federal University of Agriculture, Abeokuta, Nigeria. 2546 pp.

Jacob, D. E., Eniang, E. A. and Nelson, I. U. (2018). Impact of Nigeria National Parks on Support Zone Communities Livelihood. In: Ogunjinmi, A. A., Oyeleke, O. O., Adeyemo, A. I., Ejidike, B. N., Orimaye, J. O., Ojo, V. A., Adetola, B. O. and Arowosafe, F. C. *Achieving Sustainable Development Goals: The Role of Wildlife*. Proceedings of the 2nd Wildlife

Society of Nigeria (WISON) Conference held at the Federal University of Technology, Akure, Ondo State, Nigeria from 16th–19th September, 2018.

Jacob, D. E., Nelson, I. U., Udoakapn, U. I. and Etuk, U. B. (2015). Wildlife Poaching in Nigeria National Parks: A Case study of Cross River National Park, *International Journal of Molecular Ecology and Conservation* 5(4): 1–7 (doi: 10.5376/ijmec.2015.05.0004)

Jacob, D. E., Udeagha A. U. and Nelson, I. U. (2016). Poverty Incidence among Rural Households in Ikot Ondo Community, Nigeria. *Journal for Studies in Management and Planning*, 2(5): 80–88

Jacob, D.E. and A.U. Ogogo (2011). Community participation in protected area management: A case study of Cross River National Park. In: Popoola, L., K. Ogunsanwo and F. Idumah (eds). *Forestry in the context of the millennium development goals,* Proceedings of the 34th Annual Conference of the Forestry Association of Nigeria held in Osogbo, Osun State, Nigeria. Vol. 1, p. 412–415

Jacob, D.E., Udoakpan, U.I and Nelson, I.U. (2013). Issues in Conflict Resolution in Cross River National Park, South Eastern Nigeria. 1st International Conference on Environmental Crisis and its Solution. Scientific and Research Branch, Khouzeslan, Islamic Azad University, Kish Island, Iran, 13th–14th February, 2013. pp. 76–82.

Javed, Z.H., and A. Asif (2011). Female households and poverty: A case study of Faisalabad District. *International Journal of Peace and Development Studies* 2:37–44.

JICA (2011). Country gender profile – Nigeria: Final report, retrieved from http://www.jica.go.jp/english/operations/thematic_issues/gender/background/pdf/e10nig.pdf

Majeed, M. T. and M. N. Malik, (2014). Determinants of Household Poverty: Empirical Evidence from Pakistan. Munich Personal RePEc Archive (MPRA) Paper No. 57744, Available online at https://mpra.ub.uni-muenchen.de/57744/

Manyong, V.M., I. Okike, and T.O. Williams (2006). Effective Dimensionality and factors affecting Crop-livestock Integration in West African Savannas: A Combination of principal component analysis and Tobit approaches. *Agricultural Economics* 35:145–155.

Mohammed, S. O., Gajere, E. N., Eguaroje, E. O., Shaba, H., Ogbole, J. O., Mangut, Y. S. and Kolawole, I. S. (2013). Spatio-temporal analysis of the national parks in Nigeria using geographic information system. IFE J Sci 15(1):159–166.

Mosetlhi, B. (2012). The Influence of Chobe National Park on People's Livelihood and Conservation Behaviours. A Ph.D Dissertation presented to the Graduate School, University of Florida, 220 pp.

Murphree, M. W. (1991). *Communities as Resource Management Institutions. Gatekeeper Series No. 36*. Sustainable Agriculture and Rural Livelihoods Program: International Institute for Environment and Developments

National Bureau of Statistics (NBS) (2012). Nigeria Poverty Profile 2010. Abuja, 31 pp. http://www.nigerianstat.gov.ng/pdfuploads/Nigeria%20Poverty%20Profile%202010.pdf.

National Bureau of Statistics (NBS), (2005). Poverty in Nigeria. Abuja

National Bureau of Statistics. (NBS, 2011). NBS press briefing on Nigeria poverty profile 2010 report. www.nigerianstat.gov.ng (2011). Accessed 15th January, 2016

Nelson, I. U., E. S Udo, D.E Jacob (2018). Profitability of ornamental plant business in Akwa Ibom State, *Nigeria. Eurasian Journal of Forest Science* 6 (1), 35–43.

Nelson, I. U., E. S. Udo and D. E. Jacob (2017). Economic analysis of firewood marketing in Uyo capital city, Akwa, Ibom state, Nigeria. *Eurasian Journal of Forest Science*, 5 (2), 26–43.

Oates, J. F. (1999). Myth and reality in the rain forest: how conservation strategies are failing in West Africa, Berkeley: University of California Press, xxviii, 310 pp.

Ogogo, A. U., A. A. Nchor and D. E. Jacob (2010). Challenges of Buffer Zone Management in Cross River National Park, Southeastern Nigeria. *Journal of Research in Forestry, Wildlife and Environment* 2(2): pp. 156–163.

Ogunjinmi, A. A., Onadeko, S. A. and A. A. Adewumi (2012). An Empirical Study of the Effects of Personal Factors on Environmental Attitudes of Local Communities around Nigeria's Protected Areas. *The Journal of Transdisciplinary Environmental Studies* 11(1):40–53.

Ogwumike, F.O., and M.K. Akinnibosun. (2013). "Determinants of Poverty among Farming Households in Nigeria", *Mediterranean Journal of Social Sciences*

Ohwotemu, O.L. (2012). A distributional analysis of income in Nigeria. Unpublished M.Sc thesis. Available at: https://pdfs.semanticscholar.org/2bad/bdc649b344ac069587c2d98f67f8a021bbe5.pdf?_ga=2.82911058.515652779.1587211984-857019837.1587211984

Okojie, C. E. (2002). Gender and Education as Determinants of Household Poverty in Nigeria, Discussion Paper No. 2002/37, World Institute for Development Economics Research (WIDER).

Olaniyan O. and Bankole, A. S. (2005) Human capital, capabilities and poverty in rural Nigeria. Research Report submitted to the African Economic Research Consortium.

Olawuyi, S. O. and Adetunji, M. O. (2013), Assessment of Rural Households Poverty in Nigeria: Evidence from Ogbomoso Agricultural Zone of Oyo State, Nigeria. *Journal of Scientific Research and Reports* 2(1): 35–45.

Olayide. O. E., A. D. Alene and A. Ikpi (2009). Determinants of Fertilizer Use in Northern Nigeria. *Pakistan Journal of Social Sciences* 6(2): 91–98.

Olorunsanya, E. O. and O. A. Omotesho (2011). A Gender Analysis of Poverty Profile of Rural Farming Households in North Central, Nigeria. *International Journal of Agricultural Economics and Rural Development* 4(2): 11–27.

Oluwatusin, F. M. and A. B. Sekumade (2016). Farm Households Income Sources Diversification Behavior in Nigeria. *Journal of Natural Sciences Research* 6(4): 102–111.

Oyedele, D. (2014). Nigeria's Population Now 168m, Hits 221m in 2020. ThisDayLive.Com. Available at http://www.thisdaylive.com/articles/nigeriaspopulation-now-168m-hits-221m-in-2020/101436/. Accessed October 17, 2014.

Ruben, R and van den Berg, M. (2001). Non-farm employment and poverty alleviation of rural farm households in Honduras, *World Development*, 29(3): 549–560

Rubin, A and Babbie, E. R. (2008). *Research Methods for Social Work.* 6th ed., Belmont: Thomson and Book/Cole, pp. 459–600

Sanderson, S. (2005). Poverty and Conservation: The New Century's Peasant Question? *World Development*, 33(2):323–332.

Silver, J. J., Gray, N. J., Campbell, L. M., Fairbanks, L. W. and Gruby, R. L. (2015). Blue economy and competing discourses in international oceans governance. *Journal of Environment and Development*, 24(2): 135–160.

Terborgh, J. (1999). *Requiem for Nature.* Washington DC: Island Press / Shearwater Books.

Tumusiime, D. M. (2006). Dependence on environmental income by households around Rwenzori Mountain National Park, Western Uganda. Master's thesis, Norwegian University of Life Sciences, Department of International Environment and Development Studies, NORAGRIC.

Udeagha, A. U. (2015). Impact of Climate change on the contributions of *Irvingia* fruits and kernels to Rural Household Economy in Cross River State. Unpublished M.Sc. Dissertation, Department of Forestry and Natural Environmental Management, University of Uyo, Uyo, AkwaIbom State, Nigeria. 1–175 p.

Udeagha, A. U., Udofia, S. I. and Jacob, D. E. (2013). Cultural and socio-economic perspectives of the conservation of Asanting Ibiono Sacred Forests in Akwa Ibom State, Nigeria. *International Journal of Biodiversity and Conservation.* 5(11), pp. 696–703.

Uganda Bureau of Statistics (2002). *Uganda Bureau of Statistics: The 2002 Uganda Population and Housing Census, Economic characteristics. October 2006, Kampala, Uganda*

Wells, M., K. E. Brandon, and L. Hannah. (1992). *People and parks: Linking protected area management with local communities.* Washington, DC: World Bank.

Wilkie, D.S., Morelli, G.A., Demmer, J., Starkey, M., Telfer, P., and Matthew S. (2006). Parks and People: Assessing the Human Welfare Effects of Establishing Protected Areas for Biodiversity Conservation. *Conservation Biology,* 20 (1): 247–249.

World Bank (2002). Empowerment and Poverty Reduction: A Sourcebook, Washington, DC: The World Bank.

World Bank. 2008. *Nigeria 2008—Enterprise Surveys.* Washington, DC: World Bank.

Yunez-Naude, A., Taylor, J. E. (2001). The determinants of non-farm activities and incomes of rural households in Mexico with emphasis on education. *World Development,* 29(3) 561–572.

12

Social Responsibility as a Tool for the Human Resources Policy Development and Reducing Inequalities on Tourism Industry

Aleksander Sapiński[1], Sabina Sanetra-Półgrabi[2], Serhii Y. Kasian[3], Nataliya Rozhko[4] and Medani P. Bhandari[5]

[1]PhD Student, MSc, Lecturer at Bielsko-Biala School of Finance and Law (Poland), ORCID: 0000-0002-3326-2387
[2]PhD in political studies, Pedagogical University of Krakow (Poland), ORCID: 0000-0001-9628-5327
[3]PhD, Associate professor, Head of Marketing Department Dnipro University of Technology (Ukraine), ORCID: 0000-0002-7103-4457
[4]Associate Professor Ternopil Ivan Puluj National Technical University, (Ukraine), ORCID: 0000-0002-1400-9503
[5]Professor and Advisor of Gandaki University, Pokhara, Nepal, Prof. Akamai University, USA and Sumy State University, Ukraine
E-mail: olek.sapinski@interia.pl; sab_san@poczta.onet.pl; kasian.s.ya@nmu.one

Abstract

The aim of this theoretical chapter is to present the role and importance of corporate social responsibility (CSR) as a tool for supporting the development of innovation in the field of human resources management in the hotel industry. It is an interdisciplinary article because it deals with a research subject that can be looked at from an economic, legal, and pedagogical perspective. The very fact of hotel activity as an effect of tourism in its broadest sense may bring a new innovative approach with the use of the novelties in the field of pedagogy and education. The interdisciplinarity of tourism and hospitality makes its huge development over the last two decades a kind of viral activity, because as a virus combines and infects, tourism and hospitality combine other disciplines to implement pro-innovative solutions, making its practical part more attractive to potential customers. The tourism industry, thanks to its interdisciplinary nature, creates many jobs, which helps to combat social exclusion. Many tourist destinations are also combating many of the social inequalities that have existed to date, thanks to investment in tourism capital. Social responsibility and HRM are ideal for tourism and hospitality due to their interdisciplinary character. Today, social responsibility is a concept that permeates all areas of knowledge and economic life.

Keywords: CSR, hospitality, HRM, tourism manager, hospitality management, social impact.

12.1 Introduction

> "Social responsibility is the ideological notion that organizations should not behave unethically or function amorally and should aim (instead) to deliberately contribute to the welfare of society or societies—comprised of various communities and stakeholders—that they operate in and interact with. As such, the notion of social responsibility is effectively taken to apply to all and any organizational entities, whether a government, a corporation, and institution, or an individual, dealing with society at large when conducting core (commercial) activities. In recent decades, however, social responsibility has come to be acknowledged as particularly relevant in relation to corporate behavior, that is, in relation to the way in which businesses and managers behave and conduct themselves in societal relationships, and the extent to which these actors commit themselves to socially oriented initiatives aimed at improving the quality of life in and overall well-being of the society" (Brigitte Planken-2013- Springer-Encyclopedia of Corporate Social Responsibility— https://link.springer.com/referenceworkentry/10.1007/978-3-642-28036-8_476).
>
> Inequality—the state of not being equal, especially in status, rights, and opportunities—is a concept very much at the heart of social justice theories. However, it is prone to confusion in public debate as it tends to mean different things to different people. Some distinctions

> are common though. Many authors distinguish "economic inequality," mostly meaning "income inequality," "monetary inequality" or, more broadly, inequality in "living conditions." Others further distinguish a rights-based, legalistic approach to inequality—inequality of rights and associated obligations (e.g., when people are not equal before the law, or when people have unequal political power) (UN (2015:1)).

The issue of social responsibility has become a kind of a platform for the development of many branches and areas of life. Challenges that appear before the modern economy of each country require that managers create opportunities for their employees, not only in the area of increasing financial profit. For many years now, the ideas and directions of business development referring to social responsibility have been gaining more and more interest among managers of tourism and recreation. Implementation of the concept of responsibility and business toward the society as well as responsibility and society for business and the wider category also for the state of the economy. Promoting social responsibility among employees, as well as their families and friends, builds bridges to reduce social inequalities resulting not only from the reasons strongly described in the literature, but also from the point of view of the culture of life in a given country or a given local community. Looking at social inequalities through the dimension of the culture of living in a place shows not only material deficiencies, but also deficiencies in the system of social hierarchy and in the development of social authority. Shaping attitudes of social openness and social responsibility can bring about a change in the culture of human existence, which builds attitudes related to entrepreneurship, self-reliance, and a desire to coexist in a group and help others (Ciupka, 2020). It seems worrying, however, that some researchers have begun to treat social inequalities as a minor topic (Haber, 2012, p. 196). All the considerations presented in this article are set in the context of reducing social inequalities, including those at the micro level, which sometimes exist through intolerance, stereotypes, and resentment. A strong inclusive response from socially excluded people can build the basis for creating pro-innovative attitudes and actions in the company. Innovative business can become a key success factor for innovative economy in a given country (Zharova, 2019).

The relationship between innovation and corporate social responsibility (CSR) seems to be an interesting issue for the needs of not only the hotel industry, but also other sectors of economic life. In the case of many global corporations, social responsibility is not only a response to the needs of their customers or contractors but is a kind of business dialogue with the anticipated needs of society, and not just marketing the moment. The credible fulfilment of the mission of CSR is closely linked to the issue of sustainable development (Bhandari, 2019). Understanding the relationship between the main concepts on which the modern world is based becomes an opportunity to fill in the white spaces that provide space for creating and implementing innovative ideas in a given industry. Innovation itself is a well-known and desirable phenomenon in various disciplines of knowledge and economic life. However, the strength of innovation does not always lie in the individual, but also in the group that creates innovative links (Tidd and Bessant, 2013, p.388). Groups of people who often work in different places or even parts of the world can "peek at others" through an exchange of views, creative discussion, or industry-oriented internet forum. In this way, the principle of promoting the group's strength leads to the building of knowledge banks, which, constantly updated and created for the benefit of the industry, can also bring results in the free market economy.

The social responsibility itself, which is driven by the mix of social needs and the creative idea of their solutions, can become very effective in such an issue. The specificity of the hospitality industry requires special consideration because some general specifications will be common to the whole industry, but a large part of the industry's conditions are dictated by cultural, legal, and other considerations specific to a particular country or region of the world. Therefore, the potential impact of socially responsible activities at some point in their implementation may require adjustments that consider the specificities of the place. This way of thinking allows for the implementation of development plans in the hotel industry according to the assumptions of the concept of sustainable development. This concept treats man as a supreme good, a host of the planet, who has duties and opportunities. These opportunities will arise from responsibilities for the welfare of the general public and the surrounding world and are the basis for the development of innovation in the field of human resources in the hospitality industry. From the social point of view, the responsibility and management of the human resources in the hotel industry are updated with each tourist season. In spite of a wide range of different courses or programs of academic studies in the field of tourism and only a person who understands the need to implement innovative solutions will be able to properly perform their work in this position.

12.2 Theoretical Approach

The theoretical concept of CSR has become the subject of serious scientific deliberation for almost 30 years now. The realization of the practical dimension of this concept has given a new breath of fresh air and a certain dynamism to the understanding and updating of the very concept of the social responsiveness of business. In many definitions, the concept of CSR is minted as self-regulating (Smith and Langford, 2009).

In contemporary literature on CSR, too much attention is paid to positive phenomena concerning cooperation between business and the environment in the field of action, which in many examples, however, are only a smoke screen. The literature review shows the clear side of companies' activities in terms of their impact on society and the environment, as well as their far-reaching effects (Bartkowiak, 2011, pp.72-73). Social and environmental damage is a frequent occurrence for many companies, but the assessments and effects of social damage are often underestimated. This is because the aims and objectives of CSR are voluntary for businesses (Tamvada, 2020). The conception and implementation of the concept of socially responsible business must therefore be implemented not only in the legal acts of states or company statutes, but above all, it must be shaped in the consciences of its employees (Smith and Rhiney, 2020). An important factor that could shape and quickly implement the implementation of CSR in practice was product labels that would promote socially responsible enterprises. The choice of informed customers would be obvious in this case and would allow such companies to grow rapidly in relation to their competitors (Rodrigues and Borges, 2015). Understanding the social responsibility of a company, including its employees, is a form of taking care of the common good both for the environment and for the organization itself (Popescu, 2019). The authors believe that it is a mistake to see CSR as an issue that rests solely on the company.

The company itself can often be understood as an abstract reality for employees who do not identify themselves with the company. Therefore, an extremely important element relating to social responsibility is education and awareness-raising of workers and their families in order to achieve full effect. This is a viral action that is required to fully achieve all the objectives of sustainable development (Zharova, 2019). Many researchers believe that one of the important issues relating to social responsibility is gender equality (Nguyen and Huang, 2020), which, in the case of the tourism industry, is extremely important in order to achieve the tort objectives of CSR policy in this area. For a correct understanding of the need to implement the CSR concept in the hospitality industry, it is important to understand that there is a human being at the center of CSR, just as there is a customer at the center of the hospitality industry.

The development of the theory and definition of social responsibility can be summarized as, playing by pro-social principles (Żemigła, 2013, p. 392), an extra-horizontal concept with a strong impact on the environment (Michalaka, 2017 p.48), or a broadly understood responsibility for the company of its employees, their families, and future generations (Moczydłowska, 2010, p.218; Szołtysek, 2013, p. 174).

12.3 Materials and methods

The article is illustrative and contributes to further research in this area. The main aim of the article is to present the position that CSR can influence the development of innovation in the area of human resources. From the observations of the industry, especially over the last 20 years, once the materials available to the public can provide the direction for the future research. The co-occurring objectives are also objectives such as presenting the possibilities of combining social responsibility with didactic processes occurring in the organization but also in society, as well as combining this phenomenon with the concept of tourism potential. Incorporating the phenomena of CSR into the didactic process becomes crucial in connection with the tasks of combating the differences that cause social inequalities.

The research methodology used in this article is based on empirical research methods in tourism and recreation. The research methodology was adopted after Henryk Grabowski (Grabowski, 2013). The adopted methodology has been chosen in the intercultural and interdisciplinary trend, because the very subject of tourism as science is of such a nature. The research also included management research methods (Kostera, 2015). In the field of management sciences, the humanistic current of management has been chosen, because man has been treated subjectively and not instrumentally in the article. The main research method chosen to conduct the research was perception, i.e., ad-hoc observation. This method has been used both directly at the facilities and by means of distance testing aids. According to Roman Batko, a valuable source of support for observation is the "word" (Batko, 2015). As he continues to write, this supports the research process and allows for the discovery of a concrete meaning, which often appears during the observation of interviews or analysis of literary texts. The choice of research methods makes sense in the research process; however, the essence is constituted by research questions that set the direction for research and in a way respond to the needs of science and practice. The formulation of research questions itself is repeatedly a test of intelligence and scientific intuition for the researcher himself. This article answers the questions:

- What is the interdisciplinarity of corporate social responsibility and what discipline correlates with each other for the purpose of implementing this concept?
- Do corporate social responsibility activities have an impact on reducing social inequalities, especially social exclusion?
- What practical realizations are already operating in the hospitality industry in the area of social responsibility and how do they result in human resources development?

12.4 Discussion

Developing the impact of social responsibility on the development of innovation in the field of human resource management in the hotel industry requires looking at society as a kind of innovation incubator, whose driving force is creativity and often the need for time. It is important for the manager to capture the right moment and the factors that will determine the introduction of appropriate processes that lead to the creation of specific projects. The project combats barriers and introduces new ideas of socially excluded people to the tourism industry by reducing social differences through creating a culture of understanding and openness to others. Humility as a value of personality does not lose its importance with time. Balancing the work of responsibility-based business development can lead to the creation of jobs for socially excluded people who, through their life experiences, can not only improve the process of doing particular jobs, thus contributing to greater efficiency, but also to reducing costs. Life experience, which is often lacking in young graduates from different kinds of schools and universities, is a great asset for excluded people who want to change their life, on their way to a career (Binda, J., Sapiński and Pochopień, 2019).

In tourism and hospitality, an extremely important form of innovation is to reduce costs (Kachniewska, 2014) and reduce the negative impact of tourists on the environment. It should be stressed that in each country, or even region or Euroregion, national, political, cultural and even economic divisions are a great obstacle to real cooperation between the public and the hospitality industry (Półgrabi 2018, p. 134). Therefore, on the basis of market observations, it can be concluded that important determinants influencing positively the development of mutual cooperation between the hotel industry and society as a broadly understood entity are important:

- ability to think holistically,
- openness to generational diversity(Wziątek-Staśko, 2012),
- sense of community

Describing different kinds of theories that contain determining factors like the above ones is an important part of science, but science itself does not contribute much without a practical definition of the expected results. It is therefore extremely important to combine socially responsible actions resulting directly from processes dedicated to sustainable development into a range of actions. CSR will therefore be the responsibility of the whole team (Bartkowiak, 2011, pp.123-156) and will need to be spread among customers coming (Wziątek-Staśko, 2012, pp. 142-149) to and using the services of the hotel industry. There are examples of modern CSR solutions for the hotel industry:

1. modern examples of advanced technologies that will grant the environment, but also service and users;
2. supporting the processes of non-governmental organizations which are active in the development of the hotel industry, as well as the environment in which the facilities of this industry are located (human resources, natural resources, unique places, health promotion);
3. supporting their employees as a unique human capital that is not found in any other organization, since the value and human capital are attested to by individual human predispositions and their competences and the skills, they use in a given situation;
4. working with employees to listen to their ideas and turn them into action, instead of corporate indoctrination;
5. creating hotel policy in such a way as to prevent the illegal exploitation of children and prostitution in hotel facilities (e-hotelarz.pl);
6. education of staff and hotel guests in order to raise awareness of what materials are recycled; and
7. creating opportunities for the implementation of innovative activities that produce results in other parts of the country, the continent or the world, with regional modification—seeing good practices also beyond the horizon of their national borders.

It is important for the management to realize that responsible actions in HRM are those that will reduce the turnover of highly qualified staff. The modern working environment, especially in the hospitality industry, in its approach to innovation in the HRM area cooperating with CSR must introduce standards of respect for international culture, the determinants of which can be observed through:

1. to interpret and anticipate the behavior and expectations of representatives of other cultures;
2. creating new global standards of raising and educating children and youth;
3. the process of lifelong learning with an emphasis on overcoming social stereotypes, especially those generating xenophobic and aggressive behavior;
4. developing new forms of social responsibility based on the reduction of social inequalities through the development of new forms of work; and
5. constant awareness of their impact on others even in a blind way, which can have serious business consequences.

The essence of applying social responsibility and as a tool for the development of quasi-innovative HRM activities can be found in pro-educational theories of the philosophy of education. Two terms which play an important role are: authority (Szołtysek, 2013, p.207) and leadership. Many researchers stress that in order to achieve key objectives, sustainable management should be introduced in the organization, which in the matter under consideration becomes an extremely important set of tools (Bhandari, 2019, p. 10). Authority understood as a model and leadership as invaluable help and support become a way to introduce sustainable management of the organization, and thus to provide CSR models and improve interpersonal skills.

12.5 Conclusion and Recommendations

When you consider the possibilities of social responsibility of an organization and the possibilities to innovate in the area of HRM of the hotel industry, you can see an unprecedented variety of ideas, nationalities, trends and threats that cannot be found in other industries in such quantities. This is due to the enormous amount of human exchange that takes place in the hotel industry and the intercultural exchange that characterizes the tourist industry. It should therefore be specified for the subject matter adopted that:

1. for the proper implementation of the social responsibility policy in the tourism industry, it is important to look not only at "each other" but also at other industry players on the scale of global trends;

2. identify the actions and socially responsible indications put forward by other industries in the region relevant to the tourism operator in order to interact and integrate with other actions for the development of social capital and human resource potential in the region;
3. for the development of quasi-innovative human resources, legal changes should be made in the area of anti-competitive activities;
4. increase financial and training investment in building social links with socially excluded people, who as employees will relieve the state welfare system and increase the diversity of human resources of the company and the sector;
5. to create a platform for socially responsible human resources activities for the hospitality industry as a space for exchange of good practices at national and international level;
6. the creation of international legal standards both in the field of labor law and employers' and tourism law, which will be in line with the objectives of the United Nations' sustainable development agenda;
7. the development of human resources is dependent on supporting faculties of higher education institutions that provide courses in tourism and response or sport and education for specific regions for the specialization of university graduates;
8. to work "together" according to the rule of Saint Benedict (Bianchi, 2009, p.175), which will provide an opportunity to specialize and teach humility and openness to other people's needs, and thus to understand social needs from a local, national, or international perspective; and
9. introduction of work based on process teams for improvement of pro-social activities and implementation of improvement of self-education work and awareness of one's work by lower-level employees.

Creating new solutions to improve the situation of employees and people from the hotel industry environment will require a lot more effort due to the dynamics of the industry, which interacts with each other through global trends, often rooted in local hotel groups and their inventive employees. The orientation of employees and other participants in the industry will allow, especially in times of crisis, for self-development and a look at the environment on a micro and macro scale will be an opportunity to maintain the development dynamics of the hotel industry and, what is important, the competencies of employees in the industry. The way of creating innovation for the development of human resources in the hotel industry therefore differs from other industries in that the external human factor will play a greater role here, which will be the motivator of change and the main controller of its implementation. It should be remembered that the dynamics of change requires interaction on two planes. The first one will be the plane of real human contact, while the second plane is a virtual reality, which, especially for older workers and managers, can be the most challenging. However, the virtual plane may become the biggest ally of the hotel industry.

As you can already see, evaluating the activity of the industry through opinion-forming portals or booking services gives the most information for various recipients of the industry about their work and through the publication of the opinion about the work of competitors, which should be perceived here as another element of the industry chain.

References

Alkire, S., Foster, J., Seth, S., Santos, M. E., Roche, J. M., and Ballon, P. (2015). 'Multidimensional Poverty Measurement and Analysis', Oxford: Oxford University Press. Retrieved on 2 October 2015 from Oxford Scholarship Online: August 2015.

Bhandari, Medani P. (2020). Security and sustainability. Scientific Journal of Bielsko-Biala School of Finance and Law, 24(4), 5-8. DOI: 10.19192/wsfip.sj4.2020.1 https://asej.eu/index.php/asej/article/view/554/487 https://asej.eu/index.php/asej/article/view/554

Bhandari, Medani P, (2019). Live and let other live- the harmony with nature /living beings-in reference to sustainable development (SD)- is contemporary world's economic and social phenomena is favorable for the sustainability of the planet in reference to India, Nepal, Bangladesh, and Pakistan? Adv Agr Environ Sci. (2019);2(1): 37–57. DOI: 10.30881/aaeoa.00020 http://ologyjournals.com/aaeoa/aaeoa_00020.pdf

Bhandari Medani P. (2019). "Bashudhaiva Kutumbakam"- The entire world is our home and all living beings are our relatives. Why we need to worry about climate change, with reference to pollution problems in the major cities of India, Nepal, Bangladesh and Pakistan. Adv Agr Environ Sci. (2019);2(1): 8–35. DOI: 10.30881/aaeoa.00019 (second part) http://ologyjournals.com/aaeoa/aaeoa_00019.pdf

Bhandari, M.P. (2019) Institutional Goals of Sustainability in the Context of Higher Education-Contribution. Scientific Journal of Bielsko-Biala School of Finance and Law, 23 (4): 5-12, DOI: 10.5604/01.3001.0013.6853

Bartkowiak, G. (2011) Społeczna odpowiedzilaność biznesu w aspekcie teoretycznym i empirycznym, Warszawa: Difin.

Batko, R. (2015) Czytanie Tekstów. In. Kostera, M. (ed.).Metody badawcze w zarządzaniu humanistycznym. Warszawa: Wydawnictwo Akademickie SEDNO.

Bianchi, P.G. (2009) Duchowość i zarządzanie, TYNIEC Wydawnictwo Benedyktynów.

Binda, J., Sapiński, A., & Pochopień, J. (2019). The importance of social economy practice for the development of social capital of local self-governments in the perspective of labour market security. Journal of Scientific Papers Social Development and Security, 9(6), 3–10. https://doi.org/10.33445/sds.2019.9.6.1

Ciupka, S. (2020). The stability of the world image in creating a global sense of security. Scientific Journal of Bielsko-Biala School of Finance and Law, 24(2), 5–9. https://doi.org/10.5604/01.3001.0014.3278

Ferreira, F. H. G., Vega, J. R. M., Paes de Barros, R., and Chanduvi, J. S. (2009), 'Measuring Inequality of Opportunities in Latin America and the Caribbean', The World Bank and Palgrave Macmillan.

Friedman M. (1997) Tyrania status quo, Panta, Sosnowiec.

Grabowski, H. (2013) Wykłady z metodologii badań empirycznych, Kraków: Implus.

Haber G. (2012). Nierówności społeczne jako problem globalny w XXI wieku, ze szczególnym uwzględnieniem państw peryferyjnych w światowym systemie społeczno-ekonomicznym.M.W. Solarz (ed.), Polska geografia polityczna wobec problemów i wyzwań współczesnej. Polski i świata. Wybrane problemy, Toruń.

Kachniewska, M. (2014) Big Data Analysis jako zrodlo przewagi konkurencyjnej przedsiębiorstw i regionów turystycznych. In. Folia Turistica 32, Kraków.

Kostera, M. (red.). (2015). Metody badawcze w zarządzaniu humanistycznym. Warszawa: Wydawnictwo Akademickie SEDNO.

Michalaka J.M. (2017) Agent kultury różnorodności. Percepcja roli menedżerów w organizacjach wielokulturowych, Łódź: Wydawnictwo Społecznej Akademii Nauk w Łodzi.

Moczydłowska, J. (2010) Zarządzanie zasobami ludzkimi w organizacji, Warszawa: Difin.

Popescu, C.R.G.(2019). Corporate Social Responsibility, Corporate Governance and Business Performance: Limits and Challenges Imposed by the

Implementation of Directive 2013/34/EU in Romania. Sustainability, 11, 5146. doi:10.3390/su11195146
Rodrigues, P. and Borges, A.P. (2015), Corporate social responsibility and its impact in consumer decision-making. , Social Responsibility Journal, Vol. 11 No. 4. doi.org/10.1108/SRJ-02-2014-0026
Sen, A. K. (1999), 'Development as freedom', Anchor Books.
Stewart, F. (2002), 'Horizontal Inequalities: A Neglected Dimension of Development', UNU World Institute for Development Economics Research (UNU/WIDER), WIDER Annual Lectures 5.
Szaban, J. (2011) Zarządzanie zasobami ludzkimi w biznesie i administracji publicznej, Warszawa: Difin.
Szołtysek, A.E. (2013) Filozofia edukacji. Kształtowanie umysłu, Kraków: Impuls.
Śliwerski, B. (2015) Współczesne teorie i nurty wychowania, Kraków: Impuls.
Smith, V., & Langford, P. (2009). Evaluating the impact of corporate social responsibility programs on consumers. Journal of Management & Organization, 15(1), doi:10.5172/jmo.837.15.1.97
Smith, D., Rhiney, E.(2020). CSR commitments, perceptions of hypocrisy, and recovery. Int J Corporate Soc Responsibility 5, 1. doi.org/10.1186/s40991-019-0046-7
Tamvada, M.(2020). Corporate social responsibility and accountability: a new theoretical foundation for regulating CSR. Int J Corporate Soc Responsibility 5, 2. doi.org/10.1186/s40991-019-0045-8
Thanh Binh Nguyen and Qi-Wen Huang (2020). Impact of gender and education on corporate social responsibility: evidence from Taiwan. Problems and Perspectives in Management, 18(1), 334–344. doi:10.21511/ppm.18(1).2020.29
Tidd, Bessant (2013) Zarządzanie innowacjami, Oficyna Wolter Kluver, Warszawa
United Nations Development Programme (2013), 'Humanity Divided: Confronting Inequality in Developing Countries'.
United Nations (2015), Concepts of Inequality, Development Strategy and Policy Analysis Unit Development Policy and Analysis Division Department of Economic and Social Affairs, Development Issues No. 1, United Nations, New York https://www.un.org/en/development/desa/policy/wess/wess_dev_issues/dsp_policy_01.pdf
Wziątek-Staśko, A. (2010) Diversity management, Warszawa: Difin.

Żemigła, M. (2013) Odpowiedzialność społeczna organizacji, Warszawa: PWE.

Zharova, L.V. (2019) Sustainable development of territories in the framework of smart initiatives. Scientific Journal of Bielsko-Biala School of Finance and Law, 23 (4): DOI: 10.5604/01.3001.0013.6826

13

The Pathways to Address the Challenges of Inequality

Medani P. Bhandari

Professor and Advisor of Gandaki University, Pokhara Nepal, Prof. Akamai University, USA and Sumy State University, Ukraine

In terms of pathways to reduce the inequality, the United Nations is the key stakeholder to energize the governments, the World Bank, the Regional Development Banks, and other governmental and nongovernmental national and international organizations. In addition to the governments, UN, Development Organizations, the academic sector—high schools to universities, research institutions, media organizations, advocacy groups, the activists and even the public have been advocating to reduce the widening inequalities (economic, social, political, religious, cultural, and even inequality created, and boasted by the tradition).

> "Economic inequalities include access to and ownership of financial, human, natural resource-based and social assets. They also include inequalities in income levels and employment opportunities.
> Social inequalities include access to services like education, healthcare, housing, etc.
> Political inequalities include the distribution of political opportunities and power among groups, such as control over local, regional, and national institutions of governance, the army, and the police. They also include inequalities in people's capabilities to participate politically and express their needs.

> Cultural inequalities include disparities in the recognition and standing of the language, religion, customs, norms and practices of different groups"
>
> –(Stewart 2010:1).

However, in practical terms, the inequalities are increasing within economic, social, political, religious, and cultural domain and additionally, even in new domains created by the technology and techno-driven work environment.

Inequality is an unsolved problem, and now we need to find out whether it is unbeatable. There is still lack of knowledge on how inequality has been grounded throughout the human civilization; why society is stratified, classified, economically, politically, socially, and religiously; and why the discrimination due to gender, sexual orientations, country of origin, languages differences, immigration status, caste, race, and ethnically created inequality. This book addresses these issues of inequalities in a holistic way as well as with the case studies of various countries and tries to find out why inequality has been unbeatable and what would be the best policies to overcome from this challenge—Inequality.

The book chapters have provided the few pathways to minimize the inequality "The Governments should include Social Inclusion Component (SIC) in its developmental projects targeting the marginalized and poverty-stricken population in their capacity-building in order to realize the full benefits of developmental projects while addressing the critical issues of equity, social justice and marginalization. It is important that comprehensive policies, strategies, and practices are developed and implemented, bringing together various community-based organizations, leaders, community groups, governmental agencies, marginalized population, and other stakeholders while addressing these issues.......... There is a need for developing a large number of new laws, regulations, and policies in relation to natural resource ownerships and users' rights. In addition, evaluation of existing laws, policies, and regulations on natural resource ownerships and rights and make necessary improvement is suggested" (Chapter 2).

United Nations (2015) clearly acknowledges that inequality is one of the major challenges of contemporary world. As an evidence "Out of the 17 goals, eleven address forms of inequality, in terms of equality, equity and/or inclusion (Goals 1, 3, 4, 5, 6, 7, 8, 9, 10, 11, 16 and 17), and one goal (Goal 10) explicitly proposes to reduce various forms of inequalities" (Freistein and Mahlert 2015:7). For example, Goal 6. states, Ensure availability and sustainable management of water and sanitation for all—and target states—By 2030, achieve universal and equitable access to safe and affordable drinking water for all; By 2030, achieve access to adequate and equitable sanitation and hygiene for all and end open defecation, paying special attention to the needs of women and girls and those in vulnerable situations; Support and strengthen the participation of local communities in improving water and sanitation management. Similarly, Goal 7. states, Ensure access to affordable, reliable, sustainable, and modern energy for all—and target—By 2030, ensure universal access to affordable, reliable, and modern energy services; By 2030, expand infrastructure and upgrade technology for supplying modern and sustainable energy services for all in developing countries, in particular least developed countries, small island developing states, and landlocked developing countries, in accordance with their respective programs of support (United Nations 2015).

Importantly, these targets clearly identify the major indicators of inequality (measurable or unmeasurable)—age, sex, disability, race, ethnicity, origin, religion, or economic or other status; inequalities of outcome, including through eliminating discriminatory laws, policies, and practices; labor wage, and developing–developed countries variation (Bhandari and Shvindina 2019). However, it is already 2021, and the world is trapped, wounded within the Covid-19 regime. The chapters of the book show that, Covid-19 has even widened the inequality, whereas even the haves' world is not being able to cope with the pandemics, insofar, the middle income and low-income world have even losing the pathways of survival. The traditional discriminations on the basis of gender, sexual orientations, country of origin, languages differences, immigration status, caste, race, and ethnicity are still dominant factors of inequality; however, the unequal distribution of vaccines or even the business motive behind the vaccine supply has created unseen struggles within the countries as well as beyond the neighboring countries. The suffering of labor class due to lockdown is still undocumented; however, the webs of such pain are growing and will remain as one of the painful histories of human civilization. Book tries to document the pain of poverty as case study; however, captures the cloud of uncertainty of present and for near future. "Still, many of us are under Covid-19 radar, still getting sick, being hospitalized and even having last breath. We all, let us say who has access to information (any kind), to the news, know about the dreadful nature of Covid-19, which has already taken thousands of lives. Many of us are in virus trap or on the way to its shadow. As we already know that Covid-19 is so dangerous that its entry into our body is like entry of poison, which has insofar no direct treatment. We all are noticing that virus is super painful in various ways. First there is fear of death, that is what is happening now, and we never know how long it will be continued; secondly, it hits directly to the human civilization, because it is a pandemic so, whoever got the virus—he or she is out from the inherited family system. It breaks the chain of love–emotion system, sensitivity, and immediately, forces to think, "who is important."......... It shoots into the eternal part of the feelings and penetrates into the thinking pattern which separates us from us to I and makes each of us individuals. It is the direct threat of our social system. We all know that death is the ultimate reality; however, none of "us" or "I" are prepared or will be prepared and ready to accept (except few who might have different issues in lives). Covid-19: now we can add new dictionary meaning as "pain on humanity" (Chapter 9). The book chapter also examine the circulating pain globally linking with the globalization process—"globalization deepened it in some way since the scope of global inequality, reflecting the difference in welfare between the richest and poorest countries in the world, exceeds the inequality within a particular country" (Chapter 4).

The authors of the book chapters have shown the courage that, even in the traumatic state of mind, explored, dig out the challenges of widening inequality and have suggested the policy directives for the better future-socially inclusive society.

The first chapter by Medani P. Bhandari and Shvindina Hann briefly introduces the topic and shows why the world needs to work very hard to minimize the chronic situation of inequality and cope with the Covid-19. Book chapters are built within the framework of theories, particularly within the economic development linked with the sustainable development goals—"(1, 3, 4, 5, 6, 7, 8, 9, 10, 11, 16, and 17); which explicitly proposes to reduce various forms of inequalities" (Freistein and Mahlert 2015:7).

In this order, Chapter 2 by Durga D. Poudel, titled "Social Inclusion, Sustainable Development and ASTA-JA in Nepal" unveils the major challenges of social inequalities; economic inequality; gender inequality; healthcare inequality; education inequality; employment inequality, and spatial inequality and argues that until or unless society adopts the measures of equity, social justice, social inclusion, the goals of sustainable development cannot be obtained. Chapter elaborates about ASTA–JA framework (interconnectedness of eight natural resources)—*Ja* in Nepali letter, *Jal* (water), *Jamin* (land), *Jungle* (forest), *Jadibuti* (medicinal and aromatic plants), *Janshakti* (manpower), *Janawar* (animal), *Jarajuri* (crop plants), and *Jalabayu* (climate) and argues that the proper management of these resources could be the way to attain sustainability. Chapter shows that inequitable access to natural resources such as land, water, rangelands, forests coupled with entrenched social inequality in the society, relatively large population is marginalized, resource-deprived, and is under poverty. Due to pervasive discriminations, social exclusions and poverty, the economic growth, community resiliency, and socio-economic transformation of the nation have become very challenging. Authors affirm that, Nepal is facing multidimensional socio-economic developmental challenges including the development of infrastructures such as roads, irrigation system, hydroelectricity, hospitals, educational facilities, and industries, in the meantime overcoming serious issues of equity, inequalities, and social justice.

Chapter 3 forecasts the inequality through innovative principles. Titled "Forecasting Inequality—The Innovation Implementation Benefits for Countries with Different Economic Development Levels" by Anna Rosohata and Liubov Syhyda—investigates the inequalities of the benefits' distribution as a result of the radical and improving innovation implementation in countries with different levels of economic development; to create a forecast for further benefits' distribution; to develop ways of identified inequalities overcoming; to determine the main prospects for the economic development of developing societies (in particular, Ukraine) in terms of Industry 4.0. Positive and negative changes can occur at the international level. This can lead to situations when the active development of one country dramatically affects others. It seems that positive economic effects lead to impoverishment and poverty. This creates an imbalance and inequality. The logic of modern business processes orients on profit maximization and cost reduction. Based on it, developed countries often focus on production relocation to peripheral countries, due to cheap labor, lower levels of labor protection. This process causes the reduction of automation in developed countries. The reason is that there is a replacement of production processes in the periphery. As a result, this provides a certain positive effect on developing countries. But it should be clearly understood that this process has a controlling position from the standpoint of developed countries. And in certain cases, it can be quickly modified.

Chapter 4 provides the broader picture of inequality through cross country comparison of income inequality. Authored by Olha Kuzmenko, Anton Boyko, and Victoria Bozhenko, titled the "Impact of Income Inequality on Financial Development: A Cross-Country Analysis"—argue that financial relations are a key element in ensuring the macroeconomic stability of the state and supporting the welfare of the population. The main problem of recent decades, which is inherent in countries with different levels of financial and economic development, is the unequal income distribution, which leads to decreased well-being of citizens and increased social tension in society. The income inequality of the population should be considered as a complex dynamic system, which is influenced by different factors and spheres of influence. The level of unequal income distribution in a particular country depends on various factors, the most important ones of which, include the intensification of globalization processes, rapid introduction of innovative technologies in various spheres of society, growing financialization of the economy, development of financial inclusion (gaining full access to key financial services in the country), as well as significant imbalances in the levels of social protection and security. A canonical analysis confirmed the hypothesis about a close relationship between the state of development of the country's financial system and income distribution. It has been established that the foreign economic activity has the main influence on the differentiation of income of the population for most countries of the world, regardless of the state financial regulation model. Based on the calculated canonical weights for further research, when assessing the dependence of the change in the effective parameter (equality of income distribution) on a number of factorial features (indicators of the country's financial development), it is advisable to exclude some parameters such as foreign direct investment, net inflows market capitalization of listed domestic companies, stocks traded, and government expenditure on education. Thus, intensive lending by financial institutions, the active use of stock market instruments to attract and place temporarily free financial resources and the growth of the country's foreign trade turnover led to an increase in the income gap between the poor and the rich.

Chapter 5 tries to explore the role of United Nations to minimize the economic inequality. Authors examine the role of green investment, innovative technology to achieve the inequality issues. Whereas United Nations (2015) clearly acknowledges that inequality is one of major challenge of contemporary world. As an evidence "Out of the 17 goals, eleven address forms of inequality, in terms of equality, equity and/or inclusion (Goals 1, 3, 4, 5, 6, 7, 8, 9, 10, 11, 16, and 17), and one goal (Goal 10) explicitly proposes to reduce various forms of inequalities" (Freistein and Mahlert 2015; Bhandari and Shvindina 2019). Chapter by Olena Chygryn, Tetyana Pimonenko, and Oleksii Lyulyov, titled "Green Investment as an Economic Instrument to Achieve SDGs"—where authors argue that—increased interest in climate change and environment problem has increased attention to investments in green technologies and sustainable practices. For this reason, the important economic relationships, challenges, perspectives, and investment opportunities related to renewables and other green technologies were investigated. The findings allow making the conclusion that green investment is a very wide category, and it is being used at all levels: the investment in primary technologies and projects and also to green companies and financial assets. Green investment can be independent, a sub-set of a broader investment theme or closely related to other investment approaches such as socially responsible investing, environmental, social and governance investing, sustainable long-term investing, and others. The investors' and financial institutions' attention to climate change and environmental problems in general, has been rising in recent years and investor financial initiatives in this respect are growing also. It is important to note that energy efficiency and environment protection represent a significant largely untapped opportunity for meeting the dual goals of risk-adjusted financial return and environmental protection. Authors recommend that it is essential to develop specific recommendations for government and private investors for "green investing," which should include encouraging consideration of green standards for all levels of the investment decision-process; transparency in "green issues" and strengthening disclosure to consumers and investors; encourage capacity building and development of internal and external "green audit" as well as raising "green" awareness and education. Ukraine will need to prioritize activities in investing in renewables and green technologies by selecting its own criteria, perhaps in conjunction with potential investors of a project-by-project basis.

Chapter 6, authored by Sergij Lyeonov, Tatiana Vasylieva, Inna Tiutiunyk, and Iana Kobushko, titled "The Effect of Shadow Economy on Social Inequality: Evidence from Transition and Emerging Countries." Chapter describes the general context of the economic crisis in the country, the decline in the level of material well-being and social protection of the population, the problem of income shadowing, as the main component of reducing the level of social protection in the country, increasing the level of its economic inequality, becomes especially relevant. Nowadays, the shadow economy is one of the threats of economic development of the country and one of the main obstacles of its social development. One of the biggest negative effects of the shadow economy on the level of social development of the country is the evasion of social security contributions, which leads to a significant decrease in funding for programs and measures of social security of the population, problems with social benefits, pensions, unemployment insurance, health insurance, etc. Shadowing the economy significantly reduces the pace of reform of health care, education, social protection, etc.

In Chapter 7 titled "Cultural Dynamics and Inequality in the Turkish Sphere Concerning Refugees," by Kemal Yildirim, provides the historical account of culture building in Turkey and unveils the pain of Syrian refugees. Chapter explains the current position and role of the Turkish civil religion in historical perspective on political agenda with effects on social and cultural agenda on the society. *In the late 1970s, the Turkish political scene was characterized by a thorough ideological polarization between right-wing ultranationalists (ülkücüs—idealists) and radical left-wing groups, along with a lack of decisive authority on the part of the government. Author argues that* Turkish community hosting Syrian refugees should be encouraged to integrate. The success of this public relations approach, which aims to develop the needs, might be measured time to time for a better outcome. A normative public relation as a reference point for the outputs of the theory of Perfect Public Relations at the point of measurement should be taken into account. Providing detailed information with refugees about Turkish society as well as in return to provide detailed information about the refugees to the Turkish society would be important for both parties to be familiar with each other's. In this context, informing about the refugees to the Turkish society in a most accurate way will contribute to the integration process. On the other hand, refugees are also needed to have a strong identity with the Turkish society that hosted them for a long time as a result of this, it is therefore important that they are willing and positive toward the adaptation process.

Chapter 8 titled "Genesis and Reasons for Social Inequality in the Globalization Era" by, *Teletov Aleksandr Sergeevych*, and *Teletova Svetlana Grigorievna,* demonstrates that globalization has not reduced social differentiation between people. Moreover, globalization deepened it in some way since the scope of global inequality, reflecting the difference in welfare between the richest and poorest countries in the world, exceeds the inequality within a particular country. At the same time, inequality indices in individual countries continue to increase. There are many examples, including those that confirm the efforts of the civilized world to reduce inequality between people. In general, the problem of inequality between people, the current rate of which is a major threat to stability in the whole world, has not been solved yet. Thus, one should, first, identify the type of social system that would be acceptable if not for everyone, at least for most people. In practice, states should strive to balance their citizens' living standard by eliminating economic, social, age, gender and ethnic discrimination, regulating labor and capital markets, and introducing progressive taxation and social security policies. There is no doubt that the world will be changed after the COVID-19 pandemic: the economic crisis will be deepened, deglobalization will be observed, interstate relations will be partially limited etc. It is also an opportunity to reduce inequality between people to some extent. The main task for further research on inequality and ways to overcome it in society is to find concrete steps to achieve this objective.

The Chapter 9 titled "The Luxury Lockdown: Tackle to Covid-19, Fatal to Hunger- Widening Inequality" by Man Bahadur Bk and Medani P. Bhandari, presents the humanitarian aspects of Covid-19, leading toward the more divisive and unequal society. Covid-19 is a global crisis and the crisis always hits to the people who has no resources to alter with the situation. The Covid-19 regime is about a year old; however, it completely changed the global picture of human civilization. Firstly, it remained with the blame culture, the so-called conspiracies within nations and even among scientists and the world media big and small also played important roles to horrify to the general public. The world's economic situation and political or power greediness divided the world due to few tempered leaderships, whose role was basically to blame to other nations, or in the worse cases own scientists and society. Until, very recently, world was in unprecedented crisis, however, there are lights of hope. This chapter basically presents the opinion with the grounded factual evidence. The case of Covid 19, or any future communicable diseases is / will not be the same because we all citizen of the world are connected in a way that, if any small event occurs in one part of the world, it can spread to the globe, because we are connected through modernism (technology). Therefore, we should understand that our small negative action in the society could have global negative impact. In addition to Covid 19, the world is already facing various challenges: political, social, religious, etc., inequalities, ageing, aids, atomic energy, children, climate change, economic colonization, democracy fights, poverty, food insecurity, gender inequality, lack of access to healthcare, human rights, questions on international law and justice, migration, challenges on peace and security, population growth, refugees challenges, scarcity of water, misuse of technology, drugs, growing individualistic approach among youth, deviant behavior to attend temporary pleasure, corruption, wars, violation of social norms and values as so on. In addition to that the inequality is growing and the discriminations on the basis of race, caste, color, country of origin, religion, migratory status, gender, and sexual orientations are still in practice. The economic inequality is widening and widening, and the Covid-19 is playing a favorable role to increase inequality. Scholarly world has important role to overcome from these challenges.

Chapter 10 titled "Inequality in labor—Rural Agricultural Transformation in Nepal" by Prem Bhandari and Medani Bhandari, provides the general scenario of food insecurity and labor intensity in farming, whereas both male and females contribution is about the same. They argue that food insecurity is a global challenge. Whereas women play a triple role in agricultural households: productive, reproductive, and social (World Bank, ILO, and IFAD2009:19). This situation applies mostly to undernourished people live in developing countries and are mostly the subsistence-based smallholder farmers. Although controversies abound about the roles of green revolution technologies worldwide, their roles cannot be underestimated in increasing food production and therefore, in reducing world hunger and food insecurity. This chapter unveils this reality by using the uniquely detailed data from a rural food insecure agrarian setting of Nepal by examining the relationships between family labor availability and use of modern labor-saving technologies in agriculture among smallholder farmers. We use the labor demand framework to examine the relationships. Results from multi-nominal logistic regression revealed that the availability of family labor, both males and females, discouraged the use of such technologies in crop production net of household- and neighborhood-level factors. Moreover, the results further provide evidence that both the availability of male and female laborers are equally important in hindering the use of labor-saving farm technologies.

Similarly, Chapter 11 is the boarder of the inequality graph—which deals with the inequality in terms of uses of natural resources in the National parks buffer zone of Nigeria. Authored by Jacob, D. E., and I. N. Ufot, chapter titled "Income Inequality and Poverty Status of Households Around National Parks in Nigeria"—examines the levels of income inequality and incidence of poverty among households in the support zone communities of Nigeria National Parks. The findings depict the presence of income inequality among the various park households. The socioeconomic characteristics of the household were the major determinant of poverty incidence in the communities. Author recommends, Policies and programs should also be put in place that will help ameliorate the negative effects the park have on the households as it will go a long way in enhancing households' levels of productivity and incomes. However, such policies should be highly localized to enable it to achieve better outcome. Additionally, any poverty alleviation programme initiated in the area should effectively serve as safety nets for the households by decentralisation of its implementation to enhance resource governance which will reduce poverty.

Chapter 12 titled, "Social Responsibility as a Tool for the Human Resources Policy Development and Reducing Inequalities on Tourism Industry," by Aleksander Sapiński, Sabina Sanetra-Półgrabi, Serhii Y. Kasian, and Nataliya Rozhko, presents the role and importance of corporate social responsibility as a tool for supporting the development of innovation in the field of human resources management in the hotel industry. It is an interdisciplinary article because it deals with a research subject that can be looked at from an economic, legal, and pedagogical perspective. The very fact of hotel activity as an effect of tourism in its broadest sense may bring a new innovative approach with the use of the novelties in the field of pedagogy and education. The interdisciplinarity of tourism and hospitality makes its huge development over the last two decades a kind of viral activity, because as a virus combines and infects, tourism and hospitality combine other disciplines to implement pro-innovative solutions, making its practical part more attractive to potential customers. The tourism industry, thanks to its interdisciplinary nature, creates many jobs, which helps to combat social exclusion. Many tourist destinations are also combating many of the social inequalities that have existed to date, thanks to investment in tourism capital. Social responsibility and HRM are ideal for tourism and hospitality due to their interdisciplinary character. Today, social responsibility is a concept that permeates all areas of knowledge and economic life.

There are many publications about the challenges of inequality (Bhandari 2019, 2020; Bhandari and Shvindina 2019; Collier and Rohner 2008; Freistein and Mahlert 2015; Nygård 2017; Klarin 2018; Mair et al. 2017; Sambanis 2004; Soares et al. 2014; Stewart and Brown 2010; UNESCO 2013; United Nations 2015); however, still there is knowledge gap—incompletes, because the pain of inequality begins from personal level to community and community to nation and nation to nations. The main pain holder due to inequality is the developing world—who faces each kind of inequalities within and beyond. The haves world also have challenges however, those challenges could be minimized if proper directive measures were implemented. However, the haves' group is benefiting from have nots group, country level to individual level; therefore, they only look the economic part of the society. As a result, the rich are getting richer, and poor are getting poorer. There is a need of strong intervention were, the marginalized world's voice could reach to them who holds the key of the society. The chapters of this book are the sound of reality and hopefully, the world will listen one way or another and begin the right path of social inclusiveness.

In addition to the chapters, book contains the forwards from prominent scholars of social sciences—academia, former staff of United Nations Agencies, high ranking government officials who argue that, there is a need of transformative re-thinking whether traditional measures are still relevant for new challenges if we are serious about finding pragmatic approaches to equitable equilibrium level of wellbeing for all, and thereby reducing deep disparities and inter-social inequalities. Each of the chapters is unique and unveils societal inequalities through regional and country specific cases, mostly backed by principles of survival equally. The chapters provide in-depth problems and consequences of inequality, with the case studies of various countries of the world. The book speaks the pain of Covid 19, how it is triggering inequality; unveils pain of child labor, the discrimination on the basis of gender, sexual orientations, country of origin, languages differences, immigration status, caste, race, and ethnicity. And also, provides the general pathways to minimize inequality through implementation of socially inclusive programs and policies. The book also provides the future directions of further research on this unbeatable challenges and ways out to beat the challenges.

References

Bhandari, Medani. P. (2019a), "Vasudhaiva Kutumbakam"- The entire world is our home, and all living beings are our relatives. Why we need to worry about climate change, with reference to pollution problems in the major cities of India, Nepal, Bangladesh, and Pakistan. *Adv Agr Environ Sci.* (2019);2(1): 8–35. DOI: 10.30881/aaeoa.00019

Bhandari, Medani. P. (2019b), Live and let other live- the harmony with nature /living beings-in reference to sustainable development (SD)- is contemporary world's economic and social phenomena is favorable for the sustainability of the planet in reference to India, Nepal, Bangladesh, and Pakistan? *Adv Agr Environ Sci.* (2019);2(2): 37–57. DOI: 10.30881/aaeoa.00020

Bhandari, Medani P. (2020), The Phobia Corona (COVID 19) - What Can We Do, Scientific Journal of Bielsko-Biala School of Finance and Law, ASEJ 2020, 24(1): 1–3, GICID: 01.3001.0014.0769, https://asej.eu/resources/html/article/details?id=202946

Bhandari, Medani P. and Shvindina Hanna (2019), Reducing Inequalities Towards Sustainable Development Goals: Multilevel Approach,

River Publishers, Denmark / the Netherlands Print: 978-87-7022-126-9 E-book: 978-87-7022-125-2

Bhandari, Medani. P. (2020), In the Covid-19 Regime – What Role Intellectual Society Can Play. International Journal of Science Annals, 3(2), 5–7. doi:10.26697/ijsa.2020.2.1 https://ijsa.culturehealth.org/en/arhiv https://ekrpoch.culturehealth.org/handle/lib/71

Collier, P. Hoeffler, A. Rohner, D. (2008), Beyond Greed and Grievance: Feasibility of Civil War. Department of Economics, University of Oxford.

Freistein, K., and. Mahlert, B. (2015), The Role of Inequality in the Sustainable Development Goals, Conference Paper, University of Duisburg-Essen, See discussions, stats, and author profiles for this publication at: https://www.researchgate.net/publication/301675130

Håvard Mokleiv Nygård (2017), Achieving the sustainable development agenda: The governance – conflict nexus, International Area Studies Review, Vol. 20(1) 3–18

Klarin, Tomislav (2018), The Concept of Sustainable Development: From its Beginning to the Contemporary Issues, Zagreb International Review of Economics & Business, Vol. 21, No. 1, pp. 67–94, DOI: https://doi.org/10.2478/zireb-2018-0005 https://content.sciendo.com/view/journals/zireb/21/1/article-p67.xml

Mair, Simon, Aled Jones, Jonathan Ward, Ian Christie, Angela Druckman, and Fergus Lyon (2017), A Critical Review of the Role of Indicators in Implementing the Sustainable Development Goals in the Handbook of Sustainability Science in Leal, Walter (Edit.) https://www.researchgate.net/publication/313444041_A_Critical_Review_of_the_Role_of_Indicators_in_Implementing_the_Sustainable_Development_Goals

Sambanis, N. (2004), Poverty and the Organization of Political Violence. In: Globalization, Poverty, and Inequality. Brookings Trade Forum. Washington D.C.: Brookings Institution Press, pp. 165–211

Soares, Maria Clara Couto, Mario Scerri and Rasigan Maharajh -edits (2014), Inequality and Development Challenges, Routledge, https://prd-idrc.azureedge.net/sites/default/files/openebooks/032-9/

Stewart and Brown, cited in Stewart, F. (2010), Horizontal Inequalities as a Cause of Conflict: A Review of CRISE Findings. World Development Report 2011 Background Paper

UNESCO (2013), UNESCO's Medium-The Contribution of Creativity to Sustainable Development Term Strategy for 2014–2021, http://www.unesco.org/new/fileadmin/MULTIMEDIA/HQ/CLT/images/CreativityFinalENG.pdf

United Nations (2015), Transforming our world: the 2030 agenda for sustainable development. New York (NY): United Nations; 2015 (https://sustainabledevelopment.un.org/post2015/transformingourworld, accessed 5 October 2015).

Index

F

G

Foreword Writers' Biographies

Douglass Lee Capogrossi has established and administered a variety of formal and non-formal education programs including distance learning colleges, trade apprenticeships, correctional education programs, work experience projects, on-the-job training ventures within industry and the human services, cross-border university affiliations, distance learning training programs for industry, and adult job training through center-based programs. He holds permanent teaching credentials in the United States in commerce and social studies and lifetime certification as a counseling teacher for the emotionally disturbed. Dr. Capogrossi has ten years' experience developing and delivering successful correctional education programs for adult inmates in Hawaii prison facilities with emphasis in parenting, cognitive development, transition to work and community, and adult basic education skills. Dr. Capogrossi is an experienced community service administrator, where his expertise rests primarily with creation of NGO corporate structures, program funding, excellence of Board operation, and implementation of a wide spectrum of government and nonprofit training programs and emergency services projects. Dr. Capogrossi has held top management positions in industry, serving as General Manager of Micrographic Systems, a medical cameramanufacturing firm in Silicon Valley California for a brief period in the early 1980s and he owned and operated America Builders, a successful licensed general contracting firm. Dr. Capogrossi has extensive community service experience on NGO boards and is highly experienced with founding activities for new nonprofit ventures, especially with antipoverty programs. He has served as a member of the Board of an international quality assurance agency in higher education. Dr. Capogrossi earned his bachelor's in business administration, his Master's in Curriculum and Instruction, and his Ph.D. in Adult and Continuing Education from Cornell University, USA, where he completed an extensive dissertation investigating the effectiveness of the American education system. Among his more recent published scholarly papers, The Assurance of Academic Excellence among Nontraditional Universities was published in the Journal of Higher Education in Europe. Dr.

Capogrossi has dedicated his career in service to humanity, through his efforts to improve the human condition.

Mary Jo Bulbrook, her career expansion covers over 51 years primarily serving higher education in leadership positions in academia starting in community college education in nursing, then higher education at Texas Woman's University to University of Utah. Then she was guided to move out of country to Memorial University of St. John's Newfoundland, Canada, and from there to Edith Cohen University in Perth, Western Australia. On the side her role was a practitioner, educator, and researcher in a range of energy therapies starting with Therapeutic Touch, Healing Touch, Touch for Health, Transform Your Life through Energy Medicine (TYLEM) her own creation blending her profession as a psychiatric mental health clinical specialist with energy therapies. This expansive background of energy therapies was not only taught throughout USA and Canada but also taken to Australia, New Zealand, South Africa, and Peru where she worked with traditional healers while education professors, nurses, and other health care, and lay people. Her journey served to capture the worldwide perspective on education, health and healing to address local, national, and international ways to improve the human condition where she gained firsthand perspectives on worldwide needs, values and challenges.

The issues of "what if," "why," and "how" dominated her search to understand the broader perspectives of world strategies, cultural norms, and expectation of how to prepare individuals, families, and communities for being in the world, surviving and thriving. We are not alone and need each other standing as equals became her trademark and motto in search of how to prepare educators in different cultures regarding the blending cultural needs with innovative strategies and techniques of national heroes of health and healing with native traditions. She stood alongside those she served as equals not as "I am better than" or "have all the answers" to a "we are one," "united by our humanness as equals," and "each with different perspective on the meaning of life," living in this world in a cooperative way with a win-win perspective as her personal guiding light which serves as the philosophy in her new role leading Akamai University as their first woman president and innovator holding the light for the faculty and students to shine their light.

Vasyl Karpusha, is the Rector of the Sumy State University, He is Ph.D. in Physical and Mathematical Sciences.

Jacek P. Binda is a Rector and a researcher at Bielsko-Biała School of Finance and Law, Poland (pol. Wy?zsza Szk´oa Finans´ow i Prawa w Bielsku- Białej, Polska). Prof. Binda obtained his PhD at the Warsaw University of Technology and his Post-Doc (habilitation) at University of Zylina, Republik of Slovakia. His principal fields of research include economy and finance, problems of the high-risk financial instruments, banking, local and public finances, e-economy, local government, project management and controlling. Since 2005, he has coordinated several medium-size and large research projects focusing on border studies supported by the EU's Framework Programmes, National Research and Development Centre in Warsaw, Poland. He is an expert of the National Research and Development Centre inWarsaw, European Commission, Scientific Grant Agency of the Ministry of Education of Slovak Republic and the Academy of Science. Author or co-author of numerous monographs, over 60 national and international publications in the field of finance and information technology. He is a member of the Scientific Boards and Editorial Boards, incl.: Strategic Planning for Energy and the Environment (Journal in River Publishers), MDPI Journals, Switzerland; Sumy National Agrarian University: Sumy, UA; Scientific Journal of Bielsko- Biala School of Finance and Law (Editor-in-Chief), PL. Awarded the Gold Cross of Merit for achievements in research and teaching by President of Poland.

Stanisław Ciupka, Doctor of Humanities. Dean of the Faculty of Law and Social Sciences at Bielsko-Biała School of Finance and Law (Poland) where he works at the Department of Internal Security. Co-author of numerous scientific publications (national and international) in the field of humanities and social sciences. A former professor at the University of Economics and Humanities in Bielsko-Biała where he worked at the Department of Human Resource Management, a long-term scientific editor of the Polonia Journal, editor and publisher of a scientific series supporting local scientists in southern Poland "Beskidzkie Dziedzictwo," co-founder of the Beskidzki Institute for Human Sciences, chairman of the Bielsko-Biała branch of the Catholic Association Civitas Christiana. He has collaborated with universities in Ukraine and Slovakia.

Ambika P. Adhikari is a Principal Planner managing the long-range planning division at City of Tempe in Arizona, USA. He has over 30 years of professional and academic experience in urban and environmental planning and international development in several countries including Nepal, India, USA, Canada, Mexico, Kenya, and Fiji.

Ambika was a Village Planner and Project Manager at the City of Phoenix, and a Senior Planner at SRPMIC in Scottsdale, Arizona. He then joined Arizona State University as a Program and Portfolio Manager for Research. Prior to that, Ambika worked as a Director of international programs at DPRA Inc. in Toronto and Washington DC, where he managed international urban and environmental and economic development programs and projects in Canada, USA, Mexico, Nepal, and India.

In Nepal, he served as Nepal's Country Representative of the Switzerland based IUCN—International Union for Conservation of Nature, leading Nepal's national and some Asia regional programs in several environmental and conservation fields. He was also a member of the Government of Nepal's National Water and Energy Commission, the highest policy-making national body in this sector. Ambika was an Associate Professor of Architecture and Planning at the Institute of Engineering, Tribhuvan University, Nepal.

Ambika has authored one book, co-edited six books, published numerous reports and articles and refereed articles in journals. He regularly writes and lectures on sustainability and urban planning, international development, climate change policy, and international environmental policies and programs.

Ambika received his Doctor of Design (DDes.) degree in Urban Planning and Design from Graduate School of Design, Harvard University. He obtained his post-graduate fellowship from Massachusetts Institute of technology (MIT), M.Arch. from University of Hawaii, and B.Arch from M.S., University of Baroda. He is a member of American Institute of Certified Planners (AICP) and holds the designations of Project Management Professional (PMP) through Project Management Institute (PMI), and LEED AP (ND) through the US Green Building Council (USGBC). He is Fellow of American Society of Nepalese Engineers, where he also serves in the Advisory Council.

Ambika has been active in serving the Nepali community in the US and Canada by volunteering for various community organizations. He was the president of the NRNA National Coordination Council of US and Canada, NRNA Regional Coordinator for North America, Advisor to NRNA International Coordination Council (ICC), and Patron of NRNA ICC. He is the Chair of Board of Trustees for the Association of Nepalese in the Americas (ANA), Board Member of Asta-Ja USA, and Executive Member of Global Policy Forum Nepal (GPFN).

Keshav Bhattarai, born and raised in a rural area of Balkot (Current Chhatradev Village Palika), Arghakhanchi district of Nepal, Dr. Keshav Bhattarai has contributed hugely to bridge Nepali and US education system.

Bhattarai family comprised of four siblings (three brothers and one sister). The family lived on subsistence farming. Bhattarai completed High School (10th grade) from Mahendra Vidya Bodh High School, Balkot, Arghakhanchi district of Nepal. He earned Intermediate of Science (I.Sc.), Bachelor of Science (B. Sc.), and Bachelor of Arts (BA) from Tribhuvan University. He also earned Associate of Indian Forest College (AIFC) post-graduate Diploma from Indian Forest College, Dehradun, India, and Master of Science (M.Sc.) in Natural Resource Management from the University of Edinburgh, Scotland, U.K. He completed Ph.D. from Indiana University, Bloomington, Indiana, US in 2000 in Geography. He later joined Eastern Kentucky University as a faculty member, and in 2002, he moved to the University of Central Missouri, Warrensburg, Missouri.

His career on academia began as an Associate Instructor at the Department of Geography, Indiana University, Bloomington, Indiana in 1996. He has since taught a variety of courses at Eastern Kentucky University and University of Central Missouri. Dr Bhattarai is a member of Nepal Foresters' Association since 1983 and Association of American Geographers since 2002. A recipient of more than a dozen grants, Dr. Bhattarai has also taken the responsibility of journal article reviewer and book reviewer.

Dr. Bhattarai has been working as a Professor/ Geography at the School of Geoscience, Physics, and Safety Sciences at the University of Central Missouri since 2011. Dr. Bhattarai has also worked as a Short-term Consultant to theWorld Bank to assess Hydropower potential in Nepal. He has also worked as an Interim Chair, Department of Geography at the University of Central Missouri from 2009 to 2011.

Dr. Bhattarai bridges between the University of Central Missouri (UCM) and various universities of Nepal. Nepali universities and UCM have started 2+2 and 3+1 programs with UCM. The 2+2 refers to completing two years of undergraduate studies in Nepal and remaining 2 years in the US to complete undergraduate studies and vice-versa. The 3+1 program refers to completing three years education in Nepal and transferring to the US and vice-versa to complete an undergraduate degree.

Before embarking for his higher studies in the USA, Dr. Bhattarai worked in various capacities under the Ministry of Forest and Soil Conservation in Nepal from 1983 to 1995. Bhattarai left the Government job due to repeated political illogical impositions on him on the settlements of political cadres on forest areas without any norms and land use planning.

Dr. Bhattarai's urbanization research reveals that Nepal's urbanization is moving too quickly without needed infrastructure for the urban dwellers. The unplanned political ad-hoc decision to annex many rural areas into urban definition has increased vulnerabilities among urban dwellers since Nepal is located in between frequently moving Eurasian and Indian tectonic plates. Nepali urbanization has become *ruralopolis* (political decision to annex rural areas into municipalities). Such ad-hoc urbanization has increased the costs of lands and taxes on low-income people without giving any needed facilities to them. Dr. Bhattarai served as Fulbright Specialist at the Central Department of Geography (CDG) at Tribhuvan University, Kathmandu in the summer of 2018 to facilitate CDG to embark to active participation in urban planning through research.

Dr. Bhattarai is the recipient of Gorkha Dakhsina Bahu, award for civil servant decorated by the King Birendra for outstanding performance in forestry service and administration in 1993. He has also received Mahendra Vidyabhusan for his academic excellence in 2004. Moreover, he is also the recipient of All Round Forest Award by Indian Forest College, Dehradun, India in 1983, International Scholar Award by the University of Central Missouri in 2003 and Chairperson's award for Outstanding Academic Performance by the Department of Geography, Indiana University, Bloomington in 1999. He has also bagged half a dozen faculty research award for his academic excellence from 2007 to 2020.

He has published a number of academic papers, authored four books, coauthored one book and contributed many book chapters, journal articles, and presented his research at a number of professional meetings.

About the Authors

Dr. Medani P. Bhandari, PhD, Prof. Bhandari is a well-known humanitarian, author, editor, and co-editor of several books and authors of hundreds of scholarly papers on social and environmental sciences: a poet, essayist, environment, and social activist, etc. He is a true educator, who advocates that education should not be only for education's sake and for employment (get position, earn money, etc.) but should bring meaningful change in students' lives, which could provide the inner peace in life and be able to face any kind of challenges. Prof. Bhandari is also a motivational speaker—life skill coach (how to remain in peace and calm when situation is not in our control, with application of Basudhaiva Kutumbakam principles)—and in writings and teaching, he exemplifies how studies of scientific theories can be enjoyable and applicable in day-to-day life. As such the epistemology of knowledge which is the basis of theory building is considered not fun subject for everybody; however, Prof. Bhandari's works show that, there is always a lucid and attractive side in any difficult subject.

Dr. Medani P. Bhandari has been serving as a Professor of interdisciplinary Department of Natural Resource & Environment/Sustainability Studies, at the Akamai University, USA and Professor at the Department of Finance, Innovation and Entrepreneurship, Sumy State University (SSU), Ukraine. Dr. Bhandari is also serving as Editor in Chief of International Journal—The Strategic Planning for Energy and the Environment (SPEE), Denmark, The Netherlands, USA—ISSN: 1546-0126 (Online Version), ISSN: 1048-4236 (Print Version) https://www.journal.riverpublishers.com/index.php/SPEE/about He is also serving as managing editor at the Asia Environment Daily, (the largest environmental journal of the world)- https://asiaenvdaily.com/index.php/editorials-2/2-uncategorised/29760-about-asia-environmental-daily

He holds M.A. Anthropology (Tribhuwan University, Nepal), M.Sc. Environmental System Monitoring and Analysis (ITC—The University of Twente, the Netherlands), M.A. Sustainable International Development (Brandeis University, Massachusetts, USA), M.A. and Ph.D. in Sociology (Syracuse University, NY, USA).

Dr. Bhandari has spent most of his career focusing on the Social Innovation, Sociological Theories; Environmental Sustainability; Social Inclusion, Climate Change Mitigation and Adaptation; Environmental Health Hazard; Environmental Management; Social Innovation; Developing along the way expertise in Global and International Environmental Politics, Environmental Institutions and Natural Resources Governance; Climate Change Policy and Implementation, Environmental Justice, Sustainable Development; Theory of Natural Resources Governance; Impact Evaluation of Rural Livelihood; International Organizations; Public/ Social Policy; The Non-Profit Sector; Low Carbon Mechanism; Good Governance; Climate Adaptation; REDD Plus; Carbon Financing; Green Economy and Renewable Energy; Nature, Culture and Power.

Dr. Bhandari's major teaching and research specialties include: Sociological Theories and Practices; Environmental Health; Research Methods; Social and Environmental innovation; Social and Environmental policies; Climate Change Mitigation and Adaptation; International environmental governance; Green Economy; Sustainability and assessment of the economic, social and environmental impacts on society and nature. In brief, Prof. Bhandari has sound theoretical and practical knowledge in social science and environment science. His field experience spans across Asia, Africa, the North America, Western Europe, Australia, Japan, and the Middle East. Dr. Bhandari has published several books, and scholarly papers in international scientific journals. His recent books are *Green Web-II: Standards and Perspectives from the IUCN (2018); 2nd Edition 2020; Getting the Climate Science Facts Right: The Role of the IPCC; Reducing Inequalities Towards Sustainable Development Goals: Multilevel Approach; and Educational Transformation, Economic Inequality—Trends, Traps and Trade-offs; Social Inequality as a Global Challenge, the Unbeatable Challenges of Inequality, Perspectives on Sociological Theories, Methodological Debates and Organizational Sociology.* Additionally, in creative writing, Prof. Bhandari has published 100s of poems, essays as well as published two volumes of poetry with Prajita Bhandari.

Prem B. Bhandari is a Social Researcher (Demographer). He has a Ph.D. in Rural Sociology and Demography from the Pennsylvania State University, USA. He completed his M.Sc. in Rural Development Planning from the Asian Institute of Technology, Thailand and earned a B.Sc. degree in Agricultural Economics from Tribhuvan University, Nepal. Currently, Bhandari is working as the Evaluation Coordinator/Data Analyst for a School

Feeding Program Evaluation in Cambodia and Haiti through KonTerra Research Group, Washington, D.C. In addition, he is the Managing Director of South Asia Research Consult, Inc. He is also an Adjunct Professor at Agriculture and Forestry University in Nepal. Prior to this, he worked as a researcher (Co-Investigator) and Assistant Research Scientist at the Institute for Social Research, University of Michigan. He has over 20 years of experience in designing and managing large scale social (survey) research programs. In Nepal, he worked as the assistant professor of agricultural economics at Tribhuvan University, Nepal. His major research interests include social research methods (survey design, survey instrument design, pre-testing and finalization, survey data collection, survey management and quantitative analysis of survey data, and qualitative surveys); socio-economic and cultural inequalities and determinants of demographic behaviors (population health and migration); social inequalities; food security and nutrition; and overall rural social change. He has a strong background and experience in quantitative (and qualitative) analysis. Bhandari has published a number of articles in peer-reviewed journals and as book chapters and presented dozens of papers in national and international professional meetings. He is the editorial member of several journals and an occasional reviewer of over a dozen of peer-reviewed journals.

Anton Boiko graduated from the Ukrainian Academy of Banking of the National Bank of Ukraine (M.S in "Finance," Ph.D. in "Finance, Money and Credit," D.Sc. in "Finance, Money and Credit" and "Economics and Management of National Economy"). Currently works as an Assistant Professor at the department of Economic Cybernetics at Sumy State University (Ukraine). He is the author of more than 70 papers. Scientific interests include economic security, money-laundering risk, financial monitoring, insurance market, and systemic risk. He participated in a research funded by the state budget.

Victoria Bozhenko graduated from Ukrainian Academy of Banking of the National Bank of Ukraine (M.S in "Finance," Ph.D. in "Finance, Money and Credit"). Currently works as an Assistant Professor at the Department of Economic Cybernetics at Sumy State University (Ukraine). She is a managing editor of the scientific journal "SocioEconomic Challenges" (Ukraine). She is the author of more than 60 papers. Scientific interests include money-laundering risk, income inequality, and systemic risk. She participated in research funded by state budget.

Olena Chygryn is Associate Professor of the Department of Marketing at Sumy State University, Ukraine. She got the scientific degree of PhD in Economics in 2003.

Olena Chygryn has published more than 100 scientific papers, including 3 papers in international peer-reviewed journals. She is the Scholarship holder of International Programs (ITEC, UNDP in Ukraine) and the participant more of 10 international training and seminars. The main sphere of her scientific interests includes: Green Production, Environmental Management, Green Competitiveness.

Durga D. Poudel, the Founder of Asta-Ja Framework and Professor in School of Geosciences at University of Louisiana at Lafayette, Lafayette, USA, was born and raised in Tanahu, Nepal. He spent his childhood in Tanahu and Lamjung districts, both in the Mid-Hills of Nepal. Dr. Poudel passed his S.L.C. exam from Nirmal Vocational High School, Damauli, Tanahu in 1977 and received an I.Sc. in Agriculture at Tribhuvan University at Rampur (IAAS), Chitwan, Nepal in 1980; a B.Sc. in Agriculture (Major: Agricultural Economics) at University of Agriculture, Faisalabad, Pakistan in 1987; a M.Sc. in Natural Resource Development and Management at Asian Institute of Technology, Bangkok, Thailand in 1991; and a Ph.D. in Soil Science, University of Georgia, Athens, GA, U.S.A. in 1998. Dr. Poudel is an expert in environmental science, climate change adaptation, and sustainable agricultural development; soil physical, chemical, and mineralogical characterization; soil classification; soil and water conservation, water quality, roadside vegetation management, and natural resources conservation and development. His recent research focuses on water quality monitoring and modeling, smallholder mixed-farming system, climate change adaptation, geohazards, wildflower, and environmental soil chemistry. Dr. Poudel has published over 100 peer-reviewed journal articles, conference proceedings papers, book chapters, research briefs, and scientific abstracts. He has published numerous popular articles. He has given over 100 scientific presentations to regional, national, and international conferences. Dr. Poudel has received more than $4.48 million in external funding as a PI and more than $1.76 million in external funding as a CO-PI. He has supervised more than two dozen graduate students in their Thesis/Dissertation research.

Dr. Poudel's professional experience consists of Research Fellow at Asian Vegetable Research and Development Center, Taiwan (1991–1994); Graduate Research Assistant in Sustainable Agricultural and Natural Resource

Management Collaborative Research Support Program, University of Georgia (1994–1998); and Visiting Research Scholar, University of California Davis (1998–2000), USA. Dr. Poudel joined the University of Louisiana at Lafayette, USA, as an Assistant Professor of Soil Science in August 2000, and currently is a tenured Professor and Coordinator of Environmental Science Program of School of Geosciences, Director of Ag. Auxiliary Units (Model Sustainable Agriculture Complex (600-acre Cade Farm), Crawfish Research Center, and Ira Nelson Horticulture Center), and Board of Regents Professor in Applied Life Sciences at the University of Louisiana at Lafayette, Louisiana, USA. Dr. Poudel's professional affiliations include membership in Soil Science Society of America, American Society of Agronomy, Crop Science Society of America, Geological Society of America, American Geophysical Union, and Soil and Water Conservation Society, USA. Dr. Poudel is the life member of NRNA, Nepalese Association in Southeast America (NASeA), Asta-Ja Abhiyan Nepal, Asta-Ja USA, and Association of Nepalese Agricultural Professionals of Americas (NAPA). Dr. Poudel is the founding member of Asta-Ja Abhiyan Nepal, and the founding President of Asta-Ja Research and Development Center (Asta-Ja RDC), Kathmandu, Nepal, and Asta-Ja USA, Honolulu, Hawaii, USA. Dr. Poudel was the Chair of NRNA ICC Agriculture Promotion Committee America Region, 2017–2019. Dr. Poudel served as a Board Member of the Bayou Vermillion Preservation Association (BVPA), Lafayette, Louisiana, USA, and an Advisor of the NASeA. Dr. Poudel is a Resource Person of Louisiana Organics, and a member of the Louisiana Technical Advisory Committee, USDA-NRCS, Louisiana, USA. In 2017, in recognition of the impact and quality of one of his research papers, Dr. Poudel and his co-authors were awarded for the "2017 Best Research Paper Award for Impact and Quality Honorable Mention" by the Journal of Soil and Water Conservation. Dr. Poudel believes on providing his services to others, promoting social inclusion, integrity, goodwill, and peace, and advancing on better understanding of the world.

Imaobong Ufot Nelson is a graduate of the University of Uyo, Nigeria (Ph.D. and M.S in Forest Resources Management and Economics; and Bachelor of Forestry and Wildlife). Currently works as a Lecturer in the Department of Forestry and Wildlife at the University of Uyo, Nigeria. She is an invited reviewer for several scientific journals. She is the author of more than 40 papers. Scientific interests include Environmental Resources Management, Resources Economics, Sustainable Development, and Operational Research.

Daniel Etim Jacob graduated from the Federal University of Agriculture, Abeokuta, Nigeria (Ph.D. in "Park Planning and Recreational Development) and the University of Uyo, Nigeria (M.Sc in Wildlife Resources Management and Bachelor of Forestry and Wildlife). He is a recipient of the Federal Government Scholarship (M.Sc) and Exon-Mobil Scholarship (Undergraduate). He currently works as a Lecturer in the Department of Forestry and Wildlife at the University of Uyo, Nigeria. He is a member of IUCN-WCPA, Connectivity Specialist Group, African Forest Forum (AFF), among others. He is also a member of editorial boards of several scientific journals and an author of more than 70 papers. Scientific interests include Natural Resource Governance, Biodiversity Conservation, Resources Economics, and Eco-Modelling

Serhii Y. Kasian, PhD in economics, associate professor. The Head of Marketing Department Dnipro University of Technology, Ukraine. Member of the Public Organization "Innovation University". The member of 07 Management and Administration Sectoral Expert Councils, Specialty 075 Marketing, National Agency for Higher Education Quality Assurance in Ukraine (NAQA). Participant of the international educational and scientific project "Innovative University and Leadership": education and micro-project, Warsaw University, Faculty of "Artes Liberales," Foundation "Instytut Artes Liberales"; International Foundation for Educational Policy Research. The Expert in the field of the Internet Marketing Communications, High-Tech Enterprises Marketing, International Innovative Marketing, Interaction of Economic Agents in E-Business.

Iana Kobushko graduated from Sumy State University (M.S in "Management of Organizations, Ph.D. in "Environmental Economics and Natural Resources Protection"). Currently works as a Senior Lecturer at the Department of Management at Sumy State University (Ukraine). She is a member of the Ukrainian Association for Management Development and Business Education. She is author of more than 40 papers. Scientific interests include Team Management, Investment Potential, Mechanisms for Counteracting Income Shadowing. She is an active member at NGO "Institute Business Educations", where she organize the projects on educational empowerment for the local community.

Olha Kuzmenko graduated from Ukrainian Academy of Banking of the National Bank of Ukraine (M.S in "Economic Cybernetics," Ph.D. in "Finance, Money and Credit," D.Sc. in Mathematical Methods, Models and

Information Technologies in Economics). She is Chief Editor of the International Scientific Journal "SocioEconomic Challenges". Currently works as a Head of the Department of Economic Cybernetics at Sumy State University (Ukraine), Head of Scientific and Educational Centre for Business Analytics (Ukraine). She won international projects in the sphere of Education Quality Assurance System; international programs, and internships concerning International business conferences for IT developers, Erasmus + programs. She is the author of more than 100 papers. Scientific interests include modelling of economic processes (optimization methods and models); construction and research of dynamic econometric models; digitalization of the public sector, social and economic relations; stability and sustainability of socio-economic systems; influence of state policy on socio-economic development; economic security and the fight against corruption; actuarial calculations; formation and development of insurance and reinsurance markets; financial monitoring in banks.

Serhiy Lieonov is vice-rector of Sumy State University. He graduated from the Sumy State University in 1999. He got PhD in speciality 08.02.02—Economics and Management of Scientific and Technical Progress (2003), Associate Professor of Management Department (2006), Doctor of Economics in speciality 08.00.08 —Money, Finance and Credit (2010), Professor of Finance Department (2012). Since 2017 he is Professor at Economic Cybernetics Department. Professor Lieonov is a research leader at his department. Research interests are in the spheres of banking, investing, reproductive processes in the economy, risk management, innovation, corporate reporting and its audit, investment activity. He is the author of more than 150 scientific papers and more than 50 monographs and numerous presentations. Professor Lieonov is a Head of the Specialized Academic Committee at Sumy State University; he is a mentor for many Ph.D. students and Post-Docs. He is a contributor and a member of the Working Group of scientists who develop the forecasting of economic and social development of the Sumy region for 2018–2022 period. He is an active participant in the social projects for the local community, as well as educational events.

Oleksii Lyulyov is Head of Department of Marketing of Academic and Research Institute of business, Economics and Management at Sumy State University, Ukraine. He got the scientific degree of D.Sc. in Economics in 2019. Oleksii Lyulyov has published more than 144 scientific papers, including 40 papers in international peer-reviewed journals. He is the Scholarship holder of International Programs (ERASMUS+ KA1 (2018) and the participant more of 10 international training and seminars. The main sphere of her scientific interests includes country marketing policy, country image, macroeconomic stability, innovative development, sustainable economic development, strategy development, modeling and forecasting development trends.

Dr. Man Bahadur Bishwakarma has completed his postdoc research on food security as a Fulbright Visiting Scholar (2016/17) at Brandeis University, USA. He obtained Ph.D. on Social Inclusion in Microfinance in 2010 from Tribhuvan University, Nepal. He did a postgraduate diploma in Social Studies from the ISS, Netherlands (2004), MBA (1991) in Financial Management and MA (Economics) in Rural and Cooperative Development (2000) from Tribhuvan University, Diploma (1997) in Agro-Cooperative Management from Japan. He has long working experience with various international development agencies such as GTZ, Save the Children, UNICEF, CWS UK, UNDP. He has profound experience in conflict, peace, and the constitution building process as he led the Constitutional Dialogue Centre/UNDP during the Constitutional Assembly-I. In 2010, he rejoined the government bureaucracy as a joint secretary and served as the Director General for Labour Department, Chief District Officer in several districts, Secretary for provincial Government of Lumbeni and Karnali; and now the Principal Secretary for Bagamati province. He has been serving as a visiting faculty member for the M.Phil Program (Education) and MBA-BF Program of Tribhuvan University and written a dozen of textbooks for college students and several books on development issues, including Social Inclusion in Microfinance, Eradicating Hunger: Rebuilding Food Regime.

Tetyana Pimonenko is Deputy Director for International Activity of Academic and Research Institute of Business, Economics and Management at Sumy State University, Ukraine. She got the scientific degree of Dr.Sc. in Economics in 2019. Tetyana Pimonenko has published more than 150 scientific papers, including 44 papers in international peer-reviewed journals. She is the Scholarship holder of International Programs (Fulbright Visiting Scholar Program (2018-2019); Latvian Government (2018); ITEC Program (India, 2018); National Scholarship Programme of the Slovak Republic for the support of mobility of students, PhD students, university teachers, researchers and artists (2017) and the participant more of 12 international training and seminars. The main sphere of her scientific interests includes:

Green Marketing, Green Investment, Green Economics, Green Bonds, Alternative Energy Resources; Environmental Management and Audit in Corporate Sector of Economy, Sustainable Development and Education.

Anna Rosokhata graduated from Sumy State University (M.S. in "Marketing," Ph.D. in "Economics and management of enterprises (according to the types of economic activities)." She took part in Volunteering activities in the English-speaking environment at the EuroWeek Foundation (Wroclaw, Poland) where they were the organizational and educational activities with children, teenagers, and teachers. She has practical experience in marketing and business development: she was Manager of Marketing and Business Development at LLC "Digiline" (Sumy, Ukraine). Main areas of company activity are multiscreen OTT/IPTV solutions development, web & mobile solutions. Also, she was the Business Development Manager at LLC "SoftIndustry" (Chernihiv, Ukraine). The main company's directions are the development of Mobile, Web-solutions and other IT solutions development on demand. Currently, she works as a Senior lecturer at the Department of Marketing at Sumy State University (Ukraine). She is a project leader of 8 R&D projects. She participated in more than 30 International scientific and practical conferences. She is the author of more than 50 scientific papers. Her scientific interests include forecasting in marketing, trendwatching, branding, digital marketing, and marketing of information technologies.

Nataliya Rozhko, she is a PhD in economic sciences and an associate professor. She works in science and research at Ternopil Ivan Puluj National Technical University. Her publications are in the field of marketing, business economics, and management in general.

Sabina Sanetra-Półgrabi, PhD in humanities in the discipline of political science, assistant professor in the Department of Administrative Sciences and Public Policy at the Institute of Political Science and Administration (INPA) of the Pedagogical University in Kraków. D. thesis entitled D. dissertation "Functioning of Euroregions in the Southern Polish Borderland and Integration of the Borderland Community after 1993. Comparative analysis of three Euroregions: "Śąask Cieszyński," "Beskidy," and "Tatry" defended in 2012. Since 1 October 2013 employee of the KEN Pedagogical University in Krakow. Graduated in administration (2003), international relations (2005) and political science (2006). She also completed studies in Slavic languages and culture at the Jagiellonian University in Kraków (2008). Her research interests include issues of transboundary cooperation with particular emphasis on Euroregion in the southern Polish border area and the system of local security and crisis management, including cooperation on border rivers.

Aleksander Sapiński, PhD student in two disciplines: security studies and literary studies. Received a master's degree in international management, administration, and English language and culture studies. Currently works as an academic lecturer at the Bielsko-Biała School of Finance and Law(Poland). He is the president of the NGO Żywiecki Institute of Social Economy and president of the NGO Beskidzkie Association of Ecological Production and Tourism BES PROEKO. Deputy Editor of Scientific Journal of Bielsko-Biala School of Finance and Law, former deputy scientific editor of Polonia Journal. His research interests focus on management, security studies, ecology and social responsibility.

Hanna Shvindina graduated from Sumy State University (M.S. in "Management of Organizations, Ph.D. in "Environmental Economics and Natural Resources Protection," Dr. Sci. in "Economics & Management at the Enterprises"). She was awarded several individual scholarships and was a Post-Doc researcher in France in the US (University of Montpellier, Purdue University) in "Business Studies and Management", Fulbright Alumna (2018–2019), won ERASMUS MUNDUS scholarships (2014, 2015). Currently, she works as a Head of the department and Associate Professor at the Department of Management at Sumy State University (Ukraine). She is a member of CENA community, Researchers' excellence network (RENET), Ukrainian Association for Management Development and Business Education. She is a member of editorial boards of several scientific journals (Ukraine, Poland, Switzerland), invited reviewer for the international conferences (Poland, Germany, Ukraine). She is the author of more than 90 papers. Scientific interests include Coopetition Paradox, Strategies and Innovations, Change Management, Organizational Development, Civic Education and Leadership. She is an activist and is a CEO at NGO "Lifelong learning centre," where provides courses on Leadership and Communications, organizes the program on empowerment for the local community, and offers advisory services.

Liubov Syhyda graduated from Sumy State University (M.S. in "Marketing," Ph.D. in "Economics and Management of Enterprise (by types of economic activity)." She won the Grant of the President of Ukraine (2017) for funding her research. Currently, she works as an Associate Professor of the Department of Marketing at Sumy State University (Ukraine). She is the author of more than 100 papers. Scientific interests include Supply Chain Management, Innovative Products' Distribution, Industry 4.0, Risk Management, Marketing Attractiveness of Territories.

Teletov Aleksandr, Doctor of Economics, Professor, Professor of the Department of Public Management and Administration, Sumy National Agrarian University (Sumy, Ukraine); author of more than 50 textbooks and methodical guidelines, more than 200 scientific papers, about 400 journalistic articles.

Teletovà Svetlana Grigorievna, Ph.D., Candidate of Philological Sciences, Associate Professor, Associate Professor of the Department of Russian language, Foreign Literature and Methods of their Teaching, Sumy State A. S. Makarenko Pedagogical University (Sumy, Ukraine). She is an author of more than 20 textbooks and methodical guidelines, more than 100 scientific papers.

Inna Tiutiunyk graduated from Sumy State University (M.S in "Finance and Credit," Ph.D. in "Environmental Economics and Natural Resources Protection," D.Sc in "Economics and management of the national economy" and "Money, Finance and Credit"). Currently works as an Associate Professor at the Department of Finance and Entrepreneurship at Sumy State University (Ukraine). She is a member of editorial boards and invited reviewer of several scientific journals. She is the author of more than 60 papers. Scientific interests include Managing Tax Gap, Fiscal Policy, Shadow Economy, State Economic Policy.

Tetyana Vasilyeva is a Director of Academic and Research Institute of Business, Economics and Management, Professor (2010, Professor's degree of the Department of Management) and Doctor of Economics (2008, speciality 08.00.08—money, finance, and credit). She is the author of more than 200 scientific papers, 70 monographs, and numerous materials of the conferences. She is Editor-in-chief of the scientific journal "Business Ethics and Leadership," member of the editorial board of the journals: "Strategy and Development Review," "Carbon Accounting and Business Innovation Journal." She is a mentor for PhD students and Post-Docs and is a member of the Specialized Academic Committee at Sumy State University; one of the founders and a member of NGO "Council of Young Scientists." She was the initiator of launching the Business Support Center in Sumy region, and now she is a member of this organization that gained the support of the European Bank for Reconstruction and Development. She is a team leader, an author, and participant of numerous scientific projects that gained the highest scores in Ukraine. Research interests include banking, investment, risk management, innovation activity, social responsibility, and education. She is a leader and contributor in international projects, such as Jean Monnet projects (2014, 2015, 2017), grant projects such as “Development of Dialogue Between Banks and Civil Society in the Context of Ensuring Democratic Processes in Ukraine” (DAAD, 2016), “Enhancing Energy Security by Swiss-Ukrainian-Estonian Institutional Partnership” SCOPES IZ74Z0_160564 (Swiss National Science Foundation, 2015–2017) and many others. She is a contributor in many educational projects financed by Ministry of Economic Cooperation and Development of Germany, DVV International, DESPRO in Switzerland, American Council for Economic Education NCEE, International Foundation “Vidrodzennia” etc. Professor Vasilyeva initiated many scientific and educational events concerning the development of corporate social responsibility (e.g., winter/summer schools for youth “Economic and business values,” “School of innovations and social entrepreneurship,” “Management of socially responsible business,” etc.)

Kemal Yildirim is Professor in Comparative Politics at the European School of Law and Governance in Kosovo (ESLG) and a Senior Research Fellow in the Division of History and Civilization at EMUI EuroMed University. He earned a dual Ph.D. in comparative politics at Central University of Nicaragua and Azteca in Mexico in 2013, and has completed postdoctoral research in law, democracy, and human rights at the Institute of Human Rights at Coimbra University, Portugal, in 2018, followed by a second postdoc on immigration issues in Latin America at the University of Buenos Aires in Argentina.

He is an author of 204 books that have been translated into more than 8 languages. He has published more than 50 articles in peer-reviewed journals. Dr. Yildirim has contributed to the research environment as a reviewer for scholarly journals (see the AASCIT website) and associate editor for the World Research Journal of Political Science. He is also a film producer and director, with more than twenty feature films and documentaries under his belt. Dr. Yildirim speaks 10 languages and is the president of the Beyt Nahrin Mesopotamian Academy of Sciences and Arts (more about: www.mesopotamianacademy.org).